Electronics for Engineering

Electronics for Engineering

S. A. Knight BSc (Hons)

Newnes
An imprint of Butterworth-Heinemann
Linacre House, Jordan Hill, Oxford OX2 8DP
A division of Reed Educational and Professional Publishing Ltd

ℛ A member of the Reed Elsevier plc group

OXFORD   BOSTON   JOHANNESBURG
MELBOURNE   NEW DELHI   SINGAPORE

First published 1996

British Library Cataloguing in Publication Data
Knight, S. A.
    Electronics for Engineering – (GNVQ
    Engineering Series)
    I. Title   II. Series
    621.38

ISBN 0 7506 2301 2

Printed in Great Britain by Bath Press, Avon

Contents

Preface

This book has been written to cover the Electronic Engineering content of the Advanced GNVQ, and in particular the optional unit in Electronics. Parts of its contents will also be found useful for students taking the Intermediate GNVQ in Engineering. It will also be suitable for BTEC National and GCE A-Level students.

Particular emphasis has been placed on worked and self-assessment examples throughout the text and on graded problems at the end of each chapter. Answers to most (but not all) of these problems are provided; an occasional omission allows the reader to acquire more confidence in his or her own working. It is usually possible to assess whether the solution is 'probable' or not and it is a good philosophy to check in this way the reasonableness of an answer.

Each chapter concludes with a Summary of its contents, together with a set of Review Questions which should be treated as being there for the reader to check his or her own understanding of the preceding concepts and terminology. The use of models throughout the text simplifies the functions of diodes, transistors and transformers into linear equivalent circuits.

Electronics is a practical subject and experimental work is an essential part of it; a number of relatively simple experiments are provided at the back of the book as a basis for at least a substantial portion of the course time to be given to practical work. It is suggested that these experiments are followed in parallel with the theory where the subject matter arises.

A thorough check has been made to eliminate errors in the text but in the event of anything being unclear or ambiguous, the author would welcome constructive suggestions for possible later correction.

S. A. Knight

1

The electric circuit

THE ELECTRIC CIRCUIT

An electric circuit is a closed connecting path around which electric charges may move continuously. If such charges are to be moved from one place to another there are, in what we might call a practical circuit, four requirements: these are (a) a supply of charge carriers, (b) a source of electric force sufficient to make the charges mobile in a controlled manner, (c) a medium in or through which the charges can move relatively freely, and (d) a work element or load which will receive the supply and in some way convert the energy of the mobile charges into useful work. In *Figure 1.1* the last three of these requirements are indicated as a battery, copper connecting wires and two small electric bulbs respectively. The other requirement, the charge carriers, is not so tangible.

If we ask ourselves what electric charge really is, we have to admit that we don't really know. We know

that electrons and protons, the major constituent particles of the atom, carry such charges and that these charges are of two different kinds, negative and positive, whatever these words may actually imply. We also know that forces exist between charged particles; like charges repel and unlike charges attract. Outside of this, the nature of electric charge is something of a mystery.

However, we know a little more about the carriers of charge; those which interest us the most in electronics are electrons, though there are other kinds as we shall see later. When we talk about the movement of charge we do not usually think about the amount of charge moved; we think of the rate at which the charge flows, the rate of flow of charge. An electric current is actually a measure of the rate of flow of negative charges passing any given point in the circuit. The charges are embodied in electrons, and if 6.24×10^{18} electrons pass a given point in one second, then the current is said to be 1 ampere. In light current electrical and electronic engineering, current is measured in fractions of an ampere: the milliampere (mA) which is one-thousandth of an ampere, and the microampere (μA) which is one-millionth of an ampere.

But what agency causes the current to flow around the circuit in this way? The energy or force necessary to push the electrons around the circuit is supplied by a source of electromotive force (e.m.f.) which is usually generated by a cell or battery by chemical reactions, or by a dynamo which operates by the interaction of moving magnetic fields. Electromotive force is measured in volts or fractions of a volt, and under its influence an otherwise haphazard activity of

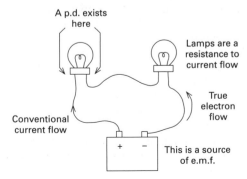

Figure 1.1

the electrons becomes a regular and ordered flow along the conducting paths.

This movement of the electrons is not, however, without incident, just as the flow of traffic is not without its own particular hazards. The hazard for the moving electrons arises as an impediment brought about by their frequent collisions with the molecules of the conducting material which introduces an opposition to their free movement. This opposition is known as resistance. Resistance may therefore be considered as that property of a material which opposes the movement of electrons through it.

So in the simple circuit of *Figure 1.1* we have the ingredients that are found in all electrical circuits, however complex: the current (I), the electromotive force (E) and the resistance (R), this last being concentrated almost wholly in the filaments of the two bulbs. An analysis of much more complicated systems can always be interpreted in terms of a series of interconnected circuits of this form.

Conductors and insulators

All material may be broadly categorized into two classes: conductors and insulators. Conductors are those materials which offer, relatively speaking, only a small resistance to the flow of current. Copper is used for most electrical wiring though silver is marginally better and is used in specialized circumstances. Both of these metals are rich in free or mobile electron carriers and large currents can be obtained under the influence of small electromotive forces. Other commonly used conductors are aluminium and (for special cases) gold. In fact, all metals are relatively good conductors, though there is one element, carbon, which is non-metallic. Basically carbon is a poor conductor but finds many uses in electronics, particularly in the deliberate making of resistance (resistors). Water, too, is a poor conductor but when contaminated with salts or acid becomes a good conductor. Such solutions are called electrolytes, and in these the carriers can be both negative and positive types.

Insulators are those materials which have practically no free electrons available as charge carriers in their make-up and so exhibit such great resistance that negligible current can flow through them, however large the electromotive force may be. Insulators are used to confine currents to the desired conducting paths; for example, most wires used in electrical equipment have copper cores which carry the current but these cores are covered with an insulating jacket as *Figure 1.2* shows. Current cannot flow through or 'escape' from this covering and so is prevented from

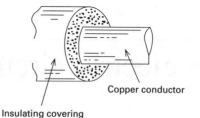

Copper conductor

Insulating covering

Figure 1.2

making contact with bare metal surfaces or other uncovered wires in the vicinity. Printed circuits are simply bare copper conducting paths glued to a base fibreglass or resin-bonded paper insulating sheet. Commonly used insulating materials are PVC and other plastics like polythene, rubber, ceramic, fibreglass, bakelite, mica and glass.

There is a third group of substances which are known as semiconductors. These substances have resistance values at ordinary temperatures which are too great to be classified as conductors, yet much too small to be classified as insulators, and so fall into a kind of no-man's-land between the other fully fledged categories.

We will encounter these materials in due course; for the immediate present we will be concerned only with good conductors.

Conventional or true?

Before the nature of electricity was properly understood (as far as it *is* understood), it was accepted that a positive current flowed from a positive terminal of the supply to the negative, that is, from a point of high potential to a point of low potential. This was analogous to the flow of water from a high to a low level as over a waterfall; on the way down, the loss of potential energy could be used to turn a wheel or drive a turbine. In the same way it was imagined that work was done by the electric current as it 'fell' from a high to a low potential level of electromotive force. This assumed direction of the current is now called the *conventional* flow.

With the discovery of the electron by J.J. Thompson early in the twentieth century, it was realized that current was actually a movement of negative particles from the negative pole of the supply to the positive pole and that the conventional interpretation so far accepted was incorrect. What is called the true or *electronic* flow of current is therefore in the direction opposite to that of conventional flow. This is illustrated in *Figure 1.1*.

The terms 'positive' and 'negative' are, however,

quite arbitrary in the sense of polarity, just as they are in the case of charges. If we take as a reference level the body of the earth and assume that this is electrically neutral, then a voltage of one kind with respect to the earth is considered positive and a voltage of the opposite kind is considered negative. Reversing these signs does not in any way affect an electrical system's function in life. In general electrical considerations therefore we work in conventional terms, that is we take the current as flowing from positive to negative; we say that the positive terminal of a source of electromotive force is at a higher *potential* than the negative terminal.

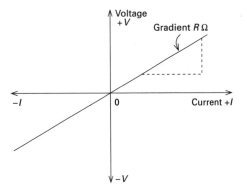

Figure 1.3

E.m.f. and p.d.

A battery or any other source of electrical pressure establishes an electromotive force within the source. However, when the source is in use, not all of this e.m.f is available to do useful work in an outside circuit, since some of it is needed to drive the current through the source itself.

It is difficult to measure accurately the precise e.m.f. acting in a circuit. Connecting a voltmeter to a battery, for instance, does not indicate the battery e.m.f. What we are actually measuring is the voltage between the terminals of the battery; this is known as the terminal potential difference or p.d. which is the electrical pressure acting between two points in a circuit when a current flows between these points. These points need not be at the terminals of a battery or any other source; a potential difference exists between *any* two points in any circuit configuration, excepting those cases where a zero p.d. is deliberately established, e.g. the Wheatstone bridge.

There is often some confusion between the concepts of e.m.f. and potential difference (or voltage drop as this is sometimes called), and because of this both e.m.f. and p.d. are often indiscriminately referred to as 'voltage'. It is true that both are measured in volts, but it is not a quibble to mention that the two concepts should be carefully differentiated in the mind, particularly when they crop up in certain questions and a distinction is necessary to clarify the answers.

Ohm's law

The current that flows in any conductor is proportional to the applied e.m.f. or to the p.d. acting between the ends of the conductor. Hence, provided the temperature remains constant, $V = kI$ where k is a constant of proportionality. This is stated to be Ohm's law, though there is a proviso which we will come

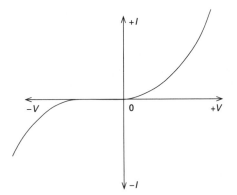

Figure 1.4

to later. By choosing the basic units of current and voltage, k can be made to provide us with the unit of resistance; we say that if the e.m.f. (or the p.d.) between two points in a circuit is 1 volt and the current is then 1 ampere, the resistance between the points is 1 ohm (Ω). Hence

$$V = IR \text{ from which } I = \frac{V}{R} \quad \text{or} \quad R = \frac{V}{I}$$

So, given any two of these quantities, we can calculate the third. For metallic conductors the proportional relationship between voltage and current may be represented graphically after the manner of *Figure 1.3*. The graph is linear and passes through the origin; it is also specified for negative values of voltage and current for reasons which will turn up in due course. It should also be appreciated that the gradient of the graph is a measure of the resistance of the conductor.

There are some circuit elements which do not obey Ohm's law; these are known as non-linear components. A diode, for example, shows a relationship between voltage and current and is drawn in *Figure 1.4*. This is decidedly non-linear and the resistance of the diode is not independent of V and I which is our strict interpretation of Ohm's law. However, the relationship $V = IR$ remains as a general definition of resistance whether or not the conducting path obeys Ohm's law; and two *related* values V and I always satisfy the Ohm's law equation.

Fundamental connections

If two or more elements are joined together so that the current flows through each of them in turn, they are said to be connected in series. So from *Figure 1.5* where the elements are the resistors R_1, R_2 and R_3, these are connected in series across a voltage source V and their total overall resistance is simply the sum of the individual resistances. Suppose the three resistors shown have an equivalent single resistance of R_e. We notice that the same current I flows in turn through each of them, so that the voltages across each resistor will be by Ohm's law IR_1, IR_2 and IR_3, respectively. Remind yourself here that these voltages are potential differences; clearly the sum of these p.d.s must equal the applied potential V, which in terms of a single equivalent resistor would be IR_e. Hence

$$IR_e = IR_1 + IR_2 + IR_3 \quad \text{or}$$

$$R_e = R_1 + R_2 + R_3 \tag{1.1}$$

and so on for any number of series-connected resistors.

Notice particularly that the applied voltage divides up into as many parts as there are resistors and that this division is in direct proportion to the resistor values.

If two or more resistors are arranged so that each forms a separate path for a part of the total current drawn from the source, they are said to be connected in parallel. This time the voltage across each resistor is the same and equal to V as *Figure 1.6* shows. What the single equivalent resistance R_e is, is not quite so easy to visualize as it was in the case of the series connection, but it is not too difficult to work it out. From *Figure 1.6* the total current I must be the sum of the separate branch currents, that is

$$I = I_1 + I_2 + I_3$$

Now since each resistor has a potential V across it, the branch currents are, from Ohm's law

$$I_1 = \frac{V}{R_1}, I_2 = \frac{V}{R_2}, I_3 = \frac{V}{R_3}$$

So, if R_e is the equivalent resistance, then $I = V/R_e$ and

$$\frac{V}{R_e} = \frac{V}{R_1} + \frac{V}{R_2} + \frac{V}{R_3}$$

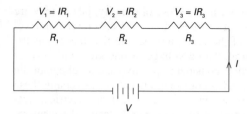

Figure 1.5

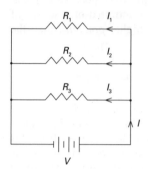

Figure 1.6

Hence

$$\frac{1}{R_e} = \frac{1}{R_1} + \frac{1}{R_2} + \frac{1}{R_3} \tag{1.2}$$

and so on for any number of paralleled resistors.

Notice here that the circuit current divides into as many parts as there are resistors. Obviously we can expect the smallest resistance to take the greatest share of the current, and the largest resistance to take the least current. This is an inverse proportion, so that if the resistance values are in the ratio 1:2:3, the current ratios will be 3:2:1.

Work, power and quantity

The charge on an electron is constant and the same for all electrons irrespective of the type of atom they belong to. This fact makes it possible to use this fundamental charge as the basis of a measure for the unit of electrical quantity. Because a single electron charge is much too small to be of practical use, the unit adopted is the coulomb (C) which is equal to 6.24×10^{18} electron charges. We have already seen that this figure corresponds to a current flow of 1 ampere, hence a current of 1 ampere carries charge from one place to another at the rate of 1 coulomb per second. Thus

Quantity (Q coulombs)
= Current (I amperes) × time (seconds)

or $Q = It$ coulombs

When current flows against resistance, electrical energy is converted to heat energy. This is because the action of any force may result in the expenditure of energy and in work being done. Different materials require different amounts of work to be done in breaking down the bonds between electrons and atoms so that the electrons become free to act as charge carriers. This results in the differing values of resistance between, say, a length of copper wire and an identical length of iron wire.

The amount of energy transferred or dissipated as heat in a resistive load is equal to the quantity of electricity multiplied by the p.d. acting across the load; the unit of energy or work done is the joule (J), so

$$\text{Energy expended} = \text{voltage } (V) \times \text{charge } (Q)$$
$$= VIt \text{ joules}$$

since $Q = It$.

Power is the *rate* of doing work or the energy used in unit time. A rate of working of 1 joule per second is 1 watt (W), hence

$$\text{Power } (P) = \text{energy expended/time}$$
$$= \frac{VIt}{t} = VI \text{ watts} \qquad (1.3)$$

In words: the power dissipated in a resistance (or the power extracted from a source) is equal to the product of the current flowing in the resistance (or from the source) and the potential difference developed across the resistance.

By noting that from Ohm's law $V = IR$, then the above power expression becomes

$$P = I^2R \text{ watts} \qquad (1.4)$$

Can you now deduce an expression for power involving V and R?

Other passive components

The passive element of resistance consumes electrical energy and generates heat. There are two other passive elements which are of importance in electronic engineering and about which certain laws and properties have been formulated from experimentally observed facts. These properties concern devices which can store electric energy and return it to its initial source when required. These elements are capacitance and inductance.

Capacitance

It is possible to add electrons to a conductor so as to give the conductor an excess of electrons; the

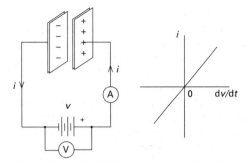

Figure 1.7

conductor is then said to be negatively charged. Also, electrons may be removed from a conductor giving it a deficit of negative charge; the conductor is then said to be positively charged. The resulting action of these two possible effects can be crudely demonstrated by rubbing a plastic pen briskly on the sleeve, when it can be observed that pieces of paper can be attracted to the pen. Such effects were well known to the Greeks. The ability of a conductor to acquire charge in this way is known as its *capacitance*.

Suppose two metal plates, separated by a small distance, the space between them being some form of insulation, are connected to a d.c. source as shown in *Figure 1.7*. A voltmeter is used to measure the voltage of this source and an ammeter is placed in series with the plates. When the circuit is completed it is noticed that a positive charge appears on the right-hand plate and a negative on the left. If the source is now disconnected, the charge persists. Such a system therefore stores electrical charge, and from its capacity to do that is given the name *capacitor*.

To accumulate this charge there must be a movement of electrons around the circuit while the charging action is taking place. In other words, electrons are displaced from the positive-pole plate and accumulate on the negative-pole plate. This momentary movement of electrons is known as a *displacement current* and is indicated on the ammeter as a brief movement of the pointer. It is important to appreciate that no current passes *through* the insulation region between the plates; the displacement is a momentary effect which persists only until the potential between the plates is equal to the source voltage. This distinguishes a displacement current from a continuous or conduction current.

It is found that the charge accumulated is directly proportional to the applied voltage, that is

$$\text{Charge } Q = CV \text{ coulomb} \qquad (1.5)$$

where C is the constant of proportionality. This constant is known as the capacity or the *capacitance* of the plate system, and is measured in farads (F) where a capacitance of 1 F will store 1 coulomb of charge under an impressed voltage of 1 V. The microfarad (μF) = 10^{-6} F and the picofarad (pF) = 10^{-12} F are the more practical units used in general electronics.

If instead of a d.c. source, the supply is derived from an alternating source, the displacement current continually flows first in one direction and then the other, and the ammeter indicates this as a constant deflection of its pointer. Under this condition, the current is now observed to be proportional, *not* to the voltage itself as it would be in a resistor, but to the *rate at which the voltage is changing*. This can be expressed by the equation

$$i = C\frac{dv}{dt} \qquad (1.6)$$

and this can be derived from (1.5) above by noting that current is rate of change of charge, or dq/dt.

To find the energy stored in a charged capacitor, we can suppose it to be charged slowly by successive additions of small equal charges, each of q coulombs. As the voltage v on the capacitor will be proportional to the charge, the voltage will increase linearly with time and at any instant

$$q = Cv = It$$

where I is the constant charging current. Then

$$I = C \times \frac{v}{t}$$

and since the average value of the voltage over a time of t seconds will be $V/2$ (V being the final voltage level) we have

$$\text{Energy stored} = \frac{V}{2} \times C \times \frac{V}{t} \times t = \frac{1}{2}CV^2 \text{ joules}$$

This energy is stored in the electric field established between the plates; it is *not* stored on the plates. When the capacitor is discharged, slightly less energy is returned to the circuit than was supplied during the charging cycle. This is mainly due to the absorption of energy in the material used between the plates and to the small but finite resistance of the plates themselves and the connecting wires which generate the inevitable heating.

A full analysis of the charge and discharge characteristics of capacitors will be given in Chapter 3.

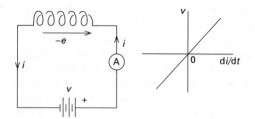

Figure 1.8

Inductance

If the plates of the elementary capacitor of *Figure 1.7* are replaced by a coil (or solenoid) of wire as shown in *Figure 1.8*, an entirely different state of affairs turns up. What this is can be derived from our knowledge of magnetic fields. A charge in motion along a wire establishes a magnetic field around the conductor; when the conductor is formed into a multi-turn coil, the magnetic field becomes highly concentrated and is large for a given current. For a d.c. source, the field once established remains constant, the current here being a conduction current, not a momentary displacement current. Energy is stored in the field so long as the current is flowing, but when the source is disconnected the field collapses and the stored energy is returned to the external circuit, usually in the form of sparking at contact points and as heating in the connecting wires. It is during the initial increase of current and during its collapse that the important features come about, for at these stages the current is *changing*.

Now Faraday's law tells us that whenever a magnetic flux linking a circuit is changing in magnitude, an e.m.f. is induced in the circuit which is proportional to the *rate* at which the flux (and hence the current) is changing. This is called the e.m.f. (e) of self-induction. Hence we can write

$$e = -L\frac{di}{dt} \text{ volts} \qquad (1.6)$$

where L is the constant of proportionality and the negative sign simply indicates that the induced e.m.f. acts to oppose the applied voltage. This is Lenz's law.

L is called the *inductance* of the coil (or whatever form the circuit actually takes) and this is measured in henries (H) where an inductance of 1 H induces an e.m.f. of 1 V when the current changes at the rate of 1 A per second. Practical units are the millihenry (mH) = 10^{-3} H and the microhenry (μH) = 10^{-6} H.

Using a similar method to that used for capacitance, the energy stored in the magnetic field of an inductor can be shown to be $\frac{1}{2}LI^2$ joules which corresponds neatly to the energy stored in a capacitor.

When the field collapses, slightly less energy is returned to the circuit than was supplied during the build-up of the field. Some of the energy is dissipated as heat in overcoming the resistance of the inductance; this is, in fact, the only expenditure of energy from the source once the field has been established, unless an iron circuit is involved, in which case there are additional eddy current and hysteresis losses when the source is alternating.

A full analysis is given in Chapter 3.

Continuity of stored energy

There is a fundamental principle in physics which states that in any system, mechanical or electrical, the energy must be a continuous function of time. This, in effect, is saying that nothing can take place in no time at all – that there is no such thing as an instantaneous change. Since power is the time rate change of energy as we saw a bit earlier on, an instantaneous change in energy would require an infinite power. Since this is an unacceptable concept to our view of physical reality, we must accept that the energy stored in any system must have a continuity in time – that it always takes some time, however small that time may be, for anything to take place.

Since the energy stored in a capacitance can be expressed as $\frac{1}{2}CV^2$ and in an inductor as $\frac{1}{2}LI^2$ we accept that in the former the voltage cannot change instantaneously and in the latter the current cannot change instantaneously.

These concepts, though seeming rather abstract at first encounter, can be used to predict the behaviour of a circuit system which undergoes an abrupt change in voltage or current for any reason, such as the closing or opening of a switch. Such abruptly changing periods are known as *transients* to distinguish them from normal continuous operations and conditions which are known as *steady state*.

EQUIVALENT CIRCUITS

Many problems can be solved by replacing the actual components in a circuit by what are known as equivalent circuits or models. This sort of thing applies particularly to the action of active devices such as transistors and other semiconductor components. The equivalent circuits obtained respond in exactly the same manner to an external stimulus as the real circuits do, but the only true similarity between behaviour of actual circuits and their equivalents lies

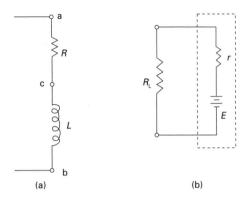

Figure 1.9

in the correspondence between the voltage–current characteristics at a particular pair of terminals.

Now, when we build a circuit out of various component parts such as wire, coils and resistors, etc., then we are actually treating these components as ideally lumped; a resistor is an embodiment of resistance alone though, particularly if it is a wire-wound component, it may possess an appreciable inductance. By this argument, any circuit component may be strictly represented by a number of elements or a combination of a number of elements. When we wire a coil into a circuit, the intention in general is to impart inductance to that circuit; we do this by referring to the component as an inductor, bearing in mind all the while that we may be introducing other circuit properties, in this case resistance. Thus an inductor is in reality a series combination of L and R, though its dominant or major circuit contribution is inductance.

In general, then, we will use the '-or' ending to describe the actual components, and the '-ance' ending will be used to describe the pure or 'untainted' elements. Thus we describe an ideal inductor in a circuit diagram as an inductance.

As we have already noted, a coil of wire has an equivalent representation to that shown in *Figure 1.9(a)* where the two terminals a and b represent the actual input points of the coil but the junction point c between the two elements does not exist as any definable point within the coil. The coil has both inductance L and resistance R but these cannot be discrete parts within the real coil although in the equivalent circuit we represent them as such. What goes on inside the equivalent diagram will have no counterpart in the real component. When current flows from the accessible terminal a to the accessible terminal b, a voltage is developed across R but this voltage is not accessible and cannot be measured as such in the actual coil. On the other hand, voltage and current conditions at the accessible terminals of the coil and of its equivalent circuit will be completely identical.

Circuit elements which may be represented in this way, as an equivalent network of resistance, inductance and capacitance, are known as passive elements and the circuits in which they act are known as linear circuits. There is a second class of circuit elements, the active elements, in which may be included sources of e.m.f. The voltage appearing at the accessible terminals of an active device is not the same in general as the e.m.f. generated. An elementary example of this is the case of a battery with internal resistance; when the battery is connected to an external load resistor R_L, as in *Figure 1.9(b)*, the terminal voltage is less than the generated e.m.f. by the drop in voltage across the internal resistance r. So, again, just as we did with passive elements above, we can represent a practical active element as an equivalent circuit made up of a combination of ideal elements.

Models very often bear no resemblance to the actual component they represent but an analysis of the model leads always to the actual functioning performance of the real component.

DECIMAL NOTATION

It is important when answering questions relating to electrical and electronic systems that you specify the value of a quantity in terms of the fundamental SI units. For this purpose it is necessary to be familiar not only with the basic SI units but also with their multiples and sub-multiples. The following table summarizes the fundamental quantities; make sure that you distinguish between the symbol and the unit abbreviation.

Quantity	Symbol	Unit (abbreviation)
Current	I	ampere (A)
Voltage	E or V	volt (V)
Resistance	R	ohm (Ω)
Charge	Q	coulomb (C)
Energy	W	joule (J)
Power	P	watt (W)
Time	t	second (s)

The range covered by the multiples and sub-multiples is very large and a special notation has been adopted to accommodate it. We have already encountered fractions of an ampere, the milliampere and the microampere; and similarly for the volt. The next table, based on the decimal powers-of-10 system, covers the essential prefixes needed for most of our work.

Multiplier	Prefix	Abbreviation
10^{12}	tera	T
10^{9}	giga	G
10^{6}	meg or mega	M
10^{3}	kilo	k
10^{-3}	milli	m
10^{-6}	micro	μ
10^{-9}	nano	n
10^{-12}	pico	p

Thus, 20 MΩ is read '20 megohms' and is equal to 20×10^6 or 20 million ohms; 30 mA is read as '30 milliamps' and is equal to 30×10^{-3} or 30-thousandths of an ampere. This might, of course, equally be written as 0.03 A.

Sometimes combinations of prefixes are used, such as kilo-megahertz for a frequency of $10^3 \times 10^6$ hertz, but it is preferable to express this as 10^9 hertz directly and use gigahertz (GHz). In the same way the old term used for capacitance, the micro-microfarad or $10^{-6} \times 10^{-6}$ farads, is replaced by the picofarad (pF) or 10^{-12} farads.

In the worked illustrative examples used throughout this book, most have their numerical quantities expressed in the fundamental SI units before being substituted into the appropriate equations. Try always to do this as far as possible; only take short cuts when you feel you are sufficiently experienced.

SUMMARY

- Ohm's law states that at a constant temperature the potential difference between the ends of a conductor is directly proportional to the current flowing through the conductor.
- For resistances in series $R_e = R_1 + R_2 + R_3 + \ldots R_n$.
- For resistances in parallel $\dfrac{1}{R_e} = \dfrac{1}{R_1} + \dfrac{1}{R_2} + \dfrac{1}{R_3} + \ldots$.
- For N equal resistors of value $R\Omega$ wired in parallel, the equivalent resistance $R_e = \dfrac{R}{N}$.
- Energy dissipation $W = VQ = VIt$ joules.
- Power P = energy dissipated/time = $VI = I^2R$ watts.

- Resistance is a measure of the ability of a device to dissipate energy irreversibly.
- Capacitance is a measure of the ability of a device to store energy in the form of separated charges or in the form of an electric field. The energy stored is $\frac{1}{2}CV^2$ joules.
- Inductance is a measure of the ability of a device to store energy in the form of a moving charge or in the form of a magnetic field. The energy stored is $\frac{1}{2}LI^2$ joules.

REVIEW QUESTIONS

1. Distinguish between electromotive force and potential difference. What are these quantities measured in? Can the e.m.f. of a battery be measured by an ordinary voltmeter?
2. Under what circumstance would the potential difference at the terminals of a battery be equal to the electromotive force?
3. Deduce that the p.d. between two points in a circuit is 1 volt if 1 joule of work is expended in moving 1 coulomb of charge between the points.
4. What is the resistance of a circuit in which a current of 1 ampere generates heat at the rate of 1 joule per second.
5. What quantities are measured in (a) joules, (b) watts, (c) amperes, (d) coulombs? Give a definition of each of these quantities.
6. What is the difference, if any, between electrical power and electrical energy?
7. For two resistors connected in parallel, show that the equivalent resistance can be expressed as

   $$\frac{\text{product of the resistances}}{\text{sum of the resistances}}$$

8. Explain the difference between a signal which travels along a cable at a speed approaching that of light and an electric charge that drifts along at only a few metres per second.
9. How can a current flow through a capacitor which has an insulator between the two plates making it up?
10. What is the significance of the negative sign appearing before the expression for induced e.m.f.?

EXERCISES AND PROBLEMS

The following problems and exercises do not only deal with the brief revisional work of this chapter but also with general basic calculations.

(1) What quantities are measured in (a) joules, (b) watts, (c) amperes, (d) coulombs? Give a definition for each of these.

(2) What is the resistance of a circuit in which a current of 1 ampere generates heat at the rate of 1 joule per second?

(3) Deduce that the p.d. between two points in a circuit is 1 volt if 1 joule of work is expended in moving 1 coulomb of charge between the points.

(4) A 12 V car light is rated at 25 W. What total charge flows through the filament in 1 minute? How many electronic charges does this involve? What is the resistance of the hot filament? Is the cold resistance greater or less than the hot resistance?

(5) A battery delivers a current of 2 A when a 1 Ω resistor is wired across its terminals. When a 3 Ω resistor is used, the current falls to 0.71 A. What is the e.m.f. and internal resistance of the battery?

(6) Find the equivalent resistance of the circuit shown in *Figure 1.10*. Find also the currents in each resistor.

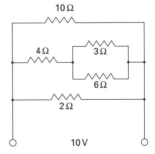

Figure 1.10

(7) Two resistors connected in parallel draw a total current of 3 A from a 100 V supply. The power dissipated in one of the resistors is 120 W. What power is dissipated in the other resistor

and what are the values of the individual resistances?

(8)   A model electric motor draws 100 mA at 6 V. If it is to be operated from a 9 V source, what value of resistance would you connect in series with it?

(9)   Two similar coils each of resistance 100 Ω are wired in series across a 20 V supply. What resistance must be connected across one of the coils to reduce the voltage across it to 8 V.

(10)   Two 100 V lamps are rated at 60 W and 200 W, respectively. They are connected in series across a 200 V supply. What current will flow through the lamps? What resistance would you place in parallel with the 60 W lamp so that each lamp gets its proper current at the proper voltage? What would be the minimum power rating for this resistor?

(11)   A potential divider of resistance 220 Ω is connected across a 100 V supply, see *Figure 1.11*. A voltmeter of internal resistance 1000 Ω is connected to the slider of the potentiometer as indicated. Plot a graph showing the variation of voltage as the contact is moved from point A to point B. (Take every 20 Ω step along the potentiometer.)

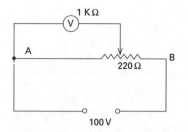

Figure 1.11

(12)   What charge is stored in a 10 μF capacitor when it is charged to a voltage of 150 V?

(13)   What will be the inductance of a coil which stores the same energy in its magnetic field as the capacitor of the previous question, when a current of 5 A flows in the coil?

(14)   A capacitor carries a charge of 0.01 C. If the energy stored is 1 J find (a) the voltage, (b) the capacitance.

(15)   Deduce, from what information you have already been given (or otherwise), that the total capacitance of a number of capacitors wired in parallel is $C_1 + C_2 + C_3 \ldots$ and of a number wired in a series is

$$\frac{1}{1/C_1 + 1/C_2 + 1/C_3} \ldots$$

2

Circuit networks and theorems

Network terminology
Ideal sources
Passive and active elements
Analysis and synthesis
Network theorems

A network may be defined as any general electrical circuit made up of generators and impedances. The generators may consist simply of batteries or other sources of direct current, or they may consist of alternators, oscillators or the outputs of electronic amplifiers. The impedances may consist simply of resistances in uncomplicated circuits, such as that sketched in *Figure 2.1(a)*, or they may consist of inductive or capacitive reactances in combination with resistive components, together with various generators, as shown in *Figure 2.1(b)*.

Don't be alarmed by the apparent complexity of this last circuit; it is simply a make-believe arrangement of bits and pieces which we will use to help us unravel the terminology associated with networks in general. We shall, in fact, restrict ourselves to networks containing only resistances, though the theo-

rems to be discussed will, in the main, apply equally well to networks containing other linear impedances.

TERMINOLOGY

There are a number of terms relating to networks which should be noted at this stage, though you are reminded that some writers define these terms in an alternative fashion. These alternatives, though consistent with what we are using here, are often the result of a pedantic approach to the subject which, at our present level of study, can be quite confusing.

Nodes   A point in a network which is common to two or more elements is called a node. For example, in *Figure 2.1(b)* the points a, b, f, g, h, j, k and m are

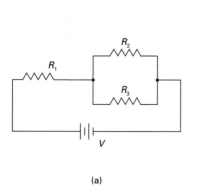

(a)

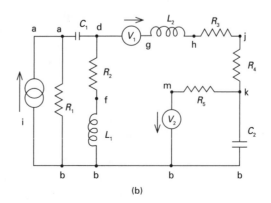

(b)

Figure 2.1

nodes. A node which is common to three or more elements is called a *junction*, and in the diagram the nodes a, b, d and k are junctions. The nodes g, h, j and m, however, are not junctions since each of them is common to two elements only.

Branch   A single element or a series of connections of elements between any two junctions is a branch of the network. In *Figure 2.1(b)* the following are examples of branches: a,b consisting of resistance R_1; a,d consisting of capacitance C_1; d,b consisting of R_2 and L_1; and d,k consisting of source V_1, L_2, R_3 and R_4. Notice that the current flowing in a branch is the same in each branch element since these elements are in series. In accordance with the definition of a branch, j,k and d,f, for instance, are not branches since they do not connect two junctions.

Loops and meshes   In some advanced work on network topology a precise differentiation is made between loops and meshes, just as there is between nodes and junctions mentioned above. But here we can use these terms in the general sense of representing the same thing; a closed path through the elements of a network. This is justified for our purposes by redrawing a network diagram so that a mesh in the old diagram can be made to become a loop in the new diagram, and conversely. We therefore talk about a loop or a mesh in connection with a specific network diagram. Strictly, a loop is a set of branches forming a closed path in a circuit such that if any branch is removed the remaining branches do not form a closed path. Two examples of loops from *Figure 2.1(b)* are a,d,f,b,a and d,g,h,j,k,m,b,f,d. A mesh on the other hand is a loop which does enclose or circle another loop and which cannot be divided into loops. In the figure, m,k,b,m is a mesh whereas the path a,d,g,h,j, k,b,a is not a mesh since it encloses the loop a,d,f,b,a. Try and find some more branches, loops and meshes from *Figure 2.1(b)*.

ANALYSIS AND SYNTHESIS

There are a number of network theorems available which enable us, being given the voltages or e.m.f.s which are applied to certain terminals in the network, to work out the voltages and currents which are established as a consequence in other parts of the network. Such calculations are known as *network analysis*. Alternatively, it may be necessary to specify a network which will exhibit desired voltages or currents

acting in certain branches within the network, and this is referred to as *network synthesis*. Whichever happens to be our concern (and we will be mainly interested in network analysis) there will be a theorem which will provide us with the approach to the problems we require.

IDEAL AND REAL SOURCES

Although every generator supplies both e.m.f. and current, it is necessary at this stage to distinguish between two general classes: those in which the voltage output is independent of the current drawn from the source, and those in which the current supplied is independent of the voltage across the source. In other words, either the voltage or the current is quite independent of the nature of the circuit to which the supply is connected. These ideal conditions are not attainable in real situations but very close approximations to them can be found or developed. A fully charged car battery, for example, will supply current up to many amperes without any appreciable change being noticed in its terminal voltage. Beyond a certain point, of course, the terminal voltage will fall because part of the chemical energy is irreversibly lost in conversion to heat, and a voltage drop occurs across the small but finite internal resistance of the battery. *Figure 2.2(a)* shows the characteristics of a real voltage generator relative to an ideal characteristic. Such an ideal source would have a perfect energy-conversion process and zero internal resistance; for the real generator we can use this idealized concept but degrade it with the addition of a small series resistance r, this internal resistance accounting for the irreversible energy losses. The symbol for this real *constant voltage* source is shown in the diagram.

The corresponding constant-current source might be represented as the collector current of a transistor, shown in *Figure 2.2(b)*, which remains sensibly constant for a range of collector voltages. The gradient of this curve is clearly indicative of a high internal resistance; for the real generator then, an ideal source having an infinite internal resistance is suggested but degraded by the parallel connection of a large but finite resistance as the diagram shows.

Figure 2.3 summarizes these classes of generators. The *ideal* voltage generator is seen in diagram (a) where the resistance seen across the output terminals (the output resistance) is zero ohms and the terminal voltage equals the generated e.m.f. for all loads R_L. The *real* voltage generator is shown at (b); here the output resistance is very small and the terminal volt-

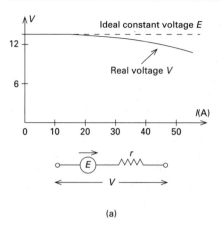

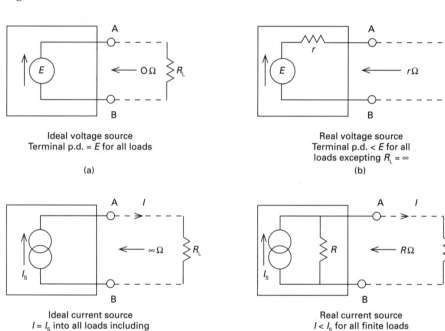

Figure 2.2

Figure 2.3

age is less than E, being equal to E only when the terminals are open-circuited. For this reason E is known as the *open-circuit voltage*.

The *ideal* current generator is shown in diagram (c); here the generated current I_s will flow into any load, including a short-circuit (for which reason it is known as the short-circuit current), and the output resistance this time is infinite. In diagram (d) we see a real *constant-current* generator where I_s will flow only into a short circuited load, the output current I being less than I_s for all other cases.

All these diagrams are known as one-port networks;

these have two terminals at which the excitation can be applied and the response measured.

This last form of generator may be a new concept to you and you may find it curious in the sense that we consider current to be the effect which follows from the cause of e.m.f. But you should keep in mind that in analysing a circuit network it is not e.m.f. or current alone which interests us but the relationship between them. So it is not necessary to treat either voltage or current as being the more fundamental quantity and the concepts of constant-voltage and constant-current generators are equally feasible. Thus, although the ideal source is not a practicable attainment in this imperfect world, the equivalent circuits are of great utility in the solution of network problems as a little experience will demonstrate.

SIGN CONVENTIONS

You will probably have already noticed that we have symbolized the flow of current in an element or elements of an electrical circuit as an arrow indicating the assumed direction of flow. This has introduced no difficulty because in assigning the direction we have taken the conventional characteristic of the flow. The same considerations have also been applied in several instances to the direction of action of circuit voltages. To establish an unambiguous base as it were, the following conventions will be followed throughout the remainder of this chapter and indeed of most of the book:

1   The head of an arrow labelling a voltage will indicate the more positive voltage, as shown in *Figure 2.4(a)*. Make a note that a voltage level of, say, -6 V, is more positive than one of -10 V.
2   A positive current flows from a more positive voltage point to a less positive voltage point, as shown in *Figure 2.4(b)*. Notice from this figure that the direction of current flow in a circuit element is opposite to the direction of the voltage across the element.

In most elementary circuits, even where a number of interconnected components are found, the directions of both flow and action are self-evident to even the most cursory inspection.

There are, however, many systems where it can prove difficult or even impossible to deduce from an inspection which way round the current will flow or the voltage act in a particular part of the system. These directions then have to be assumed, while leaving it to the algebra to sort things out for us. But we shall come to this a little later on.

NETWORK THEOREMS

Simple networks

The simple series–parallel circuit of *Figure 2.1(a)* and indeed all such circuits made up of series and parallel groups in this way, can be dealt with by first replacing all the parallel groups by their equivalent resistances. The circuit will then be reduced to a number of series resistances which can then be dealt with by elementary Ohm's law procedures, which will not only provide the total equivalent resistance of the complete circuit but also the distribution of voltages across and the currents flowing in the various branches of the network.

You may well be completely familiar with basic analysis of this sort already; if so, a little bit of additional revision may not be out of place, as these elementary procedures turn up quite regularly as 'finishing posts' after the reductions of more complicated networks by the more unusual network theorems that we will encounter in this chapter.

Here follows a little revisionary work. In a circuit made up of both series and parallel arrangements of resistive elements a logical approach to such simplification is all that is required. The following examples will illustrate each step of the implication process

> Example (1)   In the network of *Figure 2.5*, find the current I when $V = 10$ V.

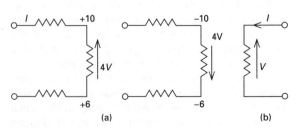

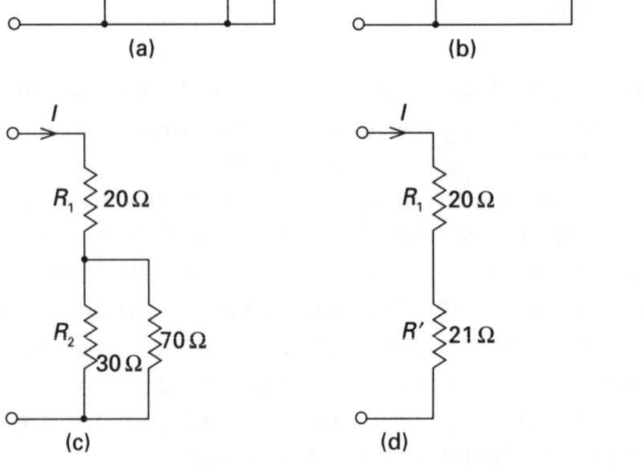

Figure 2.4

Figure 2.5

Solution   We need here to simplify the circuit into a single equivalent resistance across which the 10 V source is connected.

The general rule in cases like this is to simplify all the parallel branches first; so combining resistors R_4 and R_5 we get, using the product-over-sum rule, $R = 40 \times 60/(40 + 60) = 24\ \Omega$. We now have the circuit reduced to that of *Figure 2.5(b)*. R_3 now goes in series with 24 to give us $46 + 24 = 70\ \Omega$ and this in turn now forms a parallel arrangement with R_2 as seen in diagram (c). Combining 70 Ω with 30 Ω, therefore, we get $R' = 30 \times 70/(30 + 70) = 21\ \Omega$ and this in series with R_1 gives us diagram (d).

Finally, the 10 V supply is effectively across an equivalent resistance of 41 Ω, hence a current $I = 10/41 = 0.244$ A (or 244 mA) flows into the network.

Example (2)   In the circuit of *Figure 2.6(a)* – which is called a T-section – what value of resistor must be connected across the points C-D such that the *same* value is measured across the points A-B?

Solution   Suppose the required resistance to be R; then connecting it across points C-D leads to the circuit of diagram (b). But the resistance measured across A-B has also to be equal to R, hence we can write the equivalent resistance of the network as that derived from diagram (c) or

$$R = 40 + \frac{60(40 + R)}{60 + 40 + R}$$

using again our product-over-sum rule for the parallel branch. This reduces to

$$R(100 + R) = 40(100 + R) + 60(40 + R)$$
$$100R + R^2 = 4000 + 40R + 2400 + 60R$$
$$\therefore\quad R = \sqrt{6400} = 80\ \Omega$$

This may seem a rather curious property of a T-section network if you have not encountered them before; that the resistance connected across its 'output' can be identical to the resistance seen at the 'input'. There is only one possible resistance which leads to this result in every example of a T-section of this kind; it is known as the *characteristic resistance* of the section. T-sections are very important networks in certain electronic systems.

Now try the following problems for yourself:

(3)   Find the equivalent resistance for each of the circuits shown in *Figure 2.7*.

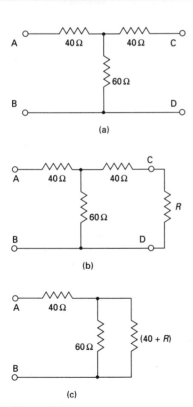

Figure 2.6

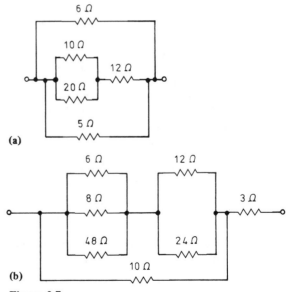

Figure 2.7

(4) In *Figure 2.8* what should be the value of resistor X for the total equivalent resistance of the network to be 18 Ω?

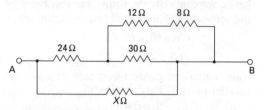

Figure 2.8

(5) A battery delivers a current of 2 A when a 1 Ω resistor is wired across its terminals. When this is replaced by a 3 Ω resistor, the current falls to 0.71 A. What is the e.m.f. and internal resistance of the battery?

Kirchhoff's laws

We have accepted a signed convention for current flow and voltage action; when we apply this to certain circuits we may find on analysis that our answers do not accord with what we have assumed, and that negative values for the currents and the voltages make their appearance. This does not mean that we have made a mistake somewhere and that our answers are incorrect; it simply implies that the true direction is the opposite to that assumed. And why do we make these assumptions anyway? Because as we noted earlier, we cannot always tell from an inspection of a network which way along a conductor a particular current is likely to flow.

It is necessary therefore to be aware of the fact that current may be positively or negatively signed in a network solution and that this will tell us whether our symbolic direction of flow is correctly marked or that the current flows contrary to that marking.

Kirchhoff's laws are extensions of Ohm's law and if you have come so far successfully, you should have no difficulties in dealing with them. The laws enable us to determine the equivalent resistance of more complex networks than we have so far encountered as well as the currents flowing in the various branches.

Kirchhoff came up with two laws: the current law and the voltage law. The current law tells us this:

At any junction in a network the total current flowing into that junction is equal to the total current flowing away from the junction.

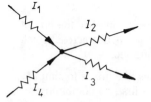

Figure 2.9

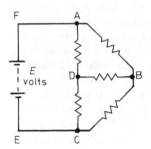

Figure 2.10

Figure 2.9 explains this. Here we have four currents acting at a junction; I_1 and I_4 are flowing towards the junction and I_2 and I_3 are flowing away from it. So by Kirchhoff's current law, $I_1 + I_4$ must be equal to $I_2 + I_3$, or what comes to the same statement

$$I_1 + I_4 - I_2 - I_3 = 0$$

That is, by assigning positive signs to currents entering the junction and negative signs to those leaving it, the law tells us that the *algebraic sum* of all the currents meeting at a point is zero. The actual signs we apply to the currents are quite arbitrary; reversing the sign order still leaves the end expression as a zero summation. Charge cannot accumulate at a junction.

We now turn to Kirchhoff's voltage law which tells us this:

In any closed loop in a network, the algebraic sum of the voltage drops taken around the loop is equal to the effective e.m.f. acting in that loop.

To illustrate this law, we will use the term loop generally, rather than introducing the term mesh in addition to this; this will avoid any ambiguities causing us confusion.

In *Figure 2.10*, for example, both routes ABDA and BCDB are closed loops. Route ABCEFA is also a closed loop. You can identify one or two more if you look carefully. The first two of these possible loops do not include the battery, so that the effective e.m.f. acting in them is zero. The third loop does contain the battery, so the effective e.m.f. acting in that loop is E volts. Kirchhoff's voltage law is simply telling us something which is intuitively obvious: that if we select any one of these loops (or any of the others that are present) and add up the *IR* voltage drops as we proceed around it, we can equate the resulting summation to the effective e.m.f. acting in the loop.

The direction in which we proceed around any loop is quite arbitrary. It is only necessary to proceed wholly in one definite direction (clockwise or anticlockwise) and to reckon (as a general rule) the *IR* products and any e.m.f. as positive when these are acting in the assumed direction of the current and negative when acting against it. The algebra will then take perfect care of everything else.

To summarize then, we apply Kirchhoff's laws to a network problem as follows:

(a) Insert current symbols, I_1, I_2, I_3 etc. (or x, y and z may be used) to represent the unknown currents whose values are asked for, assigning to them their probable directions of flow and keeping their numbers as small as you can.
(b) With these current distributions marked in, apply the voltage law to as many closed loops as there are unknown currents. A number of simultaneous equations can then be set up from which the unknown currents may be calculated.

Example (6)   Use Kirchhoff's current law to find the currents I_1, I_2 and I_3 flowing in the network of *Figure 2.11(a)*.

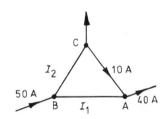

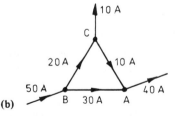

Figure 2.11

Solution   Starting at junction A we notice that 10 A is entering it along wire CA and 40 A is leaving it. Hence current I_1 must be 40 − 10 = 30 A flowing along the wire BA towards A.

Turning to junction B, we notice that 50 A is entering it and that I_1 (= 30 A) is leaving it along BA. Hence the difference, 20 A, must be flowing along wire BC towards C; call this I_2.

Finally, at junction C, 20 A is entering it along BC and 10 A is leaving it along CA. The difference must therefore be flowing out of the network from C. The full current distribution is shown in *Figure 2.11(b)*. Notice, as a check on the solution, that the total current entering the network at B, i.e. 50 A, is equal to the total current leaving it at A and C i.e. 10 + 40 = 50 A.

(7)   Find all currents, I_1, I_2 etc., flowing in the various branches of the network shown in *Figure 2.12*.

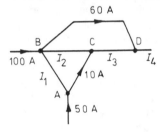

Figure 2.12

Example (8)   In the loop of *Figure 2.13*, which can be considered as part of a more complicated network, the currents x, y and z have been assigned the arbitrary directions shown. Write down an expression illustrating Kirchhoff's voltage law by working around the loop from any one of the junctions A, B or C.

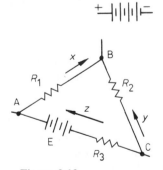

Figure 2.13

Solution   Let us proceed in a clockwise direction from point A. Going along the R_1 branch from A towards B we notice that we are proceeding along the assumed direction of current x; hence we can assign a positive sign to the voltage drop xR_1 which is developed between points A and B. From point B we proceed along the R_2 branch towards point C, noticing this time that we are going in opposition to the assumed direction of current y. Hence we assign a negative sign to the voltage drop yR_2, i.e. we write $-yR_2$. At point C we proceed along the last branch back to our starting point A. In this branch we notice that not only is there a resistance but also a source of e.m.f.

Now, in resistance R_3 our assumed direction matches that of current z, so the zR_3 voltage drop must be given a positive sign. But what about the source of e.m.f. E volts which we now encounter? Well, ask yourself in which direction this e.m.f. is acting – in the conventional sense, that is. Recall that we take the conventional flow of current to be from the positive terminal of a source, around the circuit (the loop) and back to the negative terminal. So for any source of e.m.f. which has positive and negative terminals indicated in some way, the direction in which the e.m.f. acts is from the negative to the positive, *through* the source itself.

Finally, then, we can attach a *positive* sign to the e.m.f. E as it is clearly acting in the same direction as the assumed current z. Hence by Kirchhoff's voltage law we can write an expression which tells us that the algebraic sum of the *IR* voltage drops around the loop is equal to the e.m.f. acting in the loop, that is

$$xR_1 - yR_2 + zR_3 = E$$

or

$$xR_1 - yR_2 + zR_3 - E = 0$$

You should now be able to apply Kirchhoff's laws to various circuit problems. For your guidance, three worked examples follow.

Example (9) Calculate the currents flowing in each branch of the circuit of *Figure 2.14* and determine their proper directions.

Solution It is a good plan always to mark in on a diagram all the directions in which the e.m.f.s of the sources are acting. After this, use capital letters to mark out clearly the various junction points and branches of the network.

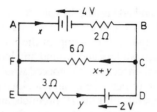

Figure 2.14

These things have been done in *Figure 2.14*. Now apply the current law to the various branches: suppose that current x flows from A to B, and that current y flows from E to D, then the sum of these currents $(x + y)$ must flow along the centre branch from C to F. Keep firmly in mind that Kirchhoff's current law must be obeyed at each junction. Notice that by adding the x and y currents in the centre branch, we saved ourselves a third unknown current z. This means that we have just two unknown currents to find instead of a possible three.

Two unknown quantities call for two independent equations for their evaluation. We obtain these by applying Kirchhoff's voltage law to any two different loops; we have available three loops from *Figure 2.14*, so choosing ABCF for starters and moving from A in a clockwise direction, we get

$$2x + 6(x + y) = -4$$
$$\therefore \quad 8x + 6y = -4$$

Notice the -4, as we have moved *against* the battery e.m.f.

Using loop FCDE and starting at E in a clockwise direction, we get

$$-6(x + y) - 3y = 2$$

or

$$-6x - 9y = 2$$

Notice again the reasons for the particular sign pattern. Solving these two equations simultaneously we find

$$x = -\frac{2}{3} \text{ A}, \, y = \frac{2}{9} \text{ A}, \, (x + y) = -\frac{4}{9} \text{ A}$$

We have negative signs attached to both current x and current $(x + y)$. This tells us that the directions we assigned to these currents were 'wrong'; x actually flows from B to A, and $(x + y)$ flows from F to C. The magnitudes of the currents, $x = 2/3$ A and $(x + y) = 4/9$ A are, of course, perfectly correct.

Example (10)   A bridge network of resistances is arranged as in *Figure 2.15*, and a battery of e.m.f. 10 V and negligible internal resistance is connected across points AC. Find the current in the centre arm of the bridge.

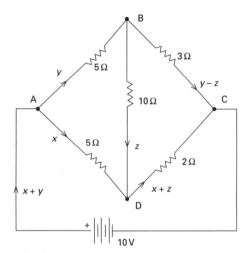

Figure 2.15

Solution   Here it is not quite so easy to make a guess as to which way the currents in the various branches will be flowing but, as the previous example revealed, what is more important than agonizing over direction is to keep the number of different currents as small as possible. In this example, a first glance might wrongly tell us that we will require five unknowns, but as the diagram shows, we have restricted ourselves to three by a careful use of Kirchhoff's current law. We shall therefore need three independent loops to form the three independent equations we need for the solution. Choosing first of all the loop ABC + battery, we have

$$5y + 3(y - z) = 10$$

or

$$8y - 3z = 10 \tag{i}$$

Similarly for loop ABDA:

$$5y + 10z - 5x = 0$$

or

$$5x - 5y - 10z = 0 \tag{ii}$$

And for loop BDCB:

$$10z + 2(x + z) - 3(y - z) = 0$$

or

$$2x - 3y + 15z = 0 \tag{iii}$$

Multiplying (ii) by 2 and (iii) by 5 and subtracting gives us

$$5y - 95z = 0$$

whence

$$y = 19z$$

Substituting this value back into (i) gives us

$$8(19z) - 3z = 10$$
$$149z = 10$$

from which

$$z = \frac{10}{149} = 0.067 \text{ A} \quad \text{or} \quad 67 \text{ mA}$$

You may, of course, use determinants for the solution of simultaneous equations if you prefer that method.

Example (11)   In the circuit of *Figure 2.16(a)* the ammeter reads 0.5 A. What will it read if the shunting resistor is changed from 2 Ω to 1.5 Ω?

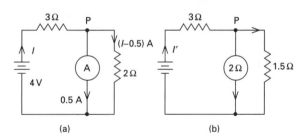

Figure 2.16

Solution   There is no problem here about the direction of current flow, so we mark in the appropriate arrows and apply Kirchhoff's current law to the junction point P; current I leaves the battery, flows through the 3 Ω resistor and divides into 0.5 A through the meter and (I − 0.5) A through the 2 Ω shunt. Now the p.d. across the meter is the same as that across the

2 Ω shunt, that is, $2(I - 0.5)$ V. But this must be equal to the applied voltage less the drop across the 3 Ω resistor. Hence

$$2(I - 0.5) = 4 - 3I$$

from which

$$I = 1 \text{ A}$$

From this we clearly see that the resistance of the ammeter itself is the same as that of the shunt, i.e. 2 Ω since equal currents flow through each, see *Figure 2.16(b)*.

When the shunt is changed to 1.5 Ω, the circuit becomes that seen in *Figure 2.16(b)*. The equivalent resistance of meter and shunt is then $2 \times 1.5/(2 + 1.5) = 3/3.5$ or 0.857 Ω. Hence, the new battery current $I' = 4/3.857$ or 1.037 A.

The p.d. across the meter is then $4 - (3 \times 1.037) = 0.89$ V and the meter current (and reading) will be $0.89/2$ or 0.445 A.

Given the current 1.037 A flowing into point P we could of course have found the ammeter current by the ratio of division method. Try this for yourself.

Now try the following problems on your own.

(12)   Two batteries A and B are connected in parallel (like poles together) across a 10 Ω resistor. Battery A has e.m.f. = 6 V and internal resistance 2 Ω; battery B has e.m.f. = 2 V and internal resistance 1.5 Ω. Calculate (a) the current flowing through each battery; (b) the terminal p.d. of each battery; (c) the power dissipated in the 10 Ω resistor.

(13)   In the circuit system of *Figure 2.17* the meter G is used to measure a resistance R Ω.

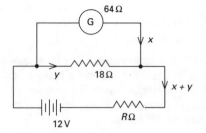

Figure 2.17

Using Kirchhoff's laws, the currents x and y amperes through the meter and the 18 Ω shunt resistor are given by the equations

$$64x - 18y = 0$$
$$18y + R(x + y) = 12$$

(a)   If $R = 84$ Ω, calculate the current through the meter.
(b)   If 50 mA gives a full-scale meter reading, calculate the lowest value of resistance that can be measured with this circuit arrangement.

Let us return now for a moment to the potentiometer or potential divider which was discussed in Chapter 1. To obtain a variable voltage V_2 from a fixed voltage V_1, a potentiometer is used with an adjustable sliding contact, point S, see *Figure 2.18(a)*. The position of S determines the ratio of the resistances R_1 and R_2 and hence the level of voltage V_2 relative to V_1. V_2 will, however, only be in direct proportion to the ratio $R_2/(R_1 + R_2)$ if no current is supplied at the point P to an external load. When such a current is supplied, problems of voltage division can be solved by a simple application of Kirchhoff's current law at the junction point.

The next worked example will illustrate this.

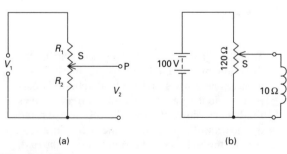
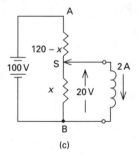

(a)   (b)   (c)

Figure 2.18

Example (14) The potentiometer of *Figure 2.18(b)* has a total resistance of 120 Ω and is connected across a 100 V supply. Find the position of the slider S so that a current of 2 A will flow through the 10 Ω load coil.

Solution The first step is to redraw the circuit and fill in all the given information, together with any unknown quantities. This has been done in *Figure 2.18(c)*. When a current of 2 A flows in the 10 Ω coil, the voltage drop across it must be 20 V, and this must also be the voltage present between the points S-B of the potentiometer. Hence the voltage across the points A-S must be $100 - 20 = 80$ V. As we are asked to find the position of the slider, we call the lower resistance portion of the potentiometer x and the remaining upper portion is then $(120 - x)$. Now applying Kirchhoff's current law to the slider junction point, we have

(a) Current entering the junction $= \dfrac{80}{120 - x}$

(b) Current leaving the junction $= \dfrac{20}{x} + 2$

$$= \frac{20 + 2x}{x}$$

These currents must be equal, so

$$\frac{80}{120 - x} = \frac{20 + 2x}{x}$$

Cross-multiplying

$$80x = (120 - x)(20 + 2x)$$

$$= 2400 + 220x - 2x^2$$

Dividing by 2 and rearranging we get

$$x^2 - 70x - 1200 = 0$$

and solving this gives $x = 84.24$ (taking the positive root).

Hence the slider position has to be 84.24 Ω along the track from its lower end point B.

In most questions involving Kirchhoff you will not be provided with a circuit diagram, so it is necessary to give yourself plenty of practice in interpreting the words of a problem into a clearly labelled diagram.

Try this out in the next two self-assessment problems.

(15) A potentiometer of total resistance 24 Ω is wired across a 50 V supply. What will be the current through a 10 Ω load when the sliding contact is exactly half-way along the potentiometer track?

(16) A potentiometer of 1000 Ω total track resistance is connected across a 25 V supply. It is required to supply a current of 10 mA to a 500 Ω load. Calculate the resistance between the position of the slider and the end to which the load is connected.

We now turn our attention to two very important network theorems in circuit analysis. The first of these is named after its originator, M.L. Thévenin, and deals with active terminal pairs.

Thévenin's theorem

This theorem tells us that in any linear active network having output terminals A and B as shown in *Figure 2.19(a)*, the circuit behaves as far as *measurements at the output terminals* are concerned as though it consisted of a single ideal voltage generator of e.m.f. E volts in series with a passive impedance Z_g as shown in *Figure 2.19(b)*. This means that if a load impedance Z_L is connected to the terminals A-B of *any* network of linear sources and impedances as seen at

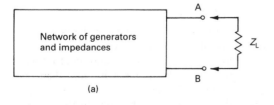

(a)

Thévenin equivalent

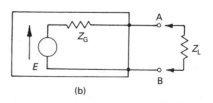

(b)

The current flowing in the load Z_L will be the same in both cases

Figure 2.19

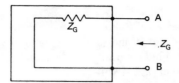

With no load connected the voltage
at terminals A and B is the same
as the generator e.m.f. E

The impedance seen at terminals
A and B with the generator
removed is equal to Z_G

Figure 2.20

(a), the current that will flow in the load will be *exactly* the same as if it was connected to the simple network shown at *(b)*.

What we have to do then is to find some way of converting the complex network into the two simple components of Thévenin's equivalent circuit. Look again at this equivalent circuit in *Figure 2.20*. If there is no load connected to terminals A-B, the voltage measured there will be equal to E since there will be no voltage drop across the internal impedance Z_G. This is then the open-circuit voltage E_{oc}, and E_{oc} (=E) is the voltage across A-B with the load impedance Z_L removed. If now we replace the generator with its own internal impedance (which is equal to zero, this being an ideal voltage generator) the impedance we measure between A and B must be Z_g. Let us illustrate this in a little more detail, working now with a network consisting of resistive elements, though keep in mind that the theorem applies equally well to linear impedances.

Let A and B be the output terminals of a network of resistances R_1 and R_2 and a source of e.m.f. E having an internal resistance r; this is seen in *Figure 2.21(a)*. When a load resistor R_L is connected across the terminals, what current will flow in this load?

What we have to do is find the Thévenin equivalent of the network to the left of the terminals; we do this by disconnecting the load resistor R_L and find the open-circuit voltage E_{oc} acting across the terminals. With R_L removed, the current through R_1 will clearly be $E/(R_1 + r)$ and the voltage drop across R_1 will be

$$\frac{R_1 \cdot E}{R_1 + r}$$

This will also be the *open circuit* voltage E_{oc} at the terminals, since there is no drop across R_2. We now have one of the equivalent circuit elements, namely E_{oc}, as shown in diagram *(b)*.

Now remove the source E and replace it by its internal resistance r as shown in diagram *(c)*. Looking back into the network from terminals A-B we see a resistance R_g where

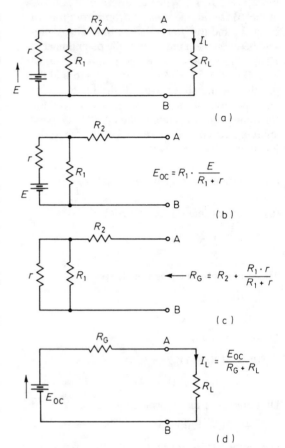

Figure 2.21

$$R_g = R_2 + \left[\frac{rR_1}{R_1 + r} \right]$$

This gives us the other component of the equivalent circuit, namely, the resistance seen at the terminals with the internal generator stopped and replaced with its internal resistance, R_g. If there is more than one generator in a network, they too can be treated separately by a theorem we shall be coming to shortly.

Hence the Thévenin equivalent of the circuit of *Figure 2.21(a)* is as shown in diagram *(d)* where the network behaves to an external load as though it were a simple series circuit made up of an ideal generator of e.m.f. E_{oc} volts and internal resistance R_g.

Now go through the following worked examples and you should then have no difficulty in dealing with the basic cases of Thévenin's theorem

Example (17)   Obtain the Thévenin equivalent circuit at terminals A-B of the network shown in *Figure 2.22*.

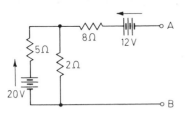

Figure 2.22

Solution   Dealing first of all with the parallel part of the circuit shown in *Figure 2.23(a)*, the open-circuit voltage at terminals C-D will be

$$E_{oc} = \frac{20 \times 2}{2 + 5} = 5.71 \text{ V}$$

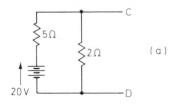

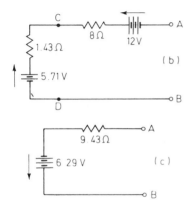

Figure 2.23

The resistance seen looking into terminals C-D with the 20 V source shorted out while leaving its internal resistance of 5 Ω in place will be

$$r_g = \frac{2 \times 5}{2 + 5} = 1.43 \text{ Ω}$$

Now this equivalent circuit can be placed in series with the remainder of the network as *Figure 2.23(b)* indicates. Clearly, this will finally reduce the circuit to that shown in diagram *(c)* which has an e.m.f. of $(12 - 5.71) = 6.29$ V and an internal resistance of $(8 + 1.43) = 9.43$ Ω. Notice that terminal B is positive in this equivalent circuit, and why.

This example illustrates the good practice of putting arrows in a diagram to indicate the direction in which batteries or other sources of e.m.f. are acting.

Example (18)   Using the circuit of *Figure 2.14* which was previously solved by Kirchhoff's laws, calculate the current through the 6 Ω resistor using Thévenin's theorem.

Solution   Referring to *Figure 2.24(a)* which repeats the previous example, we have to find the Thévenin equivalent of the remainder of the circuit by disconnecting the 6 Ω resistor (treating it as a load resistance). Let the terminals to this be A-B. Then, since the 4 V source dominates the loop, a current will flow around the loop in the direction indicated. The magnitude of this current I_1 will be the effective e.m.f. acting in the loop divided by the total resistance in the loop, that is

$$I_1 = \frac{4 - 2}{3 + 2} = \frac{2}{5} \text{ A} \quad \text{or} \quad 0.4 \text{ A}$$

The p.d. across A-B (or E_{oc}) then equals the p.d. across *either* the 3 Ω or the 2 Ω branches since these are in parallel. We can use either of these

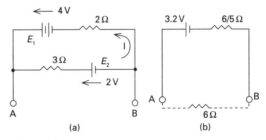

Figure 2.24

or, as a convenient check on our working, both of them.

At this point, however, pay attention to the signs appended to the two branches: in the 4 V branch the p.d. developed across the 2 Ω is opposing the e.m.f. E_1; in the 2 V branch the p.d. developed across the 3 Ω is adding to the e.m.f. E_2. This agrees with our accepted sign convention earlier on and indicates its importance. The signs in the calculation must obviously agree with this.

The open-circuit e.m.f. is therefore either

$$E_{oc} = E_1 - (I_1R_1) = 4 - (0.4 \times 2) = 3.2 \text{ V}$$

or

$$E_{oc} = E_2 - (I_1R_2) = 2 + (0.4 \times 3) = 3.2 \text{ V}$$

and these answers must, of course, agree.

With the sources now replaced by short-circuits, the resistance seen at the terminals A-B is 2 Ω in parallel with 3 Ω, giving us an equivalent 6/5 Ω and this will be in series with the open-circuit e.m.f. of 3.2 V. Hence, the equivalent Thévenin circuit is as shown in diagram (b). When the 6 Ω load is attached, the total circuit resistance becomes (6 + 6/5) or 7.2 Ω and the load current is then 3.2/7.2 = 4/9 A as in the earlier example.

Example (19)   A student is asked to do an experiment to find the e.m.f. and output resistance of a signal generator. He connects the output terminals, X and Y, of the generator to an adjustable load resistor R as shown in *Figure 2.25*, and then measures the voltage across this resistor with a voltmeter which draws negligible current. He records the relationship between the terminal voltage and the setting of the load and establishes the following table:

R (Ω)	50	100	200
V (µV)	19.2	35.7	62.5

What is the *probable* e.m.f. of the signal generator and its output (internal) resistance?

Solution   We need the Thévenin equivalent of the signal generator output circuitry. Of course the generator concerned here is not a d.c. source such as a battery but a sinusoidal oscillator, but as we have noted previously, this does not invalidate the use of Thévenin.

Figure 2.25 shows the Thévenin circuit connected to the variable load resistor R and the voltmeter at terminals X and Y. From this we have

$$V = \frac{ER}{R + r} \quad \text{or} \quad E - V = \frac{r}{R} \times V$$

From the three readings for V given in the table, we get

$$E - 19.2 = \frac{r}{50} \times 19.2 = 0.384r \qquad \text{(i)}$$

$$E - 35.7 = \frac{r}{100} \times 35.7 = 0.267r \qquad \text{(ii)}$$

$$E - 62.5 = \frac{r}{200} \times 62.5 = 0.313r \qquad \text{(iii)}$$

We have here three equations in two unknowns; strictly we need to use only two of these equations but a check is made on the experiment by using all three. Hence to find r by eliminating E:

Subtracting (ii) from (i): $16.5 = 0.027r$ or $r = 611$ Ω

Subtracting (iii) from (ii): $26.8 = 0.044r$ or $r = 609$ Ω

Subtracting (iii) from (i): $43.4 = 0.071r$ or $r = 611$ Ω

The average of these values gives us 610 Ω and substituting this into the above equations in turn gives us:

(a) $E = 19.2 + \dfrac{610}{50} \times 19.2 = 253 \text{ µV}$

(b) $E = 35.7 + \dfrac{610}{100} \times 35.7 = 253 \text{ µV}$

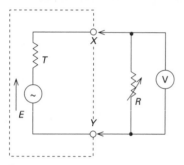

Figure 2.25

(c) $E = 62.5 + \dfrac{610}{200} \times 62.5 = 253\ \mu V$

The small variation in the value obtained for r arises from the inevitable observational errors made in recording the experimental figures. This instrument was *probably* designed to have a nominal output resistance of 600 Ω and an open-circuit e.m.f. of 250 μV.

One further example should be enough at this stage to get you fairly familiar with Thévenin's theorem.

Example (20)   In the bridge circuit of example (10) we found that the current in the centre arm of the bridge (see *Figure 2.15* again) was 67 mA, using Kirchhoff's laws. Use Thévenin's theorem to find this same current.

Solution   As the centre branch was a 10 Ω resistor and we require the current flowing through it, we treat it as the load resistance of the circuit and begin by disconnecting it. We now find the Thévenin equivalent of the remainder of the circuit. So, from *Figure 2.26(a)*, we have

p.d. between A and B $= 10 \times \dfrac{5}{5+3} = 6.25$ V

p.d. between A and D $= 10 \times \dfrac{5}{5+2} = 7.143$ V

p.d. between B and D $= 7.143 - 6.25 = 0.893$ V

point B being positive with respect to D.

Replacing the battery across A and C by its own internal resistance (zero ohms) now gives us the circuit of *Figure 2.26(b)*. The resistance of this network between B and D is then

$\dfrac{5 \times 2}{5+2} + \dfrac{5 \times 3}{5+3} + \dfrac{10}{7} + \dfrac{15}{8} = \dfrac{185}{56}$

$+\ 3.303\ \Omega$

Hence the network of *Figure 2.26(a)* is equivalent to a source having an e.m.f. of 0.893 V and internal resistance 3.303 Ω as shown in diagram *(c)*. The current through the arm B-D will therefore be

$\dfrac{0.893}{10 + 3.303} = \dfrac{0.893}{13.303} = 0.0671$ A or 67 mA

as before.

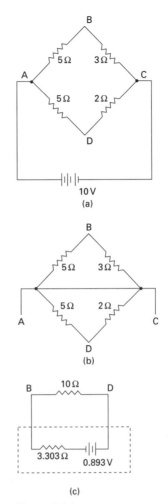

Figure 2.26

Here are two more self-assessment problems to help you master Thévenin's theorem.

(21)   Obtain the Thévenin equivalent circuits for the networks shown in *Figure 2.27*.

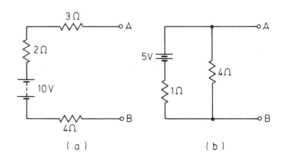

(a)   (b)

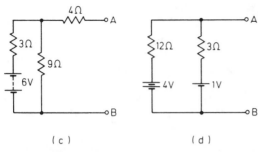

(c)

Figure 2.27

(d)

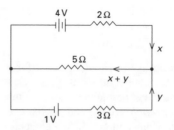

Figure 2.29

(22)   By finding Thevenin's equivalent of the circuit shown in *Figure 2.28*, determine the value of a load resistor required across terminal A-B for a current of 100 mA to flow through it.

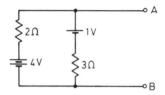

Figure 2.28

The Superposition theorem

In any network made up of linear impedances, the current flowing at any point is the sum of the currents which would flow if each generator was considered separately, all other generators then being replaced by their internal impedances. This is a statement of the Superposition theorem.

The use of this theorem permits the solution of network problems without setting up a large number of simultaneous equations as required by the direct application of Kirchhoff's laws, since only one generator at a time need be considered. This does not necessarily mean that the working will be shorter but it will in general be simpler, as each generator is reckoned to act independently of the others and its own component of current separately calculated.

Example (23)   A circuit is made up of three resistances and two sources of e.m.f. as shown in *Figure 2.29*. Calculate the current flowing in each branch of the circuit.

Solution   We have already had some practice with this kind of circuit using Kirchhoff. As for Kirchhoff we assign currents to the various branches, keeping their number as small as possible and not bothering ourselves too much about which direction the various currents will take in reality. So we assign currents x and y in the outer (battery) branches, which give us the current $(x + y)$ in the centre branch. We now proceed with the Superposition theorem.

Let the current due to the 4 V source acting alone be x_1 and let the current due to the 1 V source be y_1. Then for the 4 V source acting alone, *Figure 2.30(a)*, we have:

$$\frac{4}{[2 + (3 \times 5)/(3 + 5)]} = \frac{32}{31}$$

and of this current a proportion $3/(3 + 5) = 3/8$ flows through the 5 Ω branch.

Therefore, the current in the 5 Ω branch due to the 4 V source alone

$$= \frac{32}{31} \times \frac{3}{8} = \frac{12}{31} = 0.387 \text{ A} \qquad (i)$$

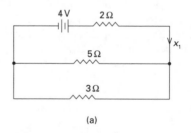

(a)

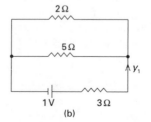

(b)

Figure 2.30

For the 1 V source alone:

$$\frac{1}{[3 + (2 \times 5)/(2 + 5)]} = \frac{7}{31} \text{ A}$$

and of this current a proportion $2/(2 + 5) = 2/7$ flows through the 5 Ω branch.

Therefore, the current in the 5 Ω branch due to the 1 V source alone

$$= \frac{7}{31} \times \frac{2}{7} = \frac{2}{31} = \frac{0}{065} \text{ A} \qquad \text{(ii)}$$

The algebraic sum of (i) and (ii) gives the required total current in the 5 Ω branch, that is, $0.387 + 0.065 = 0.452$ A.

You should now be able to show that the other branch currents are $x = 0.871$ A and $y = 0.419$ A, with the true direction of y opposite to that assumed in the diagram.

In a particularly elementary example such as this there is not a lot to choose between the Superposition and Kirchhoff and, if anything, the direct use of Kirchhoff may well be easier. Work through the above problem using both Kirchhoff and Thevenin to check that you obtain the same answer for the current in the centre branch.

When using the Superposition theorem, take careful note always of the signs (directions) of the separate currents flowing through any particular resistance; it is the *algebraic* sum we have to find in all cases. Also, and this is a general note, when *ideal* sources are removed, sources of e.m.f. are replaced by short-circuiting and ideal current sources by open-circuiting. Real sources have internal resistance and if these are indicated in the question, they must be retained in the circuit.

Example (24)   The circuit of *Figure 2.31(a)* is known as a 'summing' circuit in analogue computer systems. Use the Superposition theorem to find the voltage V across the points A-B in terms of E_1, E_2 and E_3.

Solution   Let V_1 be the p.d. across A-B when the sources E_2 and E_3 are set to zero. Then the resistance seen across A′B, diagram (b), is 0.5 Ω and hence $V_1 = 0.5E_1/(1 + 0.5) = 1/3E_1$.

From diagram (c), with sources E_1 and E_3 set to zero, the resistance across A″B is 0.4 Ω and

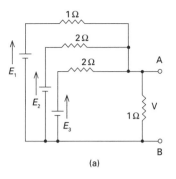

(a)

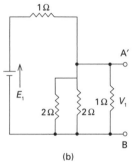

(b)

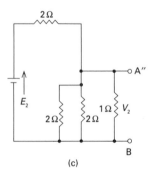

(c)

Figure 2.31

$$V_2 = \left[\frac{0.4}{2} + 0.4\right] . E_2 = \frac{1}{6}E_2$$

Similarly, we find that $V_3 = 1/6E_3$ with E_1 and E_3 set to zero.

Hence, by the Superposition theorem, the output voltage V due to the sources E_1, E_2, E_3 acting simultaneously is the algebraic sum of the outputs due to each source acting individually.

$$V = \frac{1}{3}E_1 + \frac{1}{6}E_2 + \frac{1}{6}E_3$$

As the magnitudes of these fractions are determined by the values of the resistances, it is possible to find values such that the output voltage is the true sum of the input voltages or a 'weighted' proportion of these, hence the name 'summing' circuit.

Norton's theorem

This theorem is simply a dual of Thévenin's theorem in that it enables a network to be replaced by a single generator and impedance, but in this instance the generator is of the constant-current type and the impedance is in parallel with it, whereas in Thévenin, as we have seen, the equivalent generator is of the constant-voltage type in series with the impedance. We shall here, as before, use resistive elements, but the theorem applies to all linear impedances as does Thévenin.

Norton's theorem states that the current in any impedance connected to the two output terminals of a network consisting of generators and linear impedances is the same as if a load were connected to a constant-current generator whose ideal generated current is equal to the short-circuit current measured at the terminals in question, and whose internal impedance is infinite, but being in parallel with an impedance equal to impedance of the network looking back into the terminals with all generators replaced by impedances equal to their internal impedances.

The equivalent circuits as obtained by Norton's and Thévenin's theorems give the same current in, and voltage across, any load impedance and are therefore effectively equivalent to one another. For referring to *Figure 2.32(a)*, which is the equivalent according to Thévenin, we have

$$I_L = \frac{E_{oc}}{R_G + R_L}$$

and the current on short-circuit would be

$$I_{sc} = \frac{E_{oc}}{R_G}$$

In *Figure 2.32(b)*, which is the equivalent circuit according to Norton, the load current

$$I_L = \frac{I_{sc} R_G}{R_G + R_L}$$

Combining these last two expressions, Norton's load current becomes

$$I_L = \frac{E_{oc}}{R_G + R_L}$$

which is equivalent to Thévenin's load current.

In any particular problem either theorem may be used, and the choice is usually one of convenience. One of the commonest applications of Norton's equivalent is to the parallel form of the model output circuit of a transistor.

It is important to realize, and the warning is therefore repeated here, that the equivalence afforded by both Thévenin and Norton holds for the current in the load and *not* for conditions *within* the network itself. It is the failure to recognize this limitation that incorrect and even ridiculous answers are sometimes obtained to problems when these theorems are applied to their solutions.

Since the load resistance connected to the output terminals will, in Norton, be in parallel with the internal resistance, it is often easier to treat this last element as a conductance G_G (=$1/R_G$). You will recall that current divides in a parallel circuit in direct proportion to the branch conductances.

Example (25)   Deduce the Thévenin and Norton equivalent circuits for terminals A-B in the network of *Figure 2.33*. A 10 Ω load resistor is connected across A-B. Use Thévenin's circuit to calculate the load current and Norton's circuit to calculate the load voltage. What power will be dissipated in the load?

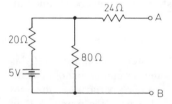

Figure 2.33

Solution   First the Thévenin: referring to the circuit given, when A-B are open-circuited, the voltage across A-B will be

$$E_{oc} = \frac{5 \times 80}{80 + 20} = 4 \text{ V}$$

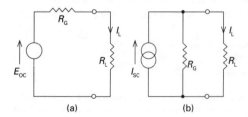

(a)   (b)

Figure 2.32

and the resistance across A-B will be

$$R_G = 24 + \frac{20 \times 80}{20 + 80} = 40\,\Omega$$

The Thévenin equivalent is then as shown in *Figure 2.34(a)*. When a 10 Ω load resistor is connected to A-B, the load current will be

$$I_L = \frac{4}{40 + 10} = 0.08\ A \quad or \quad 80\ mA$$

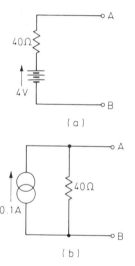

(a)

(b)

Figure 2.34

Turning now to Norton, if terminals A-B are short-circuited $I_{sc} = 4/40 = 0.1$ A or 100 mA. Hence the Norton equivalent is as shown in *Figure 2.33(b)*. If the 10 Ω load is now connected, the parallel resistances of 10 Ω and 40 Ω reduce to 8 Ω and 0.1 A, and will flow through this. Hence the voltage across the load is $V_L = 0.1 \times 8 = 0.8$ V.

Notice that this solution agrees with Thévenin's circuit conditions: there $I = 0.08$ A and this current flowing in 10 Ω will, of course, develop a p.d. of 0.8 V.

The power in the load can be evaluated by either formula $P = I^2R$ or V^2/R. Hence

$$P = \frac{0.8^2}{10} = 0.08^2 \times 10 = 0.064\ W \quad or$$

64 mW

Combining equivalent circuits

Many networks are given as, or can be reduced to, a series or a parallel combination of Thévenin and Norton equivalent circuits. It then becomes necessary to reduce such circuits to either a single Thévenin or a single Norton equivalent. What form the reduction takes depends on whether a series or a parallel arrangement is concerned. A worked example will perhaps illustrate the method best.

Example (26)   By conversion to a Norton equivalent circuit, calculate the voltage across the 10 Ω resistor shown in the circuit of *Figure 2.35*.

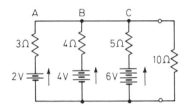

Figure 2.35

Solution   What we have here is three Thévenin equivalent circuits all in parallel. There is nothing to stop us tackling this problem by a direct approach, but by converting the branches to Norton equivalents the work is greatly eased.

For branch A, $I_{sc} = \frac{2}{3}$ A and $R = 3\,\Omega$

For branch B, $I_{sc} = 1$ A and $R = 4\,\Omega$

For branch C, $I_{sc} = \frac{6}{5}$ A and $R = 5\,\Omega$

The circuit is now transformed into three Norton equivalents which are shown in *Figure 2.36*. Here the 10 Ω load resistor remains in place. From this circuit we can see the following.

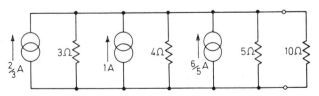

Figure 2.36

Total current $I = 2/3 + 1 + 6/5 = 43/15 = 2.87$ A and this current flows into a single equivalent resistance of 3, 4, 5 and 10 Ω in parallel, that is 1.13 Ω. The voltage across this combination is equal to the voltage across the 10 Ω load, that is $2.87 \times 1.13 = 3.24$ V.

Here are some self-assessment problems to help you master Norton's theorem.

(27)   Obtain the Norton equivalents of the circuits shown in *Figure 2.37*.

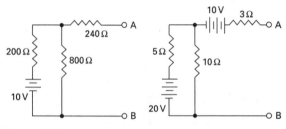

Figure 2.37

(28)   By repeated applications of Norton's theorem, or otherwise, determine the components of the equivalent Norton generator seen at the terminals A-B in *Figure 2.38*. What value of load resistance connected to these terminals will pass a current of 0.5 A?

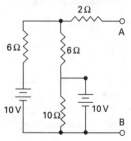

Figure 2.38

(29)   Using any appropriate network theorem or theorems, find the Norton equivalent for the circuit given in *Figure 2.39*. Hence find the current in the 3 Ω resistor. (Hint: treat the 3 Ω resistor as the load resistance.)

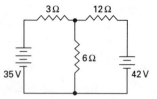

Figure 2.39

Three-terminal networks

When a network has three resistive elements, these may be arranged either in the form of a T-network (which we have already encountered) or a π-section, each of which is illustrated in *Figure 2.40(a)* and *(b)*, respectively. We shall show that it is possible to replace a T-section by a π-section and conversely. Since the T-section may be redrawn as a star, *Figure 2.40(c)* and the π-section as a mesh or delta, the conversion is generally known as the *star–delta* transformation.

Such conversions are often of considerable value in simplifying complex networks.

If a T (or star) network is to be equivalent to a π (or delta) network, the resistance seen between any pair of the terminals of *Figure 2.40(a)* must be the same as that measured between the same pair of terminals of *Figure 2.40(b)*. Thus, if we consider the terminals 1 and 3 for the two sections and equate them, we get

$$R_1 + R_2 = \frac{R_c(R_a + R_b)}{R_a + R_b + R_c}$$

Similarly, equating resistances for terminals 3 and 4 we find

$$R_2 + R_3 = \frac{R_a(R_b + R_c)}{R_a + R_b + R_c}$$

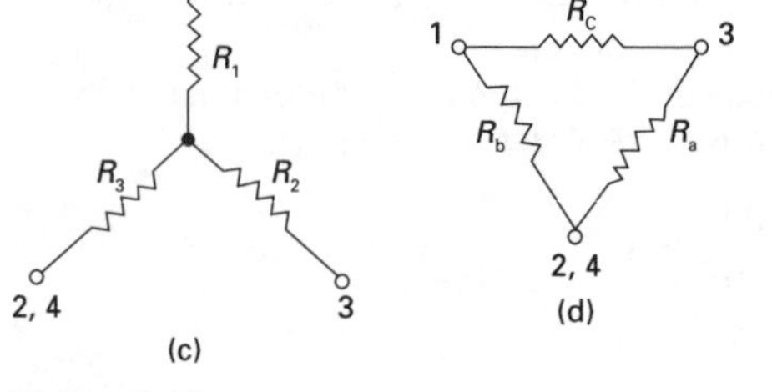

Figure 2.40

and for terminals 1 and 2

$$R_3 + R_1 = \frac{R_b(R_a + R_c)}{R_a + R_b + R_c}$$

These are three equations from which R_1, R_2 and R_3 may be evaluated in terms of R_a, R_b and R_c. Doing this we obtain the results:

$$R_1 = \frac{R_b R_c}{R_a + R_b + R_c}$$

$$R_2 = \frac{R_a R_c}{R_a + R_b + R_c}$$

$$R_3 = \frac{R_a R_b}{R_a + R_b + R_c}$$

There is a symmetry about these three expressions which should be noted: the letter subscripts in the numerators on the right-hand side are those of the resistances connected to the terminals corresponding to the numerical subscripts on the left-hand side. The equations may therefore be memorized so: the equivalent T (or star) resistance connected to a given terminal is equal to the two π (or delta) resistances connected to the same terminal, divided by the sum of the resistances.

Can you now deduce for yourself that the equivalent equations for the T to π transformations are:

$$R_a = R_2 + R_3 + \frac{R_2 R_3}{R_1}$$

$$R_b = R_1 + R_3 + \frac{R_1 R_3}{R_2}$$

$$R_c = R_1 + R_2 + \frac{R_1 R_2}{R_3}$$

It can be shown that any network of impedances, however complex, can be reduced to a T network of three impedances. This follows from the fact that the number of meshes in a network is reduced by one each time a π network is replaced by an equivalent T.

It must be pointed out that the conversion of a π to a T or a T to a π in the solution of a problem applies only with respect to the three terminals chosen. The voltage across element R which was obtained from a π has no meaning in the original circuit which included the π for example. Similarly, if a T is converted to a π, then the currents in the elements of the π have no meaning in the original circuit.

Example (30) In the circuit of *Figure 2.41(a)*, find the current I flowing from the battery.

Solution We will do this problem by a π to T conversion. We have two choices: we can convert either the π acb or the π cbd, so choosing acb we get the circuit of *Figure 2.41(b)*, where

$$R_1 = \frac{(8)(6)}{8 + 6 + 6} = \frac{48}{20} = 2.4\,\Omega$$

$$R_2 = \frac{(6)(6)}{8 + 6 + 6} = \frac{36}{20} = 1.8\,\Omega$$

$$R_3 = R_1 = 2.4\,\Omega$$

Hence, from diagram *(c)*

$$R_{ad} = 2.4 + \frac{2.8 \times 4.4}{2.8 + 4.4} = 4.12\,\Omega$$

Therefore the total resistance seen by the 6 V source $= 4.12 + 4 = 8.12\,\Omega$; then $I = 6/8.12 = 0.74$ A.

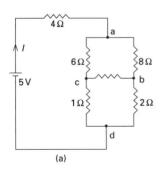

(a)

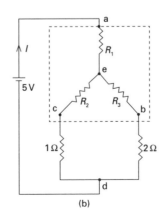

(b)

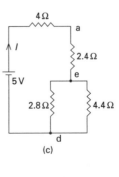

(c)

Figure 2.41

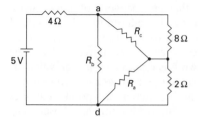

Figure 2.42

You might try this problem for yourself, making a T to π conversion. A diagram to help you in this is shown in *Figure 2.42*; now go through the rest of the procedure as necessary; your answer should be the same as that obtained above.

(31)   Transform the network of *Figure 2.43(a)* to that of *(b)* by making the appropriate π to T conversions. (Hint: replace the π network R_4, R_9, R_6, to an equivalent *T*; then replace the π network R_3, R_2, R_5, similarly. You are now almost out of the jungle!)

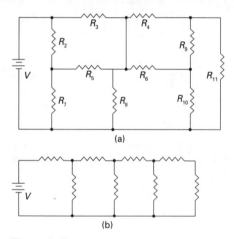

(a)

(b)

Figure 2.43

MAXIMUM POWER TRANSFER

When a generator is connected to a load, power is dissipated in the load. It is frequently necessary to ensure that the greatest possible power is dissipated in the load and not wasted in other parts of the circuit, particularly those resistances (or impedances) which are found inside the sources and so are not accessible for adjustment in the external circuit.

Since any network containing a source or sources of e.m.f. or current can be reduced to a Thévenin

equivalent, our investigation will centre on finding what value of load resistance will enable the source to deliver the greatest power to the load, restricting ourselves to those cases where the load is purely resistive.

Figure 2.44 shows a Thévenin equivalent generator with a load R.

Clearly, as R is varied, the current I will vary and hence the power dissipated in R, (I^2R) will also vary. It might appear from this that all we have to do is to maximize I, but this is an impossible answer because I can only be at its greatest when R = 0 and all the circuit power is then dissipated in the internal resistance r. As R is increased from zero, the total circuit power is distributed between R and r; this means that if R is less than r, the internal resistance is dissipating more power than is R; and if R is greater than r, the converse must be happening. But is this last statement true? Clearly, if R becomes very large, the circuit current I will be very small, and so the power dissipated in R will be very small.

It appears then that there is some intermediate point where the dissipations in R and r will be equal and this will only occur when these resistances are equal. This is the condition for the greatest power to be delivered to an external load; we can therefore state this very important rule: the maximum power in a load is obtained when the *load resistance* is equal to the *internal resistance* of the generator. If you are familiar with elementary calculus, the following proof will satisfy you on this point.

In a circuit of e.m.f. E volts, internal resistance r and load resistance R we have

$$I = \frac{E}{R} + r$$

Then the power dissipated in the load is

$$P = I^2R = \frac{E^2 \cdot R}{(R + r)^2}$$

Differentiating this expression gives us

$$\frac{dP}{dR} = E^2 \frac{\left[(R + r)^2 - 2R(R + r)\right]}{(R + r)^4}$$

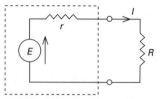

Figure 2.44

For maximum power in R, this expression will be zero, hence the numerator must be zero:

$$(R + r)^2 = 2R(R + r)$$

or

$$R + r = 2R$$

Therefore

$$R = r$$

This result will turn up in various guises throughout the remainder of this course.

MATCHING

Under the above condition for maximum power transfer from a source to a load, the efficiency of the process is 50 per cent since half the power available is dissipated (and so wasted) in the internal resistance. The process of adjusting a load to be equal to the internal resistance of a source is known as 'matching'.

Such a matching of resistances (or impedances in an a.c. system as we shall see) is not always possible and, indeed, is not desirable in many applications. No electrical appliance, for instance an electric iron, is designed to draw maximum power from a mains point; the socket, in fact, has an extremely low output impedance while the iron or whatever is simply designed to consume a definite amount of power, say 2 kW. Increasing the socket impedance to match that of the iron would only reduce the power to the appliance.

Matching resistances or impedances to source resistance or impedance is much more important in electronic instrumentation, and devices such as attenuator systems are carefully designed to ensure that the optimum loading is placed on the source output terminals whatever the setting of the attenuator.

SUMMARY

- The algebraic sum of the currents meeting at a point (node) is zero.
- The algebraic sum of the voltages acting around a closed loop or mesh is zero.
- The procedure for setting up equations for networks is:

    (a) Arbitrarily assume a consistent set of currents and voltages.
    (b) Write as many equations as there are unknown quantities by applying Kirchhoff's laws.
    (c) The resulting simultaneous equations may be solved by successive substitution or by determinants.

- Insofar as the load element is concerned, any one-port network of resistive elements and energy sources can be replaced by an ideal voltage source E_{oc} and a series resistance r (Thévenin's theorem); or by an ideal current source I_{sc} and a parallel resistance R (Norton's theorem).
- If cause and effect are linearly related, the Superposition theorem states that the total effect of several causes acting simultaneously is equal to the algebraic sum of the effects of the individual causes acting at one time.
- Maximum power is transferred to an external load resistance R when R is made equal to the internal resistance r of the source.

REVIEW QUESTIONS

1 Distinguish between 'resistor' and 'resistance'.
2 Distinguish between an ideal-voltage source and a constant-voltage source.
3 Distinguish between an ideal-current source and a constant-current source.
4 Why cannot charge accumulate at a network junction?
5 Why is a sign convention necessary when applying network theorems to circuit systems?
6 Use simple diagrams to show how Thévenin and Norton equivalent components are related.
7 Is the same power dissipated in an active circuit and the Thévenin equivalent?
8 In what way are current and voltage sources 'replaced' in network theory?

EXERCISES AND PROBLEMS

(32) In the circuit of *Figure 2.45* evaluate the currents I_1, I_2 and I_3. What powers are dissipated in the resistances?

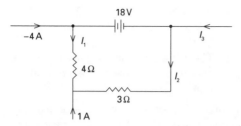

Figure 2.45

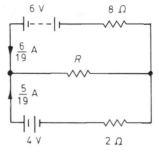

Figure 2.48

(33) Show that the currents I_1 and I_2 in *Figure 2.46* are related by the equations

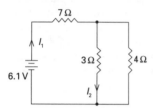

Figure 2.46

$$11I_1 - 4I_2 = 7I_1 + 3I_2 = 6.1$$

Hence calculate these currents. Convert the circuit into its Thévenin equivalent and verify your solution for I_2 by this method.

(34) In *Figure 2.47* find (a) the current in the 10 Ω resistor; (b) replace the circuit by a simple equivalent so that the current in the 1 Ω resistor can be determined by a *single* calculation, and find this current.

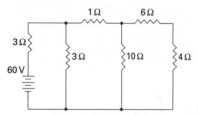

Figure 2.47

(35) In *Figure 2.48*, find the value of the load resistor R.

(36) The potential divider shown in *Figure 2.49* has a total resistance of 24 Ω and is connected to a 110 V source. What current will flow in the 40 Ω coil when the sliding contact is set at a point equivalent to 6 Ω from end B?

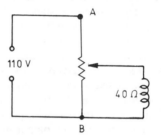

Figure 2.49

(37) Use (a) the Superposition theorem, (b) Kirchhoff's laws, to find the current in the 5 Ω resistor of *Figure 2.50*.

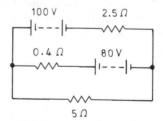

Figure 2.50

(38) Two batteries of e.m.f. 2 V and 29 V and internal resistances of 4 Ω and 3 Ω, respectively, are connected in parallel with like poles together, and a load resistance of 7 Ω is wired across the supply. Use Thévenin's theorem to find the current in the load.

(39) Two batteries of e.m.f. 6 V and 10 V have internal resistances of 0.5 Ω and 1 Ω, respectively. They are connected in parallel, like poles together, and supply current to a 9 Ω load. Find (a) the current supplied by each battery, (b) the value of a resistance which when paralleled with the load reduces the current in the 6 V battery to zero.

(40) A source is connected to an adjustable load resistor and readings are recorded of the load current I as the resistance R is varied. The results obtained are

R (Ω)	200	400	700	1200
I (mA)	12	10	8	6

Determine the Thévenin and Norton equivalent circuit for this source and hence find the maximum power available from this source.

(41) The load voltage from a power unit changes from 12.25 V to 12.23 V when the current increases from 1.55 A to 2.37 A. What is the output resistance of this unit? What would you expect the load voltage to be if the current drawn was 3 A?

(42) A signal generator has its output terminals connected to a resistor in parallel with a voltmeter which can be assumed to draw no current. The indication of the voltmeter as the resistance is varied is given in the table below. Determine the e.m.f. of the generator and its output resistance. What is the maximum power output available from this generator?

R (Ω)	25	50	100
V (μV)	33	50	67

(43) In a network the mesh currents x, y and z amperes are related by the equations

$$10y + 25(y + z) = 2$$
$$9x + 10x - 10y = 0$$
$$9x + 24(x - z) = 2$$

Calculate current x to the nearest milliampere.
  Deduce the circuit configuration responsible for the above Kirchhoff equations, showing the appropriate resistance values. (Hint: try a bridge circuit.)

(44) In *Figure 2.51* the p.d. between points A and D is 24 V and between C and B is 20 V. Find the e.m.f.s E_1 and E_2.

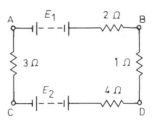

Figure 2.51

(45) In the bridge circuit of *Figure 2.52* show that the p.d. between the points B-C can be expressed as

$$\frac{(R_2R_3 - R_1R_4)}{(R_1 + R_3)(R_2 + R_4)} \text{ V}$$

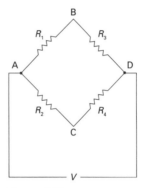

Figure 2.52

(46) In the bridge of the previous problem, let $R_1 = 3\ \Omega$, $R_2 = 1\ \Omega$, $R_3 = 6\ \Omega$ and $R_4 = 4\ \Omega$. A 10 Ω resistor is now connected between points B and C. Using the π–T transformation, or otherwise, find the resistance between points A and D.

(47) Show that the current in the 5 Ω branch of the circuit of *Figure 2.53* is 26 mA. Now,

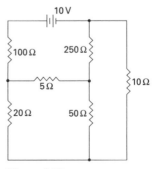

Figure 2.53

interchanging the battery and the 5 Ω resistor again find the current in the resistor. What do you notice about your answer? Can you, by taking a general linear network, demonstrate that this result is true in all cases? (This example illustrates what is known as the Reciprocity theorem.)

3

D.C. transients

The d.c. step function
Transient and steady state conditions
The exponential function
Time constant
Timing systems

THE D.C. STEP FUNCTION

When a circuit is switched from one condition to another either by a change in the applied voltage or a change in one of the circuit elements, there is a transitional period which may be extremely very short or long during which the circuit currents and voltages are adjusting themselves from their former values to new ones. After this *transitional* period, the circuit settles down to what is known as the *steady state*, and remains in this state unless there is a further change in the circuit conditions.

In this chapter we will consider what happens when we apply abrupt voltage changes to simple circuits containing resistive and reactive elements, in series or parallel configurations. Such voltage changes are known as *transients*, literally meaning a sudden transition from one particular voltage state to another.

The simplest example is that of a voltage step. This is a function which has zero magnitude prior to a time $t = 0$; it then rises instantaneously to some finite level and remains at that level (the steady state) for an indefinite period thereafter. *Figure 3.1* shows such a

voltage step function. We can clearly obtain such a voltage step from a simple battery and switch circuit as the figure shows. Before the switch is closed, the potential across terminals A and B is zero. As soon as the switch is closed the potential across A and B rises very rapidly to V volts.

The operative word in our definition of a step function is 'instantaneously', a purely theoretical concept. We cannot obtain an *instantaneous* rise in practice since this would imply that the voltage was at two different levels at the same instant, but we can, nevertheless, treat the output at the terminals as being a very good approximation to an ideal step function. Using sophisticated electronic systems, both voltage and current step functions can be generated which represent the almost perfect condition of an instantaneous rise.

FUNDAMENTAL CONSIDERATIONS

As we will now be considering voltages and currents which are no longer steady but changing during the periods of interest – the transient periods – we will use the small case symbols v, i and q to indicate that these are measures of voltage, current and quantity respectively at a particular instant of time; and similarly for other varying quantities. For constant quantities, capitals will be retained.

The purely resistive circuits dealt with so far have all had one characteristic in common: the cause and effect of all of them were time coincident. The

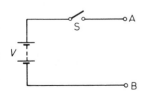

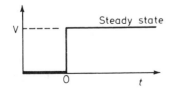

Figure 3.1

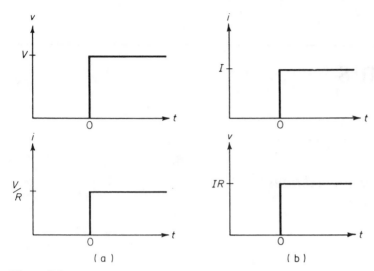

Figure 3.2

application of a voltage immediately produced a current, and a flow of current immediately generated a potential difference across the resistive elements. For a purely resistive element, current is proportional to voltage, so at any instant of time $v = iR$. Suppose a step function to be applied to a resistor; then the voltage and current waveforms will be as illustrated in *Figure 3.2* at *(a)* and *(b)*, respectively. The waveforms for current and voltage in a resistive circuit will always be *identical* with the applied voltage, the relative amplitudes being determined only by the value of the resistance.

What will happen if we apply the step function to a capacitor? As we have noted in the previous chapter, power is the time rate of change of energy, and infinite power would be necessary to cause an instantaneous change in the energy of a system.

The energy stored in a charged capacitor, $W = \frac{1}{2}CV^2$ joules, thus implies that the voltage across the capacitor cannot change instantaneously. For the same reason, the voltage across a capacitor connected to a step function cannot instantaneously rise to the level of the function.

Let a small change dQ due to a current i flowing for a time dt produce a small change of voltage dv. Now

$$dQ = C \cdot dv = i \cdot dt$$

Therefore

$$i = C \cdot \frac{dv}{dt}$$

When a step function is applied to a capacitor, the rate of change of the voltage with respect to time,

dv/dt, is infinite since we *assume* that the voltage changes from zero to V volts instantaneously. Hence the current flow will be infinite for zero time as shown in *Figure 3.3(a)*. For the current step of *Figure 3.3(b)*, consider the instant the step is applied to the capacitor; then $v = 0$ and $i = I$ and this current thereafter remains constant. Then at time t_0

$$I = C \frac{dv}{dt_0}$$

$$\frac{dv}{dt_0} = \frac{I}{C} \text{ V/s}$$

Hence, since I is constant, the capacitor voltage will increase linearly at a rate of I/C volts per second. This can be illustrated by a ramp waveform as shown in *Figure 3.3(b)*.

These two cases are, of course, ideal and theoretical. No circuit element confirms exactly and under all conditions to the ideal concepts of resistance and capacitance, though some may do so closely enough for practical purposes within a more or less restricted range of circumstances. We can, however, make a note at this point, that a capacitor which is charged from a source of *constant current* will charge linearly. This fact has a number of practical applications.

In general terms, circuits exhibit behaviour which is the result of a mixture of ideal elements (or models, as we have seen) and two of these will form our basic investigation into the transient behaviour of (a) a series arrangement of resistance and capacitance, (b) a series arrangement of resistance and inductance.

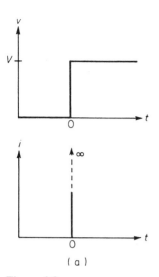

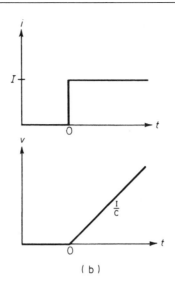

(a)

(b)

Figure 3.3

Example (1)   A 5 µF capacitor in series with a 1 MΩ resistor is charged from a d.c. source which provides a constant current of 0.5 mA. Show by means of a graph, for the voltage range 0–100 V, how the voltage across the capacitor is related to the charging time.

Solution   If the constant current of 0.5 mA flows for t seconds, then $q = 0.5 \times 10^{-3}t$ coulombs. Now

$$v_c = \frac{it}{C} = \frac{0.5 \times 10^{-3}}{5 \times 10^{-6}}t$$

$$= 100t$$

A graph of V_c against t is therefore linear with a gradient or rise rate of 100 V/s. *Figure 3.4* shows the curve.

THE SERIES *C–R* CIRCUIT

We will now investigate what happens when a voltage step is applied to a series arrangement of resistance R and capacitance C. For this purpose it is sufficient to take the step function as being generated by a simple battery and switch arrangement, this being shown in *Figure 3.5*. We understand that there is no residual charge on the capacitor before the switch is operated.

When S is closed the voltage across C does not immediately rise to the voltage level V since a movement of electric charge is necessary and the expression for charge, $Q = It$, tells us that the capacitor needs time to charge. Immediately after switch-on at $t = 0$ displacement current begins to flow around the circuit. Electrons move into the plate of the capacitor connected to the negative pole of the battery and flow out of the plate connected to the positive pole, a process which leads to the capacitor acquiring charge and hence a rise in voltage across its terminals.

The only opposition to the flow of current at the instant the switch is closed is that represented by

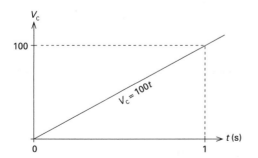

Figure 3.4

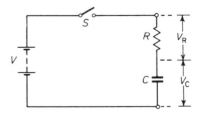

Figure 3.5

resistor R. So the instantaneous value of the current at the moment of switch-on is simply that given by the Ohm's law value of V/R. The capacitor at that instant has no charge and hence no voltage across its terminals; since the sum of v_c and v_r must at all times equal V, and v_c is instantaneously zero, all the applied voltage must appear instantaneously across R. We will indicate this initial value of the current by I_0.

This initial current does not continue unchanged as it would do if only the resistor was present. C begins to charge immediately and the voltage across its terminals rises accordingly. The voltage across R correspondingly falls by an equal amount, so that for a given value of R, $v_r = V - v_c$. Consequently at any particular time instant after switch-on, the charging current will be *less* than I_0 and will be expressed by

$$i = \frac{V - v_c}{R} = \frac{v_r}{R}$$

So, as the voltage across C rises, both the circuit current and the rate of charging falls. Hence C charges progressively more slowly as time passes. Looked at simply, as an increasing number of electrons pack into the negative-connected plate and an equal number flow out of the positive-connected plate, there is an increasing force of repulsion tending to keep out those wanting to enter the negative plate and an increasing force of attraction tending to prevent others leaving the positive plate. The process is rather like packing people into a hall. When the doors are first opened there is a comparatively rapid entry rate; as the hall fills the flow slackens, and when the hall is filled the flow stops altogether.

Likewise, with the passage of time, the capacitor must eventually 'fill up'. Its terminal voltage v_c will then be equal to V, so that v_r will be zero and the circuit current will be zero. Throughout the entire charging cycle, therefore, we can make the following observations:

At switch-on ($t = 0$):

$$v_c = 0, \; v_r = V \text{ and } I_0 = \frac{V}{R}$$

At some instant during the charge:

$$v_c > 0, \; v_r = V - v_c \text{ and } i = \frac{V - V_c}{R}$$

At the completion of the charge:

$$v_c = V, \; v_r = 0 \text{ and } i = 0$$

So the capacitor voltage has risen from zero to V volts, the resistance or voltage and circuit current have

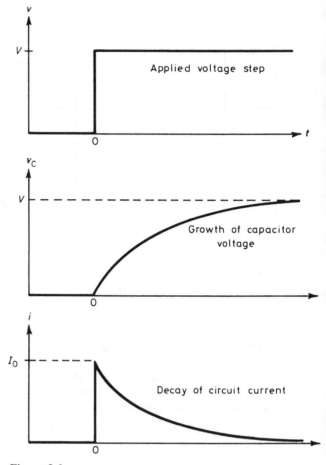

Figure 3.6

fallen from V and I_0, respectively, to zero. The kind of variation we can expect in the capacitor voltage and the circuit current after switch-on at time $t = 0$ will therefore be as shown in *Figure 3.6*. Both of the curves for v_c and i have the same mathematical form and we shall derive this a little later on. For the present we can call them the curves of exponential growth (for v_c) and decay (for i) and keep their general appearance in mind.

Follow this next worked example carefully.

Example (2)   A capacitor is connected in series with a 10 kΩ resistor and switched suddenly across a 200 V d.c. supply. What is the initial current and the voltage across C when the circuit current is 5 mA?

Solution   At switch-on, the initial current $I_0 =$ $V/R = 200/10\,000$ A $= 20$ mA.

This current decays towards zero as the capacitor charges; when it has fallen to 5 mA, the voltage across R at that instant will be

$$v_r = iR = 5 \times 10^{-3} \times 10\,000 = 50 \text{ V}$$

But at any instant

$$v_c = V - v_r$$

Therefore

$$v_c = 200 - 50 = 150 \text{ V}$$

Example (3) A fully charged capacitor is discharged through a resistor. The voltage across the capacitor decays from 200 V to zero in 5 s, and during this time the average current is 2 mA. What is the value of the capacitor?

Solution As we have seen, the discharge curve is non-linear, but if we assume an average discharge current, this is tantamount to saying that we *are* considering a linear fall in the voltage. Then for $I = 10^{-3}$ A, $V = 200$ V and $t = 5$ s we have

$$C = \frac{It}{V} = \frac{2 \times 10^{-3} \times 5}{200} = \frac{10^{-2}}{200} \text{ F}$$

$$= \frac{10^{-2}}{200 \times 10^6} = 50 \text{ }\mu\text{F}$$

This example raises an important point: although, as we can see from *Figure 3.7*, the discharge curves, whether considered as exponential *or* linear (average), have different shapes, the stored charge lost when the capacitor is discharged completely is *identical* in each case. The same is true for the charge *stored* as a capacitor is charged through a resistor.

Try the next two examples on your own:

Example (4) When a capacitor and series resistor are connected suddenly to a 50 V d.c. supply, the initial current is 100 mA. Calculate the value of the resistor and the current at the instant when the voltage across C is 20 V.

Example (5) A voltmeter whose resistance is 20 kΩ is connected in series with a capacitor and a 100 V battery. What will the voltmeter read (a) at the instant of switch-on, (b) at the instant the current is 2 mA, (c) when the capacitor is fully charged?

TIME CONSTANT

The curve of *Figure 3.8* represents the rise of voltage across the capacitor. We notice that the rate of increase of voltage is greatest at the commencement of the charge and becomes progressively smaller as the charge proceeds. This rate is indicated by the degree of steepness of the curve at any point; and this, in the usual way, is measured as the gradient of the tangent drawn to the curve at the point in question. In mathematical form the gradient is dv_c/dt, where dv_c is the small increment in the capacitor voltage over a small increment of time dt. (Note, strictly this symbolism is for the limit of the ratio $\delta v_c/\delta t$ where the changes are infinitesimally small.)

As before

$$dQ = Cdv_c = idt$$

and so

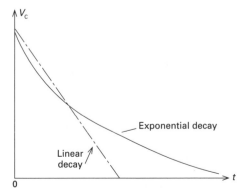

Figure 3.7

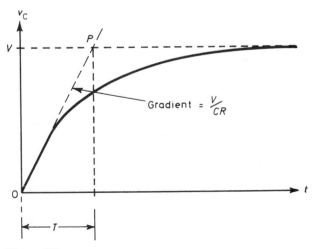

Figure 3.8

$$i = C \frac{dv_c}{dt}$$

At the instant the switch is closed, referring to *Figure 3.8*, $t = 0$ and the initial current

$$I_0 = \frac{V}{R} = C \frac{dv_c}{dt}$$

So the initial gradient of the curve is

$$\frac{dv_c}{dt_0} = \frac{V}{CR} \text{ V/s}$$

If the charge were to continue at this rate indefinitely along the broken line OP in *Figure 3.8*, the time that would elapse before v_c reached its final value V would be T seconds. But from the diagram

$$T = \frac{\text{voltage}}{\text{rate on increase of voltage}} = V \div \frac{V}{CR}$$

$$\therefore \quad T = CR \text{ sec}$$

This product is known as the time constant of the circuit and it is a very important concept.

---

(6)   Can you verify that the product CR leads to the dimension of time? (Hint: express C in terms of charge Q, and then in terms of time.)

When you use the CR product you must make sure that the correct units are employed. C must be expressed in farads and R in ohms for the answer to be in seconds. However, if C is expressed in μF, the time will be given in μs. It is often easiest to work in μF for C and MΩ for R, the answer then again being given in seconds.

Example (7)   What is the time constant of a 4 μF capacitor and a 100 kΩ resistor? Work the problem first in farads and ohms, and then in μF and MΩ.

Solution   Working in farads and ohms, we have

$$T = 4 \times 10^{-6} \times 100 \times 10^3 = 0.4 \text{ s}$$

Alternatively, working in μF and MΩ

$$T = 4 \times 0.1 = 0.4 \text{ s}$$

Example (8)   A 10 μF capacitor and a 2 MΩ resistor are switched suddenly to a 100 V supply. Find (a) the time constant, (b) the initial current, (c) the initial rate of rise of voltage across C.

Solution

(a) $T = CR = 10 \text{ } \mu\text{F} \times 2 \text{ M}\Omega = 20 \text{ s}.$

(b) $I_0 = \dfrac{V}{R} = \dfrac{100}{2 \times 10^6} \text{ A} = 0.05 \text{ mA}$

(c)   Initial rate of rise of voltage $= \dfrac{V}{CR} \text{ V/s}$

$$= \frac{100}{20} = 5 \text{ V/s}$$

Example (9)   A series circuit made up of a 50 μF capacitor and a resistance R is to have a time constant of 4 s. What should be the value of R? If the circuit is switched to a 100 V d.c. supply, find (a) the initial current, (b) the rate at which the voltage is rising when the capacitor voltage has reached 30 V.

You may perhaps have had a little difficulty with the last part of the previous problem. You have been asked to find not the initial rate of increase of capacitor voltage but the rate of increase at some later stage, when the capacitor has already acquired some of its possible charge. Let us look into this problem in a little more detail.

At any point on the charging cycle the voltage across C opposes the applied voltage, and this is a continuing process all the time the charge is taking place. Let the capacitor voltage be at some particular value v at any instant during the charge, then the current

$$i = \frac{V - v'}{R} = C \frac{dv_c}{dt}$$

and so

$$\frac{dv_c}{dt} = \frac{V - v'}{CR} \text{ V/s}$$

This is clearly less than the initial rate of increase given by V/CR. Hence the curve of capacitor voltage against time becomes progressively less steep as time goes on and will constantly approach the limiting condition of a horizontal line as $(V - v')$ approaches zero. Now, referring to *Figure 3.9* at any particular instant t_1 the capacitor has still to be charged $(V - v)$ volts. If this charging rate was then to be maintained, the time for the charge to be completed would be $(t_2 - t_1)$ seconds; then

$$t_2 - t_1 = \frac{\text{voltage available}}{\text{rate of increase of voltage}}$$

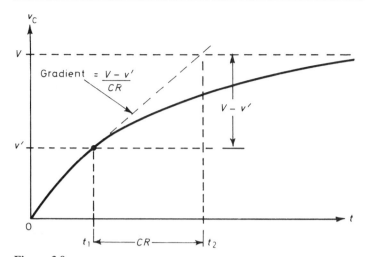

Figure 3.9

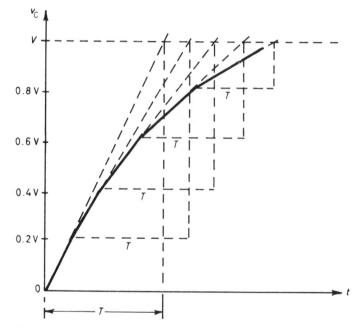

Figure 3.10

$$= \frac{V - v'}{[V - v'/CR]} = CR \text{ s}$$

Hence $t_2 - t_1 = T$, the time constant.

This argument must be true for any selected point on the charging curve. Hence at any point on the curve the time remaining to complete the charge, if the charge then continued at a constant rate, is CR s. In theory the capacitor can never be completely charged, but after a time interval equal to $5CR$ s the charge is within 1 per cent of its final value.

In *Figure 3.10* we have used this information to obtain an approximate graph of voltage against time. The time constant is the time it would take the capacitor to reach the value of the applied voltage V if the initial rate of increase could be continued. As we have seen, this statement applies to any instant in the charging cycle, but the final value of the voltage is V in each case. So we can draw a number of horizontal lines each of length equal to CR, at intervals corresponding to equal intervals of voltage. For clarity the intervals have been taken as 0.2 V, but a more accurate curve can be obtained if smaller intervals are chosen.

(10) A 10 μF capacitor is charged from a 100 V d.c. supply through a resistance of 0.5 MΩ. Use graph paper to draw an approximate curve of the rise of voltage across C for values of time $t = 0$ to $t = 30$ s employing the method outlined above. Take the voltage intervals to be 10 V. From your curve, read off the capacitor voltage after a time equal to CR s has elapsed, and express this as a percentage of the applied voltage V.

THE EQUATION OF CHARGE

To deduce the mathematical expression for the capacitor voltage at any instant calls for some advanced mathematics. If your knowledge of calculus is not up to it, you might miss out the next few lines of working and concentrate on the result we shall obtain at the end.

Let $q = v_c C$ be the instantaneous charge on C at a voltage v_c. Then from *Figure 3.11* the applied voltage V is the sum of v_r and v_c which at any instant are iR and q/C, respectively.

$$\therefore \quad iR + \frac{q}{C} = V$$

But

$$i = \frac{dq}{dt}$$

so that

$$\frac{dq}{dt} R + \frac{q}{C} = V$$

Rearranging

$$\frac{dq}{q - CV} = -\frac{dt}{CR}$$

and integrating

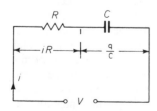

Figure 3.11

$$\log_e (q - CV) = -\frac{t}{CR} + K$$

where K is a constant.

We can evaluate K by knowing the initial conditions: when $t = 0$, $q = 0$, so $K = \log_e (-CV)$

$$\therefore \quad \log_e \left[\frac{q - CV}{-CV} \right] = -\frac{t}{CR} \text{ or } \frac{CV - q}{CV} = e^{-t/CR}$$

$$\therefore \quad CV - q = CV e^{-t/CR}$$

$$q = CV(1 - e^{-t/CR})$$

$$= Q(1 - e^{-t/CR}) \tag{3.1}$$

where Q is the final charge on the capacitor. Writing $q = v_c C$, we obtain the expression for the voltage on C at any time t:

$$v_c = V(1 - e^{-t/CR}) \tag{3.2}$$

To acquaint you with the application of this expression, here is a worked example.

Example (11) A 10 μF capacitor is charged from a 100 V supply through a series resistor of 0.5 MΩ. Calculate the voltage across C for the following intervals after the charge commences: (a) 2 s; (b) 5 s; (c) 10 s.

Solution We calculate the time constant CR: $CR = 10 \times 0.5 = 5$ s. Here $V = 100$ V, and we require to find values of v_c, given values of t.

(a) $\qquad v_c = V(1 - e^{-t/CR})$

For $t = 2$ $\quad v_c = 100(1 - e^{-2/5})$

$$= 100(1 - e^{-0.4})$$

We need now to find the value of $e^{-0.4}$. It can be proved that e is a constant given approximately by 2.7183. Values of e raised to both positive and negative powers can be found in most books of mathematical tables or they can be evaluated on most pocket calculators (you need an e^x button on your calculator; in this case, set your display to -0.4 and press the e^x button). By either of these means, we find that $e^{-0.4} = 0.67$. So

$$v_c = 100(1 - 0.67) = 100 \times 0.33$$
$$= 33 \text{ V}$$

Hence, after a time of 2 s, the capacitor voltage has grown from zero to 33 V.

(b) Here $t = 5$, so $v_c = 100(1 - e^{-5/5})$

$$= 100(1 - e^{-1})$$

We find that $e^{-1} = 0.368$, so

$$v_c = 100(1 - 0.368) = 100 \times 0.632$$

$$= 63.2 \text{ V}$$

We shall refer back to this particular answer a little further on.

(c)   Here $t = 10$, so $v_c = 100(1 - e^{-10/5})$

$$= 100(1 - e^{-2})$$

This time we find that $e^{-2} = 0.135$, so

$$v_c = 100(1 - 0.135) = 100 \times 0.865$$

$$= 86.5 \text{ V}$$

You will agree that, even if this is the first time you have had to work with an exponential equation, the calculations are not very difficult.

The result obtained for part (b) of this last example is of very particular interest to us. Notice that the value for time t was equal to the time constant of the circuit, that is, 5 s. If in any example we set $t = CR$ seconds, the term in e always evaluates to e^{-1}. Hence we draw an important conclusion: when the capacitor has charged for a time equal to the time constant, the voltage across it will always have reached 63.2 per cent of the final value V. This gives us another interpretation of time constant and one which particularly emphasizes the choice of words.

So far we have concentrated only on the exponential growth equation. If you glance back at *Figure 3.6* you will see that the circuit current follows an exponential decay curve, and quite clearly so does the fall of voltage across the resistor. We can easily deduce the mathematical expressions for the circuit current and the resistor voltage at any instant of time from our knowledge of the growth equation (3.2) above. At any instant

$$v_r = V - v_c$$

then

$$v_r = V - V(1 - e^{-t/CR})$$

$$= V \cdot e^{-t/CR} \tag{3.3}$$

Also the current at any instant in the resistor, and hence in the circuit as a whole is

$$i = \frac{v_r}{R} = \frac{V}{R} e^{-t/CR}$$

$$\therefore \quad i = I_0 e^{-t/CR} \tag{3.4}$$

When $t = 0$, $e^0 = 1$ and so $i = I_0$ the initial current.

THE DISCHARGE OF A CAPACITOR

Consider *Figure 3.12* where the capacitor is charged through resistor R by the switch connection being made as shown. When the charge is complete the switch is moved to position 2 and the series combination of C and R is short-circuited. The capacitor will then discharge through R and the circuit current will flow in the direction indicated.

By Kirchhoff, the e.m.f. acting in the closed loop is now zero, so the sum of v_c and v_r is at all times zero. At the instant the short-circuit is applied, the capacitor voltage is equal to V and the voltage across the resistor is $-V$. As the capacitor discharges, both of these voltages must decay towards zero, and so the curves of capacitor and resistor voltage will be as shown in *Figure 3.13*.

Take special note of the fact that although the resistor voltage looks rather like an exponential growth curve, it is a decay curve, the voltage falling from its initial value of $-V$ to the zero line. We should expect, therefore, that both curves will have equations of the form given in (3.3) and (3.4) above. This is indeed so, and if the necessary calculations are made, we find that

$$v_c = Ve^{-t/CR} \text{ for the capacitor}$$

$$v_r = Ve^{-t/CR} \text{ for the resistor}$$

The current in the circuit also decays and is easily calculated since $i = v_r/R$. So

$$i = -\frac{V}{R} e^{-t/CR} = I_0 e^{-t/CR}$$

where I_0 is the initial current. The negative sign tells us that the current during discharge flows in the opposite direction to that which flowed into the capacitor during the charge. As *Figure 3.12* shows, if the capacitor (or resistor) voltage continued to fall at its initial rate, the discharge would be completed in a time equal to CR seconds.

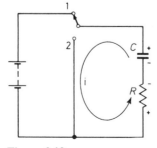

Figure 3.12

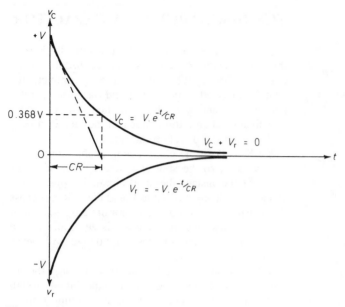

Figure 3.13

For any of the decay equations already mentioned, notice that when $t = CR$, the term of $e^{-t/CR}$ comes to e^{-1}, hence the curve, whether of voltage or current, decays away to $e^{-1} = 0.368$ of its initial value.

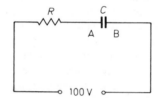

Figure 3.14

Example (12)   The circuit of *Figure 3.14* is part of an electronic system such that when the voltage across the terminals AB reaches 62.3 V the capacitor C is instantly discharged, the applied 100 V supply then being allowed to charge it through R as before. Sketch a graph showing the variation of (a) capacitor voltage, (b) circuit current throughout several cycles of the above sequence.

Calculate

(a)   the initial current;
(b)   the current at the instant before the capacitor is discharged, showing this on the graph.

Solution   The circuit time constant is 2 MΩ × 5 μF = 10 s.

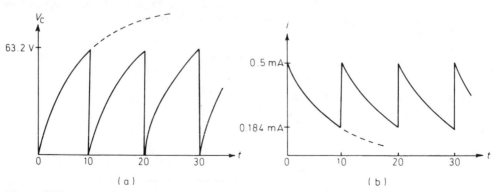

Figure 3.15

(a) As the applied voltage is 100 V and the capacitor charges to 63.2 V, the capacitor charges for 10 s before being discharged (since in a time equal to CR the capacitor voltage is always 63.2 per cent of the applied voltage). The graph of capacitor voltage against time is shown in *Figure 3.15(a)*.

(b) The circuit current will decay when the capacitor discharges, but will be restored to its initial value $I_o = V/R$ each time the capacitor is discharged. So

$$I_o = \frac{100}{2 \times 10^6} \, A = 0.5 \, \text{mA}$$

At the instant the capacitor is about to be discharged, the voltage across $R = 100 - 63.2 = 36.8$ V, hence the current at this instant will be

$$i = \frac{36.8}{2 \times 10^6} \, A = 0.184 \, \text{mA}$$

The graph of current against time is shown in *Figure 3.15(b)*. Both of the curves are examples of a particular kind of sawtooth waveform.

Example (13) In *Figure 3.13* the gradient of the line representing the initial rate of decrease of v_c is $-V/CR$. Prove that this is so.

Example (14) Sketch, on the same voltage/time axes, three curves representing the rise of voltage across a capacitor C when R has a low, medium and high value of resistance.

Example (15) Sketch on the same charge/time axes, three curves representing the increase of charge q on a capacitor C when R has a fixed value but C has a low, medium and high value of capacitance.

THE SERIES *L–R* CIRCUIT

We turn now to the series circuit made up of inductance in series with resistance. The resistance may be a separate component or it may be the actual ohmic resistance of the wire making up the inductance. When, in the circuit of *Figure 3.16*, the battery is suddenly switched to the inductance, the current will build up

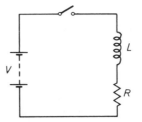

Figure 3.16

to its final possible limit of V/R but it will not do this without the passage of time. We recall that when the current is increasing towards its V/R value, the changing magnetic field induces an e.m.f. in the coil which opposes the increase, and this opposition depends on the rate of increase of the current. Hence time is required for the current to flow in an inductive circuit.

Consider some instant t sec after the voltage is applied; the current has risen to i amp and the voltage across R is then iR volts. The remaining part of V, $(V - iR)$ volts, is available to increase the circuit current and the rate at which it will increase will depend on this available voltage. The induced e.m.f. will be equal to the voltage at all times, hence

$$e = (V - iR) = L\frac{di}{dt}$$

or

$$V = iR + L\frac{di}{dt}$$

At the instant the voltage is applied, $i = 0$ and $V = L \, di/dt$

$$\therefore \quad \frac{di}{dt} = \frac{V}{L} \, \text{A/s}$$

This is the initial rate of increase of the current. If the current were to continue at this rate unopposed, it would reach its final value $I = V/R$ in a time given by

$$T = \text{final current/rate of change of current} = \frac{V/R}{V/L}$$

$$= \frac{L}{R} \, \text{s}$$

This situation is analogous to the rise of voltage across a capacitor; hence the quotient L/R is the time constant of an inductive circuit.

Example (16) Can you verify that L/R leads to the dimension of time? (Hint: express L in terms of voltage and time by Faraday's law.)

Example (17)   What is the time constant of an inductance of 10 H which has a resistance of 100 Ω?

At any instant during the rise of current in an inductance, the back-induced e.m.f. opposes the supply voltage and this is a continuing process all the time the current is rising. The rate of rise of the current gets less and less as time goes on, and the graph of current against time, instead of following the initial rate line OP in *Figure 3.17* bends over to become more and more horizontal as the final limit of current is approached. Let the current be i, then at any instant

$$L \frac{di}{dt} = V - iR$$

and

$$\frac{di}{dt} = \frac{V - iR}{L} \text{ A/s}$$

This is clearly less than the original rate of increase V/L A/s proved above.

At any instant the current rise still required is $(V - IR)/R$, and the rate of rise is $(V - iR)/L$. Hence at any instant, the time for the current rise to be completed, if it were to continue at a constant rate, is

$$\frac{V - iR}{R} \times \frac{V - iR}{L} = \frac{L}{R} \text{ s}$$

This is the same time as that found for completion at the initial rate of rise and the argument used must hold for any instant of time. Hence, at any point on the current–time curve, the time remaining for the completion of the rise of current, if the rise then continued at a constant rate, is L/R s. You will have realized by now that the situation and the form of the curve are completely analogous to the charging of a capacitor.

Current in an inductive circuit, like capacitor voltage in a CR circuit, rises gradually to its final steady-state value, the rise being rapid if L is small but slow if L is large, assuming a constant value of R. Conversely to the capacitor case, an increase in resistance reduces the time constant of the LR circuit. In theory, the current never reaches its final value since a time equal to L/R seconds always remains outstanding to complete the rise, but in practice a time equal to five times L/R brings the current to within 1 per cent of its final steady value.

It is clear that the equation expressing the rise of current in an inductance will be of exponential form and similar in pattern to that derived for the CR circuit. If the necessary calculations are carried out, we find that the current i_L at any instant is given in terms of the final current I, inductance L and resistance R by

$$i_L = I(1 - e^{-Rt/L}) \tag{3.5}$$

If now we set $t = L/R$ seconds, the equation becomes

$$i_L = I(1 - e^{-1}) = I(1 - 0.368)$$

$$= 0.632I$$

So in a time equal to the time constant, the current will have risen always to 63.2 per cent of its final value.

If you have mastered problems associated with CR

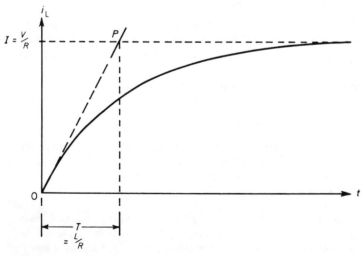

Figure 3.17

circuits you should have no difficulty in coping with the present examples. Here are two worked examples to show you the way.

Example (18)   When a coil is connected to a 100 V d.c. supply the final steady current is measured as 0.5 A. At the instant when the current was changing at the rate of 10 A/s the current was 0.2 A. What is the resistance and inductance of the coil?

What was the current in the coil at the instant $t = 0.05$ s?

Solution   Final current $I = V/R$

$\therefore$ $R = V/I = 100/0.5 = 200\ \Omega$

also

$$\frac{di}{dt} = \frac{V - iR}{L}$$

$\therefore$ $10 = \dfrac{100 - 0.2 \times 200}{L}$

$\therefore$ $L = \dfrac{100 - 40}{10} = 6$ H

For the second part of the question we require the time constant:

$$T = \frac{L}{R} = \frac{6}{200} = 0.03\ \text{s}$$

then

$$\frac{Rt}{L} = \frac{0.05}{0.03} = 1.67 \ (\text{for } t = 0.05\ \text{s})$$

so

$i = I(1 - e^{-1.67}) = 0.5(1 - e^{-1.67})$

$\quad = 0.5(1 - 0.189)$

$\quad = 0.405$ A

Example (19)   A relay coil having a resistance of 20 Ω and an inductance 0.5 H is switched across a d.c. supply of 200 V. Calculate: (a) the time constant; (b) the initial rate of rise of current; (c) the current at a time equal to the time constant; (d) the current after 0.05 s; (e) the energy stored in the field when the current is steady.

Solution

(a) $T = \dfrac{L}{R} = \dfrac{0.5}{20} = 0.025$ s

(b)   Initial rate of rise of current $= \dfrac{V}{L} = \dfrac{200}{0.5}$

$\quad = 400$ A/s

(c)   The current rises to 63.2 per cent of its final value in a time equal to the time constant.

Final current $I = \dfrac{V}{R} = \dfrac{200}{20} = 10$ A

$\therefore$ $i_L = 0.632 \times 10 = 6.32$ A

(d)   After time $t = 0.05$ s, $\dfrac{Rt}{L} = \dfrac{20 \times 0.005}{0.5}$

$\quad = 2$

$i_L = 10(1 - e^{-2}) = 10(1 - 0.135)$

$\quad = 8.65$ A

(e)   Energy stored $= \frac{1}{2}LI^2$ joules

$\quad = \frac{1}{2} \times 0.5 \times 10^2 = 25$ J

DECAY OF CURRENT

The last part of the previous example dealt with the energy stored in the magnetic field when the steady state has been reached.

Suppose a steady current I to be flowing in an inductor; now let the applied voltage V be removed and at the same instant let the coil be short-circuited. The current in the coil does not fall immediately to zero, for the collapse is opposed by the self-induced e.m.f. and this is now in such a direction that it tends to maintain the current at its original value. Let the current be i A at a time t seconds after the collapse has begun. Then

$$iR + L\frac{di}{dt} = 0$$

since there is no external voltage acting on the coil.

When $t = 0$, $i = V/R$ (its steady value at the instant of short-circuit), and so

$$IR + L\frac{di}{dt} = 0$$

$$\frac{di}{dt} = -\frac{IR}{L} = -\frac{V}{L}\ \text{A/s}$$

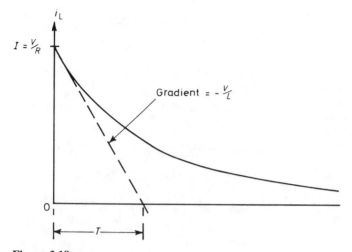

Figure 3.18

This is the initial rate of change of current, the negative sign simply indicating that the current is decreasing. Apart from this sign change, the value is the same as that of the initial rate of rise of current discussed earlier. Solving the equation above in a manner similar to that already used for the charge of a capacitor, we find that the current at any instant t during the decay cycle is

$$i_L = Ie^{-Rt/L} \qquad (3.6)$$

The curve of this equation is shown in *Figure 3.18* and is the same as the curve of current growth but turned upside down. If the fall continued at the initial rate the decay would be completed in a time equal to L/R s, the time constant again. Setting $t = L/R$ in equation (3.6) above we get

$$i_L = Ie^{-1} = 0.368I$$

Hence in a time equal to the time constant, the current in the inductor has decayed to 36.8 per cent of its initial value.

When an inductive circuit is suddenly interrupted, the induced e.m.f. may be many times greater than the applied voltage and is often sufficient to create an arc at the switch contacts. This arcing is destructive to the contacts and a spark-quench circuit is often wired across the switch terminals. This circuit is made up of a capacitor and a low value resistor in series: the energy of the collapsing magnetic field is then transferred to the electric field of the capacitor, thereafter being rapidly dissipated in the resistive part of the arrangement.

Some devices, i.e. ignition coils, operate on the principle of a high induced e.m.f. across the coil when the current is suddenly interrupted.

The next worked example brings together inductive and capacitive circuits and is possibly a little more brain teasing than those examples which have gone before. Follow the working and the comments carefully.

Example (20)   A d.c. source of 100 V e.m.f. is switched across the circuit shown in *Figure 3.19*. Sketch a graph indicating the variation with time of (a) the current in the capacitor branch, (b) the current in the inductive branch, (c) the current flowing from the source.

When steady state conditions are reached, the switch is opened to disconnect the source. Describe the subsequent behaviour of the circuit.

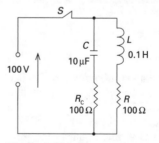

Figure 3.19

Solution   A quick survey of the situation first. Current flows into each branch of the circuit as soon as the source is connected. The capacitor will be charged by way of its $100\,\Omega$ series resistor R_c and the inductor will establish a magnetic field with its accompanying back-e.m.f. of self-induction.

The initial charging current for the capacitor will be $V/R_c = 100/100 = 1$ A, and the final current will be zero. The time constant (you will recall) is $CR = 10 \times 10^{-6} \times 100 = 0.001$ s or 1 ms.

The initial current through the inductor will be zero and the final current will be $V/R = 100/100 = 1$ A. The time constant this time will be $L/R = 0.1/100 = 0.001$ s or 1 ms. Notice that the time constant for each branch is the same and that in the steady state condition the energy stored in the capacitor, $\frac{1}{2}CV^2 = \frac{1}{2} \times 10 \times 10^{-6} \times 100^2 = 0.05$ J, is the same as the energy stored in the magnetic field of the inductor, $\frac{1}{2}LI^2 = \frac{1}{2} \times 0.1 \times 1^2 = 0.05$ J. We can now sketch the required graphs.

(a) The capacitor charging current will *decay* exponentially from its initial value of 1 A to zero, and will have the time constant $T = 1$ ms.

(b) The inductor current will *rise* exponentially from its initial value of zero to its final value of 1 A, and will have the time constant $T_L = 1$ ms also.

(c) The source current will be the sum of the current in the two branches and, as *Figure 3.20* shows, will be constant at 1 A throughout the transient period as well as the subsequent steady state condition.

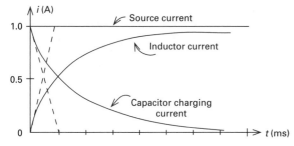

Figure 3.20

At the instant the switch is opened, a current of 1 A is flowing in the inductor branch. This same current must therefore flow through resistor R_c associated with the capacitor; hence the voltage drop across the two resistors, now in series, will be $100 + 100 = 200$ V. Since the voltage across C is only 100 V (its total charge level), it follows that at this instant an induced e.m.f. of $200 - 100 = 100$ V must exist across the inductor.

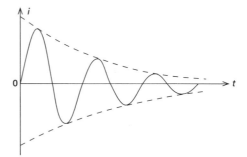

Figure 3.21

If you have got this far you will have done well enough, but what happens next? A more advanced treatment than we need for the purposes of this book is called for. Fundamentally, it all depends on what the total circuit resistance is. In this example, the current will decay exponentially from the 1 A level we found it in when the switch was opened; in other words, the energy stored in the fields of the capacitor and inductor is all rapidly dissipated in the circuit resistance and falls away to zero in a very short time. For a total resistance below a certain value (actually $2\sqrt{(L/C)}$ Ω), however, the stored energy is 'trapped' for a time within the loop of the circuit and can interchange between the electric and magnetic fields, alternately being stored first in one field and then the other. If there were no resistance, this interchange would continue indefinitely, but some resistance is inevitable and the process of *oscillation* diminishes, this taking place also along an exponential decay curve. *Figure 3.21* shows this effect which we will briefly return to in a later chapter.

DIFFERENTIATING AND INTEGRATING CIRCUITS

Suppose that a voltage pulse having a time variation as shown in *Figure 3.22(a)* is applied to a series CR circuit. There are two possibilities of interest to consider: the time constant of the CR circuit may be short relative to the pulse duration P or it may be long. Suppose firstly that the product CR is very short compared with P.

Then the charging of the capacitor through R will be completed in a fraction of the time P and thereafter V_C will remain at the peak level of the pulse amplitude, V. When the pulse ends, the capacitor will discharge rapidly through R and V_C will return to zero. *Figure 3.22(b)* shows the voltage pulse as it appears across C. Since $V_R + V_C$ must at all times equal $\hat{V}$, the voltage across R rises instantly to $\hat{V}$ at the onset

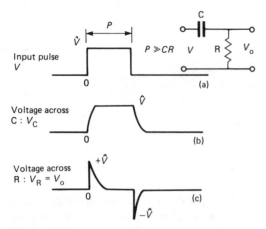

Figure 3.22

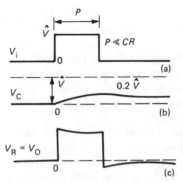

Figure 3.23

of the pulse and thereafter decays rapidly to zero as V_C rises to $\hat{V}$. It then remains at zero for the duration of the pulse.

When the pulse ends, the capacitor discharges through R; the voltage across R instantly follows the fall (which is now in the opposite sense to the charge rise) and decays from its negative level back to zero, see *Figure 3.22(c)*. The points of interest to notice here are:

(a)   the voltage output across C is a very close approximation to the shape of the input pulse, in fact, the shorter the product CR becomes, the better the approximation.

(b)   the output across R consists of two very narrow voltage pulses of opposite sign, one coinciding with the start and the other with the end of the input pulse.

This circuit, in which the product CR is very much shorter than P and the output is derived from R, is called a differentiating circuit. The narrow pulses developed across R are approximately proportional to the time rate of change of the input voltage. Suppose now that we change over the positions of C and R and make the product CR very large compared with the pulse duration P, as indicated in *Figure 3.23(a)*. At the onset of the pulse the voltage across C will now rise very slowly and will have reached a level that is only a fraction of V by the time the pulse ends, *Figure 3.23(b)*. From this point on, C will discharge through R, but as the rate of discharge is governed by the voltage available across R (and this is a lot less than V), the discharge rate will be different from the charge rate. Hence the discharge curve will exhibit a long decay tail as *Figure 3.23(b)* shows. The corresponding variation in the voltage across R is given in diagram (c). The points of interest this time are:

(a)   the voltage across R is a close approximation to the shape of the input pulse. In fact, the greater the CR product, the better the approximation;

(b)   the output voltage across C is a low amplitude pulse many times the length of the input pulse.

This form of circuit, in which the product CR is very much longer than P, is called an integrating circuit. The extended output pulse developed across C is approximately proportional to the time integral of the input voltage.

Example (20)   What would the true differentiated version of a rectangular voltage pulse look like, and how closely does a practical differentiating circuit approximate to it?

Solution   For the rectangular pulse shown in *Figure 3.24*, the instantaneous rate of change (or gradient) dV/dt is either infinite or zero. It is infinite on the leading and trailing edges of the pulse where the waveform is vertical, and zero elsewhere where the waveform is horizon-

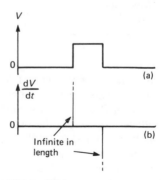

Figure 3.24

tal and the gradient zero. A true picture of the differentiated form of the rectangular pulse consequently is as shown in the figure.

The output from a practical circuit has been shown in *Figure 3.22(c)* and although this consists of narrow spikes of voltage, the similarity to the true (and unobtainable) differentiated voltage is clear. For an extremely short time constant the similarity becomes very marked and the term 'differentiating circuit' is fully justified.

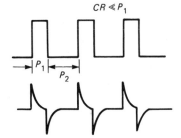

Figure 3.25

APPLICATION TO PULSE TRAINS

Suppose that a succession of rectangular pulses is applied to differentiating and integrating circuits in turn. Let P_1 be the duration of each pulse and P_2 the interval between the pulses.

The result of applying such a pulse train to a differentiating circuit where $CR < P_1$ is shown in *Figure 3.25*. The voltage waveforms developed across C and R are similar to those already described, and the output across R is a succession of alternate positive and negative spikes. If the negative spikes are removed by a simple diode connected across R (make a sketch of how you would do this) and the positive spikes are 'clipped' to a predetermined level (how might you do this?), a train of very short duration pulses may be obtained.

Some new points of discussion arise when we turn to the integrating circuit. Again, the general shapes of the voltage developed across C and R are similar to those already described, but this time the long decay tail of the discharging capacitor is not permitted to fall away to zero. As *Figure 3.26(b)* shows, the mean level of voltage across C tends to rise with each successive pulse because the capacitor has barely started to discharge before the following pulse turns up. After a number of pulses have passed, a steady state is reached in which the charge acquired by the capacitor during pulse period P_1 is exactly equal to the charge lost during interval P_2. This means that the mean current flowing into C through R is zero when equilibrium has been reached, so that the mean value of V_R is also zero.

The graph at *(c)* shows the voltage developed across R and at equilibrium the area (representing voltage × time) enclosed above the mean (zero) line during the time P_1 is equal to the area below this line during the time P_2.

The mean output across C, shown as a broken line

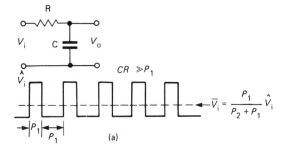

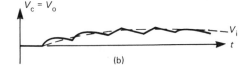

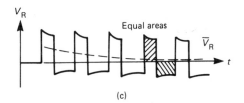

Figure 3.26

in diagram (b), is seen to be a long single pulse, albeit with a serrated edge; the circuit has integrated (or summed up) all the separate input pulses into a single pulse. The output across R is a good replica of the input voltage train, but the waveform is balanced about a zero level, unlike the input which is positioned wholly above the zero line. This effect comes about because the presence of the capacitor removes the d.c. component of the input signal. The succession of positive input pulses can be considered to be made up of a steady (d.c.) component of voltage and a purely alternating component of voltage, and only the alternating component appears across R.

In many circuit applications it is necessary to reinsert the steady component if the output has to be taken from R. This process is known as d.c. restoration.

Example (21) The input to the circuit of *Figure 3.27(a)* is the rectangular waveform of *Figure 3.27(b)*. Sketch the output waveform, marking the high and low voltage levels attained and indicating the new zero level.

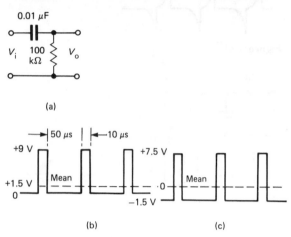

(a)

(b)   (c)

Figure 3.27

Solution In the circuit shown, the time constant $CR = 0.01 \times 0.1 = 0.001$ s or 1000 μs, which is long compared with the period between the pulses, 50 μs. Taking the output from across the resistor where the shape of the input waveform will be practically unaffected, the average value of the input is clearly

$$V_{mean} = \frac{P_1}{P_1 + P_2} \quad V = \frac{10}{60} \times 9 = 1.5 \text{ V}$$

and the output wave must then 'balance' itself so that its mean value is zero, that is, the wave must therefore vary between -1.5 V and $+7.5$ V, as shown in *Figure 3.27(c)*.

ACTIVE CIRCUITS

The C–R circuits just discussed are passive differentiators or integrators. Active forms of these circuits can be made using operational amplifiers, and these are used extensively in automatic control systems and (in the case of integrators) analogue computers.

Figure 3.28(a) shows the essential arrangement of an operational amplifier used as a differentiator.

From operational amplifier theory we have a virtual earth at the inverting input terminal and negligible current flows into the amplifier because of its very high input impedance. Hence the input voltage v_i may

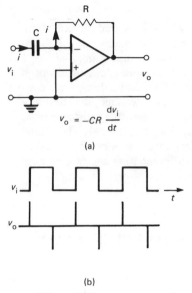

(a)

(b)

Figure 3.28

be considered to be developed across the capacitor C, the output voltage v_o across the resistor R and the input current flows through C and R effectively in series. The charge on the capacitor is

$$q = Cv_i$$

The charging current, i, is the time rate of change of charge or

$$i = \frac{dq}{dt} = C\frac{dv_i}{dt}$$

At the output, voltage v_o must obey Ohm's law and be expressed as $-iR$, hence

$$v_o = -iR = -CR\frac{dv_i}{dt}$$

The negative sign appears because the amplifier is used in the inverting mode and the output is antiphase to the input. Thus we see that the output voltage is proportional to the time derivative of the input, and with a 'gain' magnitude of CR. This means that the more rapid the rate of change of the input, the higher the output voltage. A rectangular wave input will give very sharp output pulses corresponding to the rising and falling edges where the input is changing, *Figure 3.28(b)*.

For the active integrator, capacitor C and resistor R are interchanged, *Figure 3.29(a)*. The input voltage V_1 will now attempt to charge the capacitor plate connected to the input through the resistor. Because of the phase inversion taking place in the amplifier, however, the output voltage V_o will try to charge the other capacitor plate in a direction which tends to neutralize the charge on the input plate. In other words,

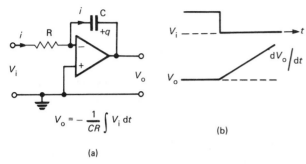

$$V_o = -\frac{1}{CR} \int V_i \, dt$$

(a)

(b)

Figure 3.29

as the capacitor charges, the output from the amplifier must go more and more negative to maintain the charging current constant. But when a capacitor is charged from a constant-current source, the rise in capacitor voltage is linear, hence with a positive input voltage, the output of the integrator will be a negative-going linear ramp, and conversely.

As for differentiation, negligible current flows into the amplifier and terminal and

$$\frac{V_i}{R} = -\frac{dq}{dt}$$

But

$$dq = C \,.\, dV_o$$

Therefore

$$\frac{V_i}{R} = -C \frac{dV_o}{dt}$$

and

$$dV_o = -\frac{V_i \,.\, dt}{CR}$$

Integrating, we get

$$V_o = -\frac{1}{CR} \int V_i \, dt$$

Therefore the output voltage of this circuit is the integral of the input voltage with a gain magnitude of $1/CR$. The response of the integrator to a negative-going voltage step is shown in *Figure 3.29(b)*.

Both the active differentiator and integrator are important circuit elements.

SUMMARY

- The voltage on a capacitor cannot change instantaneously.

- The current in an inductor cannot change instantaneously.
- Time constant T is a measure of the rate of growth or decay.
- For a capacitor $T = CR$, for an inductor
$$T = \frac{L}{R}.$$
- If a capacitor is charged from a constant current source, the rise of voltage is linear.
- The general growth exponential is $a = A(1 - e^{-t/\tau})$. For $t = T$, $a = 0.632$ A.
- The general decay exponential is $a = A \,.\, e^{-t/\tau}$. For $t = T$, $a = 0.368$ A.
- In resistance, i is proportional to v. In a capacitor i is proportional to rate of change of v.
- In an inductor, v is proportional to rate of change of i.

REVIEW QUESTIONS

1   Distinguish between transient and steady state conditions.
2   Why is the concept of time constant useful?
3   What, approximately, is the transient period for a capacitor charging through a resistor?
4   What, theoretically, is the exact transient period?
5   How can an exponential curve be approximated by the use of a series of linear parts?
6   Write down the equation of an exponentially increasing function; of an exponentially decreasing function.
7   If the *CR* product for a series arrangement of *C* and *R* is reduced, what effect does this have on v_c and v_r?
8   In theory, the charging of a capacitor through a resistor takes an infinite time. Does this imply that an infinite charge would finally be established on the capacitor?
9   Explain briefly the action of a differentiator, an integrator.
10 Two series *C–R* circuits, each having $C = 2 \,\mu\text{F}$, $R = 1 \,\text{M}\Omega$, are themselves joined in series. What is the time constant of this combination? Can you suggest an easy deductive way of answering this question without doing a calculation?

EXERCISES AND PROBLEMS

(22)   A capacitor connected in series with a 5 kΩ resistor is switched suddenly across a 200 V d.c supply. Find the initial current and the voltage across the capacitor when the current is 20 mA.

(23)   When a capacitor is connected to the 25 V d.c. supply, the initial current is limited to 100 mA by a series resistor. Calculate the value of this resistor and find the charging current at the instant the voltage across C reaches 15 V.

(24)   A 10 µF capacitor is to be charged through a 100 kΩ resistor from a d.c. supply. Find (a) the circuit time constant, (b) the additional series resistance required to increase the time constant to 4 s, (c) the additional parallel resistance required to reduce the time constant to 0.25 s.

(25)   A 150 V battery is switched suddenly across a circuit consisting of a 20 µF capacitor in series with a 0.1 MΩ resistor. Write down an expression for the subsequent voltage/time relationship. Calculate for this circuit (a) the time constant, (b) the initial current, (c) the final current.

(26)   A welding machine is controlled by a timing device which depends on the charging of an 8 µF capacitor. If the time constant of the circuit is to be variable between 0.5 s and 25 s, find the limits of the value of the resistance to be wired in a series with the capacitor.

(27)   A voltmeter whose resistance is 10 kΩ is in series with an 80 µF capacitor. The combination is switched abruptly across a 100 V d.c. source. What will the voltmeter read (a) at the instant of switch-on, (b) at a time $t = 400$ ms? What will be the circuit current at the instant the voltmeter reads 40 V?

(28)   In *Figure 3.30*, $C = 2$ µF and $R = 0.5$ MΩ. What is the potential at terminals A-B (a) be-

fore the switch is closed, (b) at the instant the switch is closed, (c) 2 s after the switch is closed?

(29)   The capacitor and resistance of *Figure 3.30* are interchanged. Calculate the potential at terminals A-B for each of the three instants given in the previous problem.

(30)   What is the time constant of a 0.5 H coil which has a resistance of 40 Ω?

(31)   Write down the time constants of the following inductive circuits: (a) 10 H, 300 Ω; (b) 0.6 H, 25 Ω; (c) 150 mH, 10 Ω; (d) 10 µH, 0.1 Ω; (e) 5 µH, 0.02 Ω.

(32)   A coil of resistance 10 Ω and inductance 100 mH is switched to a 20 V d.c. supply. What is the initial rate of rise of current, the final current and the time constant? What energy is finally stored in the magnetic field of this coil?

(33)   What happens to the stored energy in the field of the coil of the previous problem when the coil is suddenly short-circuited? What other forms of energy is it converted into?

(34)   A current flowing through a 30 mH coil is switched off and falls to zero in 40 ms. If the power dissipated is 15 W, what current was originally flowing in the coil?

(35)   The armature of a relay working on a 120 V supply does not operate until the current in the relay coil reaches 200 mA. This occurs 5 ms after the d.c. is applied. Given that the time constant is also 5 ms, what is the inductance and resistance of the coil? At what rate does the current initially rise?

(36)   A relay coil of resistance 200 Ω and inductance 8 H is connected to a 60 V d.c. source which has an internal resistance of 100 Ω. The relay closes when the coil current reaches 31.5 mA; how long does the relay take to operate after the source is connected? What energy is stored in the relay field?

(37)   What percentage of the final value of current is reached in 0.001 s, given that $L = 1 H$, $R = 6$ kΩ?

(38)   A 0.5 H coil of resistance 50 Ω is connected in series with a circuit in which the current varies in the following way: (i) the current

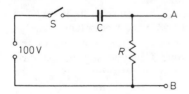

Figure 3.30

rises linearly from zero at a rate of 10 A/s for a period of 0.25 s; (ii) the current then remains steady for 3 s; (iii) the coil terminals are now shorted together and the current falls to zero.

Draw a scaled graph showing coil current against time throughout the sequence and hence determine (a) the coil current during stage (ii); (b) the energy dissipated in stage (ii); (c) the energy dissipated in stage (iii). (Hint: take particular care over part (b).)

(39)   The rate of increase of current in a certain inductive circuit is expressed by

$$i = 10(1 - e^{-40t})$$

(a) what is the time constant of this circuit, (b) the current value at time (a), (c) current when time is 0.05 s, (d) the value of the inductance, given that the circuit resistance is 200 Ω?

The circuit is now suddenly short-circuited after the steady state has been reached. Find (a) the equation of current as a function of time, (b) the rate at which the current begins to decrease, (c) the current at the instant 0.025 s after the decrease begins.

(40)   In *Figure 3.31*, C_2 is uncharged and switch S has been in the position shown for a long time. (a) Under steady state conditions, what current flows in C_1, in R_1? What is the voltage across C_1? (b) Switch S is thrown to the opposite position at time $t = 0$: write down the equation relating current i to time t; (c) Assuming that $C_1 = C_2 = 1$ μF, find the total charge stored and the voltage across each capacitor.

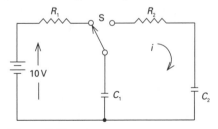

Figure 3.31

(41)   In *Figure 3.32* the circuit is in a steady state. Calculate (a) the potential between terminals A and B, (b) the voltage across the capacitor, (c) the charge on this capacitor, (d) the energy stored, (e) the current in the 1 MΩ resis-

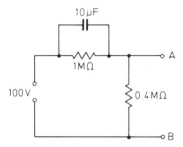

Figure 3.32

tor. The 100 V supply is now suddenly removed and replaced by a short-circuit; in what time will the capacitor discharge down to 10 V?

(42)   In *Figure 3.33* calculate the steady state values of the current in the coil and in the parallel resistor. If switch S is now suddenly opened, what will be the current in the 30 Ω resistor 0.1 s later?

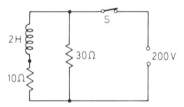

Figure 3.33

(43)   If a square wave oscillating between 0 V and −6 V at a frequency of 500 kHz is applied at input A of the circuit of *Figure 3.34* sketch a graph of the voltage waveform at B for the following values of C and R: (a) 0.001 μF, 500 Ω, (b) 0.001 μF, 2000 Ω, (c) 0.001 μF, 1 MΩ.

Figure 3.34

(44)   A telephone dial may operate between 7 and 12 pulses per second with a break percentage of 63–72 per cent. What is the duration of (a) the minimum *break* pulse, (b) the minimum *make* pulse?

4

Alternating quantities 1: waveforms

Wave characteristics
The sinusoidal waveform
Measurement of a.c. quantities
Phasor representation
Phase shift
Analysis of a.c. waveforms

An alternating electrical quantity, either voltage or current, is one which periodically goes through a definite *cycle* of variation. Over this cycle, the quantity increases to a maximum or peak value in one direction before falling to zero; it then increases to a peak value in the reverse direction before again falling to zero. By a change in 'direction' we mean a change in polarity, though the peak values reached in either of these directions are not necessarily the same. If the quantity does not reverse its direction at any time, it is not an alternating quantity.

What in some ways we might express as the ideal alternating quantity is the *sinusoidal* or simple harmonic wave which is shown in *Figure 4.1(a)*. This waveform is very important in electrical and electronic engineering studies, as it is in many other natural phenomena: the vibration of a violin string and the projection of a satellite on the rotating earth, the swing of a pendulum and the variation in the tides. For our particular purpose it is in the generation and transmission of electrical power and communication by radio links that the sinusoid has its major interest. Also, a sinusoidal source always sets up sinusoidal responses in linear circuits which is not true for other waveforms.

Not all alternating quantities are sinusoidal, however; in general terms, any wave which departs from this ideal is non-sinusoidal and, for instance, may well look like the waveforms of diagrams *(b)* and *(c)*. This fact does not affect the periodic nature of the wave or a number of those other characteristics which we will now briefly revise.

WAVE CHARACTERISTICS

Each complete cycle of an alternating quantity consists of two half-cycles, during one of which the voltage or current acts in an arbitrary 'positive' direction around a circuit, and during the other acts in the reverse or 'negative' direction. Each cycle is completed when the quantity concerned returns to its 'starting' point and is moving in the same direction.

This may be any point along the waveform diagram though it is often most convenient to take it as a zero level where the polarity changes over.

In electronics, the number of cycles occurring every second is called the *frequency*; for the domestic mains

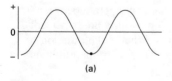
(a)

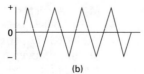
(b)

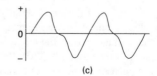
(c)

Figure 4.1

supply this is 50, but for radio transmissions it may run into many millions. Frequency (f) is measured in hertz (Hz) where 1 Hz is equivalent to 1 cycle per second. Multiples of this unit are kilohertz (kHz) = 1000 Hz, and megahertz (MHz) = 1 million hertz.

The time T taken to complete a cycle is the *period* or periodic time of the alternation; clearly, the relationship between frequency and period is

$$f \times T = 1 \ \text{ or } f = \frac{1}{T}$$

Example (1)   An audio signal has a frequency of 450 Hz. What is its periodic time?

Solution   Here $f = 450$

$$T = \frac{1}{f} = \frac{1}{450} = 0.0022\,\text{s}$$

$$= \underline{2.2\,\text{ms}}$$

Example (2)   A signal travels along a cable at a speed of 250×10^6 m/s. What is the 'length' of each cycle (the wavelength) of the transmission if the frequency is 20 kHz?

Solution   From this example we see that a wave may be defined by its frequency or its wavelength (λ). From *Figure 4.2* the relationship between these and the velocity of propagation (v) is:

$\therefore \ \lambda$ (metres)

$$= \frac{\text{distance travelled in 1 sec}}{\text{number of cycles in 1 sec}}$$

$$= \text{velocity (m/s)/frequency (Hz)}$$

$$\therefore \ \lambda \ \text{(metres)} = \frac{250 \times 10^6}{20 \times 10^3} = \frac{25 \times 10^3}{2}$$

$$= 12500\,\text{m} \quad \text{or} \quad \underline{12.5\,\text{km}}$$

In the case of radio waves, the velocity of propagation is equal to the speed of light, 300×10^6 m/s. Wave velocity along transmission lines can fall to quite small fractions of this.

---

(3)   A radio wave has a velocity $c = 300 \times 10^6$ m/s. What is the frequency of a broadcast having a wavelength of 1500 m?

MEASUREMENT OF ALTERNATING QUANTITIES

As the value of an alternating quantity is continually changing, giving a different value at every instant of time, there is no way in which such instantaneous values, as these are called, can be used as a basis for measurement.

The *peak* value might provide something of measurable value; for a sinusoidal waveform, the voltage excursions between peaks is twice the peak value since the waveform is symmetrical about the zero axis, as *Figure 4.3* indicates. Peak values (indicated as $\hat{V}$ or $\hat{I}$) are, however, themselves instantaneous values though we can, perhaps, find measurable values in terms of them.

From an examination of a sinewave cycle we see that by symmetry the *average* or *mean* value of voltage or current is zero. Over one-half cycle, however, there will be some average level; this can be shown to be $2/\pi$ or 0.637 of the peak value. Now although the average value of the wave is zero, there *must* be power dissipated when we connect a source of sinusoidal voltage across a resistor. This follows because the direction of the current in the resistor does not influence the power dissipated, hence power is provided by both halves of each sinusoidal cycle. This power dissipation must consequently be defined in

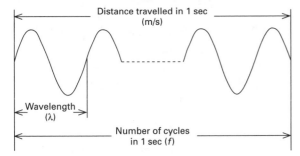

Figure 4.2

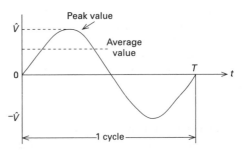

Figure 4.3

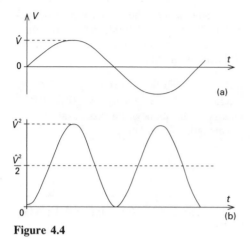

Figure 4.4

terms of the peak value so that over a complete cycle the *same* power will be dissipated as that arising from an *equivalent d.c.* voltage. This level is the r.m.s. (root-mean-square) value of the sine wave.

For a resistive load $P = I \times V = V^2/R$, therefore squaring the sine wave voltage of *Figure 4.4(a)* gives the wholly positive curve of diagram *(b)*, and the average of this is clearly $\hat{V}^2/2$. Hence the average power dissipated in a resistor will be

$$\frac{V(\text{r.m.s.})}{R} = \frac{\hat{V}^2}{2R}$$

Therefore

$$V(\text{r.m.s.}) = \frac{\hat{V}}{\sqrt{2}} = 0.707 \, \hat{V}$$

or $V = 1.414 \, V(\text{r.m.s.})$

RMS values are indicated by plain capitals, as V or I, because they represent the same power dissipation from sinusoidal waveforms in resistive circuits as their d.c. equivalents.

Voltmeters and ammeters for use in alternating circuits are normally scaled in r.m.s. values. In the commonplace d.c. moving-coil meter the current flowing in the movement coil suspended in a magnetic field produces a torque opposed by return springs. Although this torque is directly proportional to the instantaneous current, the inertia of the system prevents any rapid changes in the coil position, and the needle deflection is proportional to the average current, $0.637 \, \hat{I}$, provided the input waveform is rectified to give a wholly positive series of half-waves. The

meter scale, however, is multiplied by the factor r.m.s./ average = 1.11, so indicating r.m.s. values.

Example (4) An alternating voltage of peak value 15 V feeds a small heating element. What battery voltage would replace the alternating source and deliver the same heating effect?

Solution The r.m.s. value of the 15 V peak wave is $0.707 \times 15 = 10.6$ V. A battery of terminal p.d. 10.6 V would, therefore, supply the element with the same heating power as the alternating current.

Example (5) The diagram of *Figure 4.5* shows two cycles of a varying voltage generated by an electronic switching system. What is the frequency of this voltage? If this waveform is applied to a 15 Ω resistor, what will be the average current in the resistor? Would you consider this waveform to represent an alternating quantity?

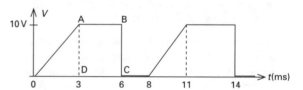

Figure 4.5

Solution The periodic time here is 8 ms, *not* 6 ms as might at first be thought. The frequency of the variation is therefore $1/8 \times 10^{-3} = \underline{125 \text{ Hz}}$.

Considering the period $t = 0$ to $t = 8$ ms, the average value of the voltage is given by

$$\frac{\text{area under the waveform}}{\text{period}}$$

$$= \frac{\text{area } \triangle OAD + \text{area } \square ABCD}{8}$$

$$= 1/2[(3 \times 10) + (3 \times 10)]/8 = 45/8 = 5.625 \text{ V}$$

The average current in the resistor is then $5.625/15 = \underline{0.375A}$. This waveform is *not* an alternating quantity; it is a varying d.c. quantity since it does not reverse direction at any time.

Example (6) A voltage wave has the rectangular form shown in *Figure 4.6*. What is the

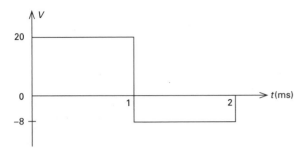

Figure 4.6

frequency of this wave? If this voltage is applied to a resistor of 35 Ω, what will be (a) the average, (b) the r.m.s. current in the resistor?

Solution   Here the single cycle shown occurs in a time 2 ms; the frequency is therefore $1/2 \times 10^{-3}$ Hz = <u>500 Hz</u>.

(a) The average must be taken over the complete cycle here as the positive and negative half-cycles are not identical. As the waveform is constant over each half-cycle, we have

$$\text{average value} = \frac{20 + (-8)}{2} = 6^{\text{H}}\text{V}$$

By Ohm's Law, the average current in the 35 Ω resistor is 6/35 = 0.171 A or <u>171 mA</u>.

(b) The r.m.s. value for the whole cycle will be equivalent to a succession of direct currents in the resistor. The mean value of the squares of the voltages is

$$\frac{20^2 + (-8)^2}{2} = 232 \text{ V}$$

and the r.m.s. voltage = $\sqrt{232}$ = 15.23 V. The r.m.s. current in the resistor is therefore 15.23/35 = 0.435 A or <u>435 mA</u>.

Example (7)   The element of a heater having a resistance of 5 Ω has direct currents of 2, 4, 6, 8 and 10 A passed through it in successive 1 m in periods. What is the total energy input? What alternating current, if passed through the element for 10 mins, will produce the same amount of heat?

Solution   Total energy input $W = (I_1^2 + I_2^2 + I_3^2 + \ldots)\ Rt$ joules
   Here $R = 5$ Ω and $t = 60$ s. Therefore $W = (2^2 + 4^2 + 6^2 + 8^2 + 10^2) \times 5 \times 60 = 66 \times 10^3$ J.

The same energy supplied by the alternating current will produce the same amount of heat. Then for a steady *direct* current I passing for 10 mins (= 600 s), we have

$$66 \times 10^3 = I^2 \times 5 \times 600$$

from which

$$I^2 = \frac{66 \times 10^3}{5 \times 600} = 22 \text{ A}$$

$$\therefore \quad I = 4.7 \text{ A}$$

An r.m.s. current of 4.7 A would therefore produce an equivalent heat to the direct current of the same level.

(8)   A moving-coil type voltmeter of f.s.d. 100 V is wired across a sinusoidal source of r.m.s. voltage 50 V and frequency 100 Hz. What will the voltmeter read? Explain your answer.

(9)   Deduce, using an intuitive method, that the r.m.s. value of the square wave shown in *Figure 4.7* is equal to the peak value.

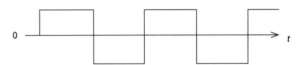

Figure 4.7

PHASOR REPRESENTATION

Clearly we need an efficient technique for analysing circuits where sinusoidal quantities are concerned. The first requirement is a convenient method of representing sinusoidal waves not as they have been drawn in the foregoing figures, which would be a problem in itself, but as a method which is the transformation of functions of time into constant quantities. This technique is known as phasor representation.

In phasor representation we associate a sinusoidal variation with the projection of a radial line of length equal to the peak value of the current or voltage, rotating with a constant angular frequency of ω radians per second. If A is a point moving anticlockwise in a circular path as in *Figure 4.8(a)*, the vertical projection of the line AB varies as the angle ω*t* varies, hence, since AB/OA = sin ω*t*, the projection of OA = A sin ω*T*, as shown in diagram *(b)*. That is, the

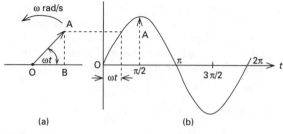

Figure 4.8

projection of the phasor on the vertical axis provides the *instantaneous* value of the quantity.

The base line of the generated sine wave is divided to represent the angular rotation of the phasor OA from the horizontal position OB. Since 2π rads = 360° which corresponds to one rotation or cycle, ω may be expressed as $2\pi f$, where f is the frequency in Hz. Hence the base line may be divided in terms of time or angle.

PHASE DIFFERENCE

The phase difference between two alternating quantities of the same frequency (we cannot compare the phases of waveforms of different frequencies) can be found by comparing the 'starting' points of the two waveforms, as seen in *Figure 4.9(a)*. Here the quantities are represented at a given instant by the phasors OA and OB; the angle between these is *phase angle* ϕ. OA is in 'front' of OB timewise and *leads* OB, while OB *lags* behind OA by the same angle. The projections forming the wave diagram of *(b)* show the relative displacements of the curves along the time axis resulting from the phase difference. Notice that the leading phasor is that which passes through its zero or its peak value first.

We can find the general equation for the phase-shifted curve easily enough, for if OB = $\hat{V}$, then Ob = $v = \hat{V} \cdot \sin(\omega t - \phi)$. To cover all possibilities, a sinusoidally varying quantity can be expressed as $v = \hat{V} \cdot \sin(\omega t \pm \phi)$ where the sign of ϕ will indicate a lead or lag relative to $\hat{V} \cdot \sin \omega t$.

The same sinusoid can also be expressed by the cosine wave function $v = \hat{V} \cdot \cos(\omega t \pm \phi - \pi/2)$ since $\sin x = \cos(x - \pi/2)$ for all values of x. Note that a phase difference of 2π radians or 360° corresponds to zero phase difference, and a phase difference of 180° corresponds to inversion or an *antiphase* condition.

We could, of course, express the phase difference in terms of time, but time in this case is inconvenient because the lead of one wave upon another would then depend on frequency; for if the speed of rotation of the phasors were doubled, for example, the lead of the OA phasor upon the OB phasor would, referred to time, be halved. The lead of the OA phasor in terms of angle, however, does not change whatever the speed of rotation of the phasors, i.e. the frequency, and so angle is always used to express the phase difference between two sine waves.

Summarizing briefly, it is important to remember when dealing with phasors in this way, that although drawn in one fixed position the phasors are rotating at a constant angular velocity ω rad/sec, and that the instantaneous values are the vertical projections of the phasors at any moment in time. We are, in effect, stopping the rotation while we make a study of conditions. It is rather like taking a film of a rapidly moving object, and then selecting one particular frame of the film for detailed study. In the case of our rotating phasors, since their angular velocities are all the same, it does not particularly matter which frame of the film we choose for study. But it is conventional to consider our instant of investigation as being at time $t = 0$ so that the origin of axes then represents zero time as well as zero angle; and for phase displacements to be taken from the horizontal left to right reference axis, positive angular displacements (a leading phase) being anticlockwise from this axis.

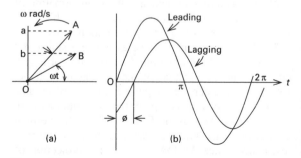

(a)   **(b)**

Figure 4.9

> **Example (10)** Analyse the equation $i = 5 \cdot \sin 314t$.
>
> *Solution* This equation concerns a sinusoidal current wave. Relating this to the standard expression $i = \hat{I} \cdot \sin \omega t$, we find that the peak value of the current is 5 A, so making the r.m.s. value equal to $0.707 \times 5 = 3.8$ A. Also $\omega = 2\pi f = 314$, so that $f = 314/2\pi = 50$ Hz.

Example (11)   Two sine waves are represented by the equations $v = 71 \cdot \sin(314t - 30°)$ and $i = 10 \cdot \sin(314t + 45°)$. Find the peak and r.m.s. values of each wave and the angle of phase difference.

Solution   As written here, the bracketed terms are hybrids, involving both radian and degree measure. This is acceptable practice but the degree symbol must be clearly indicated if you use this symbolism.

   For the voltage wave, $\hat{V} = 71$ V, hence the r.m.s. value is $0.707 \times 71 = \underline{50\ V}$. For the current wave, $\hat{I} = 10$ A, hence the r.m.s. value is $\underline{7.07\ A}$.

   The phase difference is $45° + 30° = \underline{75°}$ (or $5\pi/12$ rad); see *Figure 4.10*, and i leads v by this angle.

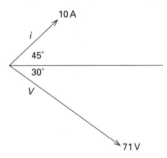

Figure 4.10

Example (12)   A sinusoidal voltage having a frequency of 50 Hz reaches a maximum excursion of 100 V at a time $t = 3$ ms. What is the equation of voltage as a function of time?

Solution   From our standard equation $v = \hat{V} \cdot \sin(\omega t + \phi)$ we have $\hat{V} = 100$, and $\omega = 2\pi f = 100\pi$ rad/s $= 314$ rad/s. The positive peak is reached when $(\omega t + \phi) = \pi/2$ or any multiple of 2π.

   Then $\phi = \pi/2 - 314 \times 3 \times 10^{-3} = 1.571 - 0.942 = 0.629$ rad, or, in degrees, $0.629 \times 360/2\pi = 36°$.

   The voltage equation is therefore $\underline{v = 100 \cdot \sin(314t + 36°)}$.

Example (13)   Two sinusodial voltages are represented by $v_1 = 100 \cdot \sin 754t$ and $v_2 = 50 \cdot \cos(754t - \pi/6)$. What is the frequency of these voltages? Use a phasor diagram to find the re-

sultant sum of these two waves and by scaling or otherwise find the phase angle between them. If the resultant is applied to a circuit of resistance 68 Ω find the r.m.s. current and the power dissipated in the resistor.

Solution   Here the frequency $f = 754/2\pi = \underline{120\ Hz}$

   We notice that v_1 has a cosine expression and this can be written as $v_2 = 50 \cdot \sin(754t + \pi/6 + \pi/2)$, since a sine wave is $\pi/2$ rad (or $90°$) in advance of a cosine wave. This kind of conversion is not strictly necessary, but it does often avoid confusion when drawing the phasor diagram. So we have that $v_2 = 50 \cdot \sin(754t + 2\pi/3)$.

   The diagram is shown in *Figure 4.11*, where v_2 leads v_1 by $2\pi/3$ rad ($120°$) and the resultant sum V_r leads v_1 by some angle ϕ. At this stage, trigonometry can be used, but phasor quantities can be added by resolving each phasor into its horizontal and vertical components – though this is only valid when phasors having the same units are involved. So we have

Horizontal component $OA = V_r \cos\phi$

$$= v_1 - v_2 \cos 60°$$

$$= 100 - (50 \times 0.5)$$

$$= 75v$$

Vertical component $AB = v_r \sin\phi = v_2 \sin 60°$

$$= 50 \times 0.866 = 43.4v$$

Hence, by Pythagoras $v_r = \sqrt{(75^2 + 43.3^2)}$

$$= \underline{86.55\ V}$$

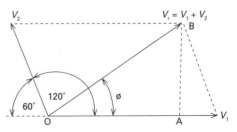

Figure 4.11

The r.m.s. value of this resultant $= 86.55 \times 0.707$

$$= 61.2 \text{ V}$$

The r.m.s. current is then $I = \dfrac{V}{R} = \dfrac{61.2}{68}$

$$= \underline{0.9 \text{ A}}$$

The power dissipated $W = I^2 R = (0.9)^2 \times 10$

$$= \underline{8.1 \text{ W}}$$

We can calculate ϕ by using the cosine rule, but by drawing a phasor diagram to a suitable scale, say 1 cm = 20 V, we can check on the above result for V_r as well as finding ϕ. If you do this, you will find $\phi = 30°$. The equation of the resultant is therefore

$$v_r = 86.55 \,.\, \sin (7554t + 30°)$$

Example (14) Determine the r.m.s. value of an alternating symmetrical triangular wave having a maximum level of 10 V and a periodic time of 16 ms.

Solution The r.m.s. value is the same for all half-cycles so only a positive half-cycle need be considered. Since the half period is 8 ms, the time axis is conveniently divided into eight equal intervals and ordinates can be erected within the waveform as *Figure 4.12* shows. We need first to establish the (voltage)^2 curve, then find its mean value, finally taking the square root of this mean value.

From the construction of *Figure 4.12*, the or-

dinates are symmetrical about the central AB and have, in order, the following lengths

0   2.5   5   7.5   10   7.5   5   2.5   0

Hence the values of V^2 at the same points along the time axis are

0   6.25   25   56.25   100   56.25   25   6.25   0

This enables the curve of V^2 to be drawn (to a new scale) as the diagram shows. Although this curve is not strictly called for or even necessary for a solution, it does illustrate the process we are using here, and also makes us aware that the shape of the V^2 curve is not just another isosceles triangle, a conclusion we might otherwise carelessly draw. The mid-ordinates can now be added to the v^2 curve.

By the mid-ordinate rule, the area under the v^2 curve can be determined as being the width of each of the strips multiplied by the sum of the mid-ordinate lengths. The values of v at the mid-ordinates are, reading along the time axis, to be more accurate than using the graph

1.25   3.75   6.25   8.75   8.75   6.25
3.75   1.25

Hence, taking the time intervals to be a unit of time, the area under the v^2 curve is

$$1 \times (1.563 + 14.06 + 39.06 + 76.54 + 76.54 + 39.06 + 14.06 + 1.563)$$

$$= 262.48$$

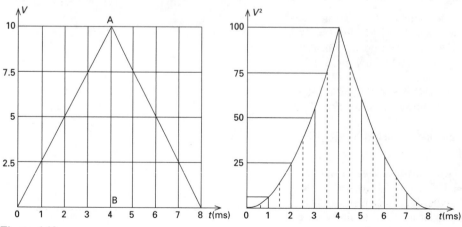

Figure 4.12

Hence

$$\text{Mean } v_2 = \frac{262.48}{8} = 32.81$$

and

$$\text{r.m.s. } V = \sqrt{32.81} = \underline{5.73 \text{ V}}$$

It can be shown that the true value of the r.m.s. level in any triangular waveform of the type seen in this example is given by (peak)/$\sqrt{3}$. The discrepancy from the true value of 5.77 V which appears in the above solution results from the small inaccuracies arising from the use of the mid-ordinate rule and the decimal approximations made in finding the v^2 values. Those who know calculus might like to verify the true result.

Subtraction of phasors

Subtraction of currents or voltages may be obtained from phasor diagrams provided the phasor for the quantity being subtracted is rotated 180°, that is, drawn in the opposite direction. *Figure 4.13* illustrates the procedure, and the resultant then follows from the application of the parallelogram rule.

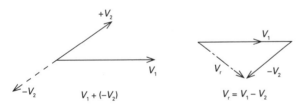

Figure 4.13

Example (15) Two alternating voltages are given by $v_1 = 20 \cdot \sin \omega t$ and $v_2 = 30 \cdot \sin (\omega t - \pi/2)$. Find from a scaled diagram the resultant of $v_1 - v_2$ and the angle between the quantities.

Solution The diagram is drawn in *Figure 4.14* where a scale of 1 cm to 10 V is used. From measurements, resultant v_r is measured as 36 V and angle ϕ as 56°. These figures are easily checked (and indeed found with greatest accu-racy) by simple trigonometry. The equation of the resultant is therefore

$$v_r = v_1 - v_2 = 36 \cdot \sin (\omega t + 56°)$$

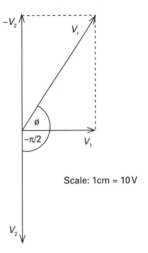

Scale: 1cm = 10 V

Figure 4.14

COMPLEX WAVEFORMS

Any repetitive waveform of frequency f can be resolved into the sum of a number of pure sinusoidal waveforms having frequencies f, $2f$, $3f$, etc., the number of these component waves being finite or infinite. By using some higher mathematics, known as Fourier's theorem, any repetitive wave, however complicated its shape may appear, can be resolved into its components sine waves. We shall not pursue this theorem any further here because it is beyond the scope of this book; but we mention it because it is a way of stating that any complex waveform can be split up into a fundamental frequency and 'harmonic' frequencies. The fundamental has frequency f, the second harmonic has frequency $2f$, the third harmonic has frequency $3f$, and so on. Sound waves produced by the voice or musical instruments always contain a large number of harmonics, and it is these additional sinusoidal components which give musical notes their particular quality or 'timbre'. What is true of sound waves is also true of electrical waves, so if a complex wave is acting in a circuit we can treat it as a series of sinusoidal alternating quantities, each acting independently of the others for purposes of analysis.

The subject is best dealt with here by way of worked examples, with the necessary discussion as we go on.

Example (16)   A certain complex wave can be expressed as

$$i = 282 \sin 250\pi t + 141 \sin 500\pi t \text{ mA}$$

This current flows in a resistance of 200 Ω. Deduce an expression for the waveform of the voltage across this resistance, and sketch on common axes the current and voltage waveforms, showing the time scales and the maximum values.

By ordinate addition derive the form of the complex resultant of which these two currents are component parts.

Solution   The current contains two separate sinusoidal components, the second being twice the frequency of the first. (How do we know this?) So, we have a fundamental plus a second harmonic. If we compare each part of the complex expression with our standard equation $i = \hat{I} \sin \omega t = \sqrt{2} . I \sin \omega t$, where I is the r.m.s. value, we notice, for the first part of the complex wave, $282 \sin 250\pi t$, that the peak value of the current is 282 mA (which is equal to 200 mA r.m.s.), and the ω term corresponds to 250π, whence

$$\omega = 2\pi f = 250\pi$$

$$\therefore \quad f = 125 \text{ Hz}$$

So this part represents a current component wave of peak amplitude 282 mA and frequency 125 Hz.

Similarly, for the second component wave, we find that the peak value is 141 mA (which is equal to 100 mA r.m.s.) and that

$$\omega = 2\pi f = 500\pi$$

$$\therefore \quad f = 250 \text{ Hz}$$

Hence this component has half the peak amplitude and twice the frequency of the first or fundamental component.

Now in a resistance the current and voltage are in phase, so that the instantaneous voltage across the resistance is the product of the instantaneous current (in amperes) and the resistance (in ohms).

$$v_r = 200(0.282 \sin 250\pi t + 0.141 \sin 500\pi t)$$

$$= 56.4 \sin 250\pi t + 28.2 \sin 500\pi t \text{ volts}$$

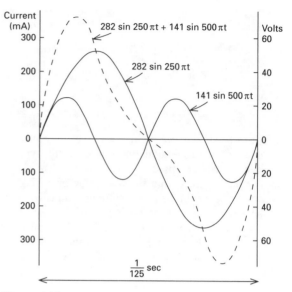

Figure 4.15

The current and voltage waveforms for the two components can be sketched in the usual way, to vertical axes of current and voltage and a common horizontal axis of time t. *Figure 4.15* shows the waveforms for one complete cycle of the fundamental. The resultant waveform actually present in the circuit is, of course, the sum of these two individual components; using ordinate addition this resultant is shown in broken lines. This is the waveform of a sine wave with the additional of a second harmonic; the distortion from a true sinusoidal wave is clearly evident, and in this example is known as second-harmonic distortion.

See now if you can show that the power dissipated in the resistance is 10 W.

SUMMARY

- The frequency of an alternating quantity is the number of complete cycles occurring in 1 second.
- The periodic time is the time taken to complete 1 cycle.
- For a sinusoidal wave the half-cycle average = 0.636 × peak value.

- The r.m.s. value of a sinusoidal wave = 0.707 × peak value.
- Any sinusoidal alternating quantity may be represented as a rotating phasor whose length is equal to the peak (or the r.m.s.) value of the sinusoidal quantity.
- Any number of sinusoidal quantities of the same frequency may be added together by adding the phasor representing them; the phasors being drawn so that the angles between them are the angles of phase difference that they represent.
- Phasor diagrams follow the rules of vector diagrams (e.g. force or velocity diagrams) and can provide approximate answers by suitable scaling, quick checking or a clear visualization of a circuit's behaviour. The procedure is (a) transform the functions of time (the waveforms) into constant phasors, (b) perform the required operations on the phasors using the phasor diagrams, (c) transform the resultant phasors into functions of time.

REVIEW QUESTIONS

1 What is meant by alternating current? Explain what is meant by (a) peak value, (b) instantaneous value, (c) average value, (d) r.m.s. value.
2 Explain why an alternating quantity is normally expressed in terms of its r.m.s. value. Why are a.c. ammeters and voltmeters calibrated to read r.m.s. values?
3 Given that a phasor is a constant quantity, how can it represent a variable function of time?
4 What is the reading of a d.c. ammeter which carries a current $i = 5 \sin 314t$ A?
5 An alternating voltage is expressed as $v = 8 \sin (\omega t + 30)$ volts. Explain exactly what each part of this expression means.
6 The expression given in the previous question is said to be 'hybrid'. What does this mean? What would the expression be if it was 'not hybrid'?
7 Given $v_1 = 10 \sin 2000t$ and $v_2 = 5$

$\cos (1000t + \pi/3)$, explain why the phasor method of adding $v_1 + v_2$ cannot be used.
8 Write down an expression for the instantaneous value of a sinusoidal current of 12 A r.m.s. and frequency 400 Hz.
9 What is the wavelength of a transmission line signal whose frequency is 1 kHz and velocity half that of light?
10 What do you understand by a complex wave? How are harmonic components responsible for the complex form of the wave?

EXERCISES AND PROBLEMS

(17) An audio signal has a frequency of 3.4 Hz. What is its periodic time?

(18) A radio programme is transmitted at a frequency of 950 kHz. What is the wavelength of this transmission?

(19) The graph of *Figure 4.16* shows one cycle of an alternating current of frequency 50 Hz. Find the average and r.m.s. values of this waveform.

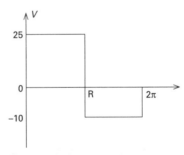

Figure 4.16

(20) The voltage in a circuit remains at 5 V for 40 ms and then to complete the cycle acts at 2 V in the opposite direction for 60 ms. Find the mean value of this voltage, and calculate the r.m.s. current which would flow in a resistor of 27 Ω connected across the circuit.

(21) Sketch the following signal waveforms to a suitable time axis: (a) a step of +10 V at a

time $t = 1$ s; (b) a triangular wave of peak value 25 V and frequency 20 Hz; (c) a train of rectangular pulses of amplitude 5 V, duration 1 ms and frequency 100 Hz.

(22)   A varying current with a periodic waveform follows the pattern shown in *Figure 4.17*. Find the mean and r.m.s. values of this current and calculate the heat dissipated in a 10 Ω resistor wired across the supply over each millisecond of operation.

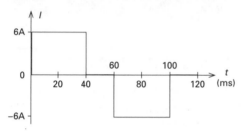

Figure 4.17

(23)   A sinusoidal waveform of peak amplitude 3 V is superimposed on a steady d.c. voltage of 5 V, see *Figure 4.18*. What is the r.m.s. value of the combined waveform?

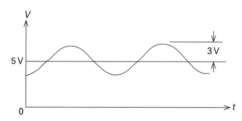

Figure 4.18

(24)   An alternating voltage has a peak value of 24 V and a triangular waveform that is in the form of an isosceles triangle. By using eight time intervals in a half-cycle and the mid-ordinate rule, find the r.m.s. value of the voltage.

(25)   An alternating voltage has a semi-circular waveform of peak value 20 V and a frequency of 25 Hz. Calculate the r.m.s. and mean level of this waveform.

(26)   What is meant by phase angle in an a.c. circuit? Explain the terms 'leading' and 'lagging'.

(27)   When a phasor diagram is being used to analyse an a.c. circuit system, what basic assumptions must be made about the a.c. voltages or currents acting in the circuit?

(28)   A voltage wave is represented by the equation $v = 10 \cdot \sin 251t$ volts. What does v actually represent? What are the peak, r.m.s. and average values of this wave, and at what frequency is it operating?

(29)   Express the following sine wave signals as specific functions of time: (a) a current of peak amplitude 20 mA and frequency 100 Hz; (b) a voltage of r.m.s. value 240 V and frequency 50 Hz; (c) a voltage of peak amplitude 20 V and frequency 15 kHz.

(30)   Three currents I_1, I_2 and I_3 have r.m.s. magnitudes of 5 A, 7 A and 10 A, respectively. I_2 leads I_1 by 40° and I_3 leads I_2 by 50°. Use a phasor diagram drawn to a suitable scale to find the r.m.s. value of the resultant current I and the phase angle between I and I_1

(31)   Sketch any three phasors and explain the graphical construction to obtain $v_1 + v_2 - v_3$.

(32)   Given that $v_1 = 5 \cdot \cos (100t + \pi/2)$ and $v_2 = 10 \cdot \cos (200t)$ explain why the phasor method cannot be used to find $v_1 + v_2$.

(33)   Find the phasor sum of the following current waves and express the sum as (a) a sine function, (b) a cosine function: $v_1 = 120 \cdot \sin (\omega t + \pi/3)$, $v_2 = 50 \cdot \cos (\omega t - \pi/2)$.

(34)   Draw phasors to represent the following voltages: $v_1 = 6 \cdot \sin \omega t$, $v_2 = 4 \cdot \sin (\omega t - \pi/3)$, $v_3 = 10 \cdot \sin (\omega t + \pi/4)$. Find the resultant sum $v_1 + v_2 + v_3$ in the form $v = \hat{V} \cdot \sin (\omega t + \phi)$.

(35)   If $i_1 = 35 \cdot \cos (\omega t + 2\pi/3)$, $i_2 = 50 \cdot \cos (\omega t + 5\pi/12)$, find $i_1 - i_2$.

(36)   At what time after $t = 0$ does a sinusoidal current $i = 20 \cdot \cos (377t - 43°)$ A reach its positive peak value?

(37)   A sinusoidal current has an r.m.s. value of 6 A and a frequency of 50 Hz. Write down the equation of this current as a function of time. Find (a) the time to increase from zero to 4 A, and (b) the level of the current 1.9 ms after passing through zero positively.

(38)   Write down an expression representing a current of 300 mA r.m.s. and frequency 10 kHz, together with its third harmonic having half the amplitude of the fundamental, the two being in step at zero time.

(39)   The fundamental component of a complex wave is expressed by $v = 40 \sin 314t$ volts.

Each harmonic in the complex wave has an amplitude one-fourth that in the preceding harmonic. Write down the expression for (a) the third harmonic, (b) the fifth harmonic. Which harmonic would have an amplitude less than one-twenty-fifth of the amplitude of the fundamental?

5

Alternating quantities 2: series and parallel circuits

BASIC RELATIONSHIPS

Resistance is that property of an electric circuit which opposes the flow of current by absorbing energy. This energy is generally dissipated as heat and the rate of dissipation is proportional to the product I^2R. Reactance is that property which opposes any change in existing current or voltage conditions in a circuit and is that quantity which determines the r.m.s. current I which flows when an r.m.s. voltage is applied. Reactance is possessed by inductors and capacitors.

In an inductance, a change in the magnitude or direction of the current changes the magnetic field associated with the current and hence changes the flux linking with the turns of the coil; an e.m.f. is then induced in the coil which opposes the change producing it. This opposition to the change in current is known as the inductive reactance, denoted by X_L and measured in ohms. When an alternating sinusoidal current flows in an inductance, there is a continual change in both magnitude and direction and the reactive opposition increases as the frequency increases, since the rate of change of current is then greater and consequently the self-induced back-e.m.f. is greater. We recall that inductive reactance is expressed as

$X_L = 2\pi fL$

where L is in henrys and f is the frequency in hertz.

When a capacitor is connected to an alternating supply, it undergoes a periodic process of charge and discharge, thus appearing to conduct an alternating current. The charge and discharge cycle constitutes an opposition to the flow of current and this constitutes the capacitive reactance of the capacitor, denoted by X_c and again measured in ohms. Capacitive reactance decreases as the frequency increases. We recall that capacitive reactance is expressed as

$X_c = 1/2\pi fC$

where C is in farads and f is in hertz.

Reactance resists the passage of an alternating current without the dissipation of energy. In purely reactive elements, all the energy supplied to the magnetic or the electrostatic fields during alternate quarter cycles of the input wave is returned to the generator during the succeeding quarter cycles. Hence the total energy supplied and the power dissipated is, in both cases, zero.

Phase

In a purely resistive circuit, the current at any instant is directly proportional to the voltage. Then, if the applied voltage is a sine wave, the waveforms of voltage and current being zero at the same instant are at their peak values at the same instant, hence voltage

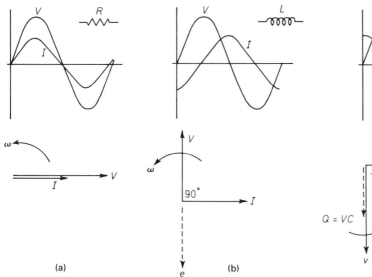

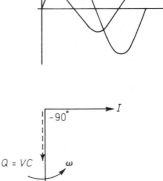

(a)   (b)   (c)

Figure 5.1

and current are in phase. *Figure 5.1(a)* shows the phase relationship and the phasor diagram.

In a purely inductive circuit, the self-induced e.m.f. is, by Lenz's law, in opposition to the applied voltage and proportional to the rate of change of current. So

$$e = -L\frac{di}{dt}$$

where the negative sign simply indicates the effect of opposition. But if the current is

$$i = \hat{I}\sin \omega t$$

the rate of change of current

$$\frac{di}{dt} = \omega\hat{I}\cos \omega t$$

so that the back-e.m.f. becomes $e = -L\,\omega\hat{I}\cos \omega t$. But this e.m.f. is equal and opposite to the supply voltage. Hence the supply voltage $v = L\,\omega\hat{I}\cos \omega t$, being a cosine wave, leads the current, which is a sine wave, by 90°. Or, as it is more usually remembered: in a purely inductive circuit the current lags on the voltage by 90°. *Figure 5.1(b)* shows the relationship and the phasor diagram.

In a purely capacitive circuit, the instantaneous charge on the capacitor $q = Cv$ where v is the instantaneous voltage. Let

$$v = \hat{V}\sin \omega t$$

then

$$q = Cv = \hat{V}C\sin \omega t$$

but current is rate of change of charge, and so

$$i = \frac{dq}{dt} = \omega C\hat{V}\cos \omega t$$

Compared with the voltage wave, the current, being a cosine wave, has a lead on the voltage of 90°. Hence, in a purely capacitive circuit the current leads the voltage by 90°. The phasor diagram for the capacitive circuit is shown in *Figure 5.1(c)*.

It is essential that you keep these basic principles in mind at all times. To refresh your memory on them, work the following assignment problems before going any further.

(1) Calculate the reactance of a coil on inductance 100 mH when it is connected to a 500 Hz supply.

(2) A coil of inductance 0.28 H has a reactance of 4400 Ω when connected to an a.c. supply. What is the frequency of the supply?

(3) Calculate the reactance of a 1000 pf capacitor when connected to a 50 V 20 kHz supply. What current will flow in the circuit?

(4) A capacitor takes a current of 5 A from a 230 V 50 Hz supply. Calculate the capacitance of the capacitor.

Impedance

The voltage–current relationship for any a.c. circuit containing inductance and capacitance obeys Ohm's law exactly as it does for d.c. circuits where only resistance is concerned. For this reason such circuits are known as linear or passive circuits to distinguish them from circuits containing, for example, diodes or transistors, which do not, in general terms, obey Ohm's law.

In a.c. circuits, in addition to resistance, we have reactance, and voltages and currents are measured by their r.m.s. values. In real circuits, reactance is never divorced from resistance, particularly so in the case of large inductances where coils with many hundreds of turns of wire may be concerned. So in such circuits there is a combined opposition, reactive and resistive, to the flow of current and this opposition is known as the impedance of the circuit. Impedance is denoted by the symbol Z and is measured in ohms.

Also, the phase angle between voltage and current is no longer exactly zero as it is for pure resistance nor exactly 90° as it is for pure reactance, but lies between these limits, the actual value depending on the relative magnitudes of the reactive and the resistive elements. Further, as the reactive element depends on frequency, the phase angle will also be a function of frequency.

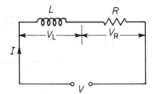

Figure 5.2

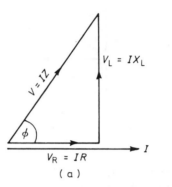
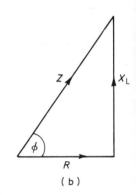

Figure 5.3

A.C. CIRCUIT PROBLEMS

The solution of a.c. circuit problems, both for series and parallel circuits, usually comes down to finding the magnitude and phase angle of the resultant phasor of a number of other phasors set up to represent the circuit conditions. There are several approaches to problems of this sort, but we are concerned with only two of these:

(a) by making a scaled phasor diagram of the circuit currents, voltages or impedances from which the required solution may be obtained by direct measurement of the resultant phasor both as regards magnitude (length) and phase (angle); and

(b) by the application of simple trigonometry to a phasor sketch diagram of the circuit quantities.

We shall concentrate mainly on methods of calculation by trigonometry but it will be wise to verify the solutions by scaled diagrams wherever possible.

Inductance and resistance in series

In the circuit of *Figure 5.2* a resistance $R\,\Omega$ is in series with an inductor of reactance $X_L\,(= \omega L)\,\Omega$. The voltage across the resistance, V_R, is in phase with the current I, and the voltage across the inductor, V_L, leads the current by 90°. The supply voltage V is the phasor sum of V_R and V_L since we are dealing with a series circuit. Also the current is common to all the component parts of a series circuit, so we can use current as our reference phasor and relate all the voltage conditions to it.

Figure 5.3(a) shows the phasor diagram of voltages; angle ϕ is the phase difference between V and I, and I lags V by this angle. Clearly for such a combination of resistance and inductance ϕ must be positive and lie between 0° and +90°. By Pythagoras we have

$$V = \sqrt{(V_R^2 + V_L^2)}$$

and

$$\tan \phi = \frac{V_L}{V_R}$$

It is usually best to work in terms of resistance, reactance and impedance rather than voltage. Look at the triangle of *Figure 5.3(a)* and note that its sides are made up of the phasors $V_R = IR$, $V_L = IX_L$ and $V = IZ$. Since the current I is a common factor to all these, we can draw a similar triangle strictly in terms of R, X_L and Z. This has been done in *Figure 5.3(b)* and gives us what is known as the impedance triangle. The angle ϕ is, as before, the phase angle between current and voltage. Again, since this is a right-angled triangle, we can write

$$Z = \sqrt{(R^2 + X_L^2)}$$

so that the current

$$I = \frac{V}{\sqrt{(R^2 + X_L^2)}}$$

and

$$\tan \phi = \frac{X_L}{R}, \cos \phi = \frac{R}{Z}, \sin \phi = \frac{X_L}{Z}$$

The impedance triangle is a very useful diagram in the solution of problems in alternating current circuits. See how it, and the voltage phasor diagram, is used in the following worked problems.

Example (5)   A 250 V 50 Hz supply is connected to an inductive circuit. A current of 2A lagging on the voltage by 30° flows in the circuit. Calculate the impedance, reactance, resistance and inductance of the circuit.

Solution   *Figure 5.4(a)* shows the circuit with the voltage and impedance triangles at (*b*). In drawing the voltage triangle, *I* is the reference phasor, V_R is in phase with *I* and V_L leads *I* by 90°.

$$\text{Circuit impedance } Z = \frac{V}{I} = \frac{250}{2} = 125 \ \Omega$$

From the impedance triangle

$$\cos \phi = \frac{R}{Z} \text{ and } \sin \phi = \frac{X}{Z}$$

for $\phi = 30°$ $\cos \phi = 0.8660$ and $\sin \phi = 0.5$.

Then

$$\text{resistance } R = Z \cos \phi = 125 \times 0.8660$$
$$= 108.25 \ \Omega$$

$$\text{reactance } X_L = Z \sin \phi = 125 \times 0.5 = 62.5 \ \Omega$$

But

$$X_L = 2\pi f L$$

$$\therefore \quad L = \frac{X_L}{2\pi f} = \frac{62.5}{2\pi \times 50} = 0.2 \text{ H}$$

Example (6)   In a series a.c. circuit a pure inductance is placed in series with a 150 Ω resistor. If the current flowing is 0.5 A when the supply is 100 V at 800 Hz, find with the aid of a phasor diagram (a) the p.d. across the inductor, (b) the inductance of the coil, (c) the phase angle between voltage and current.

Solution   *Figure 5.5(a)* shows the circuit with the voltage phasor diagram at (*b*). Using the voltages this time, we have $V^2 = V_R^2 + V_L^2$.

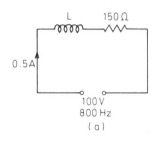

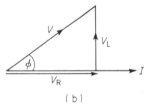

(b)

Figure 5.5

Now $V = 100$ V and $V_R = 0.5 \times 150 = 75$ V

$$\therefore \quad V_L = \sqrt{(100^2 - 75^2)} = 66.14 \text{ V}$$

Notice that the ordinary algebraic sum of V_R and V_L does not come to 100 V.

(b) $X_L = \dfrac{V_L}{I} = \dfrac{66.14}{0.5} = 132.28 \ \Omega$

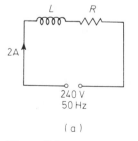

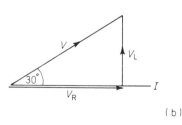

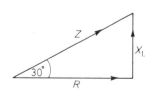

(a)   (b)

Figure 5.4

But

$$X_L = 2\pi f L$$

$$\therefore \quad L = \frac{X_L}{2\pi f} = \frac{132.28}{2\pi \times 800} = 0.026 \text{ H}$$

(c) $\cos \phi = \dfrac{V_R}{V} = \dfrac{75}{100} = 0.75$

$$\therefore \quad \phi = 41.4°$$

Try the next problem on your own.

(7)   A series circuit having a ratio of resistance to reactance at 100 Hz of 4 to 1 draws a current of 1.25 A when connected to a 50 V 100 Hz supply. Calculate the resistance and inductance of the circuit. Find the frequency at which this circuit would take 2 A from a 200 V supply.

Resistance and capacitance in series

As in the case of resistance and inductance in series, the current is limited by the combined effect of resistance and reactance, this being the a.c. impedance of the circuit. Again, the current is common to both circuit elements. The voltage across the resistor, V_R, is in phase with I and the voltage across the capacitor, V_c, lags the current by 90°. The phasor diagram of *Figure 5.6(a)* shows the addition of V_R and V_C to obtain the supply voltage V. From this triangle the phasor magnitudes are $V_R = IR$, $V_C = IX_C$ and $V = IZ$. Since the current is common, the impedance triangle of *Figure 5.6(b)* can be drawn.

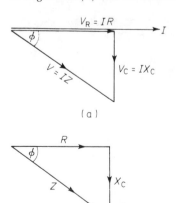

(a)

(b)

Figure 5.6

You should notice that this triangle is similar to that obtained for resistance and inductance in series except that the capacitive reactance X_C is drawn in the direction opposite to that given to X_L. Clearly for such a combination of resistance and capacitance ϕ will be negative, lying between 0° and −90°. By Pythagoras we have

$$Z = \sqrt{(R^2 + X_c^2)}$$

and so

$$I = \frac{V}{\sqrt{(R^2 + X_c^2)}}$$

also

$$\tan \phi = \frac{X_c}{R}, \cos \phi = \frac{R}{Z}, \sin \phi = \frac{X_c}{R}$$

Do not attempt to memorize these relationships or those which have gone before. Simply keep in mind the phase relationships between voltage and current for the three basic circuit elements and this, together with Ohm's law, will enable you to work out each problem on its own merits.

Here are two more worked examples to follow through to make sure that you thoroughly understand these very basic principles.

Example (8)   A resistor of 40 Ω is in series with a 15 μF capacitor across a 24 V 400 Hz supply. Calculate (a) the circuit impedance, (b) the circuit current, (c) the phase angle, (d) the voltages across C and R.

Solution   (a) We need to know reactance X_C.

$$X_c = {}^1\!/_2 \pi f C = \frac{10^6}{2\pi \times 400 \times 15} = 26.53 \ \Omega$$

then

$$Z = \sqrt{(R^2 + X_c^2)} = \sqrt{(40^2 + 26.53^2)} = 48 \ \Omega$$

(b) $I = \dfrac{V}{Z} = \dfrac{24}{48} = 0.5 \text{ A}$

(c) $\tan \phi = \dfrac{X_c}{R} = \dfrac{26.53}{40} = 0.6633$

$$\therefore \quad \phi = -33.6°$$

(d)   Volts across $R = IR = 0.5 \times 40 = 20$ V

Volts across $C = IX_C = 0.5 \times 26.53$

$$= 13.26 \text{ V}$$

Notice again that the ordinary algebraic sum of V_R and V_C does not come to 24 V.

Example (9)   A circuit comprising a 500 Ω resistor in series with a capacitor is supplied from an a.c. source whose frequency and output voltage may be independently adjusted. When the frequency is set to 1 kHz, the output voltage is adjusted until the p.d. across the capacitor is 7 V and the current drawn is 20 mA. Draw a scaled phasor diagram to find the output voltage of the source for this condition.

The frequency is now adjusted to 500 Hz. What output voltage will now be needed to give the same circuit current as before?

Solution   Again the reference phasor is current I.

Volts across $R = IR = 20 \times 10^{-3} \times 500$

$$= 10 \text{ V}$$

Volts across $C = IX_C = 7$ V and this lags I by 90°

The applied voltage $V = IZ$

Figure 5.7 shows the circuit and the scaled phasor diagram. By measurement

$V = 12.2$ V, $\phi = 35°$ lagging I

When the frequency is halved the capacitive reactance is doubled. Hence, if the circuit current is to be the same at 20 mA, the voltage across the capacitor will be twice what it was originally, i.e. $2 \times 7 = 14$ V. The voltage across R will be unchanged at 10 V. Hence the new output voltage from the source will be

$$V = \sqrt{(V_R^2 + V_L^2)} = \sqrt{(10^2 + 14^2)} = 17.2 \text{ V}$$

(10)   A circuit comprising a 1 µF capacitor in series with a 300 Ω resistor is connected to an a.c. source of 15 V. If a current of 25 mA flows in the circuit, calculate the supply frequency.

L, C AND *R* IN SERIES

As for the series circuits so far discussed, when a circuit is made up of resistive, inductive and capacitive elements all wired in series, the total applied voltage is the phasor sum of the voltages developed across the individual elements. We can perform this addition in the usual way using phasor diagrams and some elementary trigonometry. Also the circuit impedance will be the phasor sum of the separate impedances and since the current flowing is common to all parts of the circuit, the applied voltage is equal to the product of this current and the phasor sum of all the impedances.

Now that we have all three basic elements in the circuit, the phase angle may be positive or negative according to whether the supply voltage leads or lags on the current, and this in turn is dependent on which reactive element is dominant at the particular frequency concerned. Referring to *Figure 5.8*, we have

$$Z = \sqrt{\left[R^2 + (X_L - X_C)^2\right]}$$

and

$$\tan \phi = \frac{X_L - X_C}{R}$$

The direct subtraction of one reactance from the other may not at first look like our customary procedure for dealing with phasors, but as the diagram shows, X_L and X_C are in direct opposition to each other and their phasor resultant is clearly the algebraic difference between them.

In the diagram, we have assumed that X_L is greater

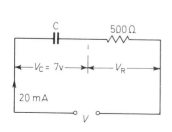

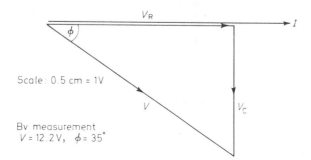

Scale: 0.5 cm = 1V

By measurement
$V = 12.2$ V, $\phi = 35°$

Figure 5.7

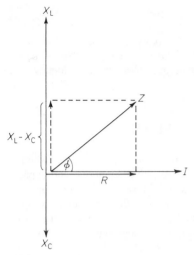

Figure 5.8

than X_C but this situation could, of course, be reversed. In the expression for the circuit impedance Z, therefore, our reactive term is represented by this difference. Since the term is squared, it is not important which reactance is subtracted from which. In the expression for tan ϕ, however, the sign of the subtraction is important if we are to avoid confusion over whether ϕ is leading or lagging: if X_L is greater than X_C, ϕ will be positive and V will lead I, if X_C is greater than X_L, ϕ will be negative and V will lag I.

Usually it is a matter of common sense to decide which way the angle is acting: if the inductive reactance predominates, clearly I will lag V, and if the capacitive reactance predominates, I will lead V. A particular case occurs when $X_L = X_C$. Here the reactive component disappears completely and the circuit phase angle becomes zero. We will investigate this situation very shortly.

Example (11)   A coil of resistance 20 Ω and inductance 0.01 H is connected in series with a capacitance of 4 μF across a 100 V 1000 Hz supply. Calculate (a) the circuit impedance, (b) the circuit current, (c) the phase angle.

Solution   We calculate first of all the circuit reactances:

For the inductor: $X_L = 2\pi fL$

$$= 2\pi \times 1000 \times 0.01$$

$$= 62.8 \ \Omega$$

For the capacitor: $X_C = \frac{1}{2\pi fC}$

$$= \frac{10^6}{2\pi \times 1000 \times 4}$$

$$= 39.8 \ \Omega$$

Obviously the inductive reactance predominates, so we can expect the circuit current to lag on the voltage, that is, ϕ will be positive. However, we now find the circuit impedance:

(a) $Z = \sqrt{[R^2 + (X_L - X_C)^2]}$

$$= \sqrt{[20^2 + (62.8 - 39.8)^2]}$$

$$= \sqrt{[400 + 529]}$$

$$= 30.5 \ \Omega$$

(b)   Circuit current $I = \dfrac{V}{Z}$

$$= \frac{100}{30.5} = 3.28 \ A$$

(c)   Phase angle: $\tan \phi = \dfrac{X_L - X_C}{R}$

$$= \frac{62.8 - 39.8}{20}$$

$$= \frac{23}{20} = 1.15$$

$\therefore$ $\phi = 49°$ leading

This result is illustrated in the phasor diagram of *Figure 5.9*. The diagram is once again an

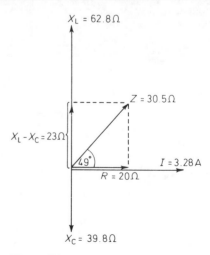

Figure 5.9

impedance diagram, the applied voltage V being the product of the current I and the total circuit impedance Z.

SERIES OR VOLTAGE RESONANCE

In the LCR circuit just discussed, the phase angle between current and voltage depended on the relative magnitudes of the inductive and capacitive reactances at a particular frequency. The voltage across R is IR volts in phase with the reference current. The voltage across L is IX_L volts leading the current by 90°, and the voltage across C is IX_C lagging by 90°.

The total voltage applied to the circuit V is the phasor sum of these three separate voltages and is represented by a resultant which may lead on, lag on or be in phase with the current, depending, respectively, on whether IX_L is greater than, less than or equal to IX_C.

The three possible cases are illustrated in *Figure 5.10*. At (a) the frequency ω is considered to be small, hence $X_L = \omega L$ is small but $X_C = 1/\omega C$ is large. Hence $(X_C - X_L)$ is large capacitively and phase angle ϕ is large and negative. At (b) ω is considered to be very large, hence ωL is large but $1/\omega C$ is small. Hence $(X_L - X_C)$ is large inductively and phase angle ϕ is large and positive. At (c) we consider ω to be such that $\omega L = 1/\omega C$, hence $(X_C - X_L)$ is zero and ϕ is zero. In this circumstance, the voltages developed across the reactances are equal in magnitude and opposite in sign and so cancel out. So the circuit current and voltage are now in phase and so the system behaves as a pure resistance.

When the circuit becomes purely resistive, it is said to be in a state of resonance. The frequency at which resonance occurs is called the resonant frequency. The phenomenon of resonance occurs in circuits containing opposite kinds of reactance because both inductive and capacitive elements can store energy during one quarter cycle of the applied alternating voltage and return it to the generator during the following quarter cycle. Since the inductance is storing energy when the capacitor is returning energy, and vice versa, it is possible for the elements to transfer energy from one to the other successively.

The effect can be compared to familiar mechanical systems such as a swinging pendulum, a bouncing ball or a vibrating string. All these systems have a common property: energy of position or state, known as potential energy, and energy of motion, known as kinetic energy. The pendulum has potential energy at the extremes of its oscillations, but no kinetic energy, the motion being instantaneously halted. As the pendulum descends, the potential energy is converted to kinetic energy, the conversion being complete as the pendulum passes its lowest point, for there the velocity is momentarily greatest. Some energy is lost because of air resistance and friction and as a result the pendulum eventually comes to rest. The energy loss in electrical circuits takes place in the resistance of the components and the connecting wires.

Let the frequency of resonance of a series circuit be f_o. Then, since at this frequency $\omega L = 1/\omega C$, we have

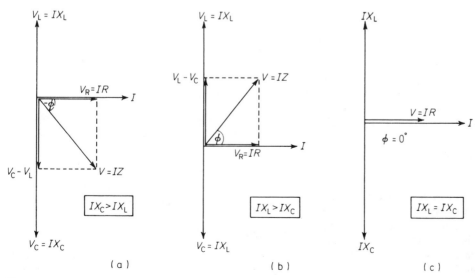

(a)   (b)   (c)

Figure 5.10

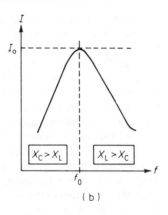

Figure 5.11

$$\omega^2 = \frac{1}{LC} \text{ and } \omega = \frac{1}{\sqrt{LC}}$$

but $\omega = 2\pi f_o$

$$\therefore \quad f_o = \frac{1}{2\pi\sqrt{LC}}$$

We could define resonance as that frequency at which the circuit impedance is a minimum, for since

$$Z = \sqrt{\left[R^2 + (X_L - X_C)^2\right]}$$

$$Z = \sqrt{R^2} = R \text{ when } X_L = X_C$$

and this must be the smallest possible value for Z.

It is, however, strictly correct to say that resonance occurs when the voltage and current are in phase.

Since $Z = R$ at resonance, the current I_o at resonance must be at its greatest value:

$$I_o = \frac{V}{R}$$

Figures 5.11(a) and *(b)* show, respectively, the variation of impedance Z and current I as the frequency is increased through resonance for a series circuit. Below resonance $X_C > X_L$ and the impedance is predominantly capacitive. Above resonance $X_L > X_C$ and the impedance is predominantly inductive and increases with frequency. The circuit is known as an 'acceptor' circuit because it presents the lowest impedance to the flow of current when it is resonant.

Example (12) A coil of resistance $10\,\Omega$ and inductance $100\,mH$ is wired in series with a capacitor of $2\,\mu F$ to a $50\,V$, variable frequency supply. Calculate the resonant frequency. What are the voltages across (a) the capacitor and (b) the coil at resonance?

Solution $f_o = \dfrac{1}{2\pi\sqrt{(LC)}}$

$$= \frac{1}{2\pi\sqrt{(0.01 \times 2 \times 10^{-6})}}$$

$$= \frac{10^4}{2\pi\sqrt{2}} = \frac{10^4}{8.886} = 1125 \text{ Hz}$$

At resonance

$$I_o = \frac{V}{R} = \frac{50}{10} = 5 \text{ A}$$

(a)   Voltage across C:

$$V_C = IX_C = 5 \times \frac{1}{2\pi fC}$$

$$= 5 \times \frac{10^6}{2\pi \times 1125 \times 2}$$

$$= 5 \times \frac{10^6}{14\,137} = 354 \text{ V}$$

(b)   Voltage across L:

$$V_L = IZ_L = I\sqrt{(R^2 + X_L^2)}$$

Notice that will use the coil impedance Z_L not the coil reactance X_L here. Remember that the resistance is an integral part of the coil, and cannot be separated out from the inductance.

Then

$$X_L = 2\pi fL = 2\pi \times 1125 \times 0.01$$

$$= 70.7 \;\Omega$$

and

$$Z_L = \sqrt{(10^2 + 70.7^2)} = 71.4 \;\Omega$$

Hence

$$V_L = 5 \times 71.4 = 357 \text{ V}$$

which for all practical purposes is the same as the voltage across C.

What this example shows us is that the voltages developed across the reactive elements at resonance may be much greater than the supply voltage.

Voltage magnification

Is this last statement always true? Let the resonant voltage across the inductor be V_L and across the capacitor be V_C. Then

$$V_L = I\sqrt{(R^2 + \omega^2 L^2)}$$

$$= \frac{V}{R}\sqrt{R^2 + \frac{L}{C}} \text{ since } \omega^2 = \frac{1}{LC}$$

$$= V\sqrt{R^2 + \frac{L}{CR^2}}$$

Hence

$$V_L > V \text{ for all values of } L \text{ except } L = 0$$

Now

$$V_c = \frac{I}{\omega C} = \frac{I\sqrt{LC}}{C} = \frac{V}{R}\sqrt{\frac{L}{C}} = V\sqrt{\frac{L}{CR^2}}$$

Hence

$$V_c > V \text{ if } L > CR^2$$

These are the conditions for V_L and V_c to be greater than the source voltage V.

At resonance the circuit current $I_o = V/R$ and the voltage across the inductor and capacitor is $I_o\omega L$ and $I_o/\omega C$, respectively. Since $\omega L = 1/\omega C$, these voltages are equal. So

$$V_L = \frac{\omega L V}{R} \text{ and } V_C = \frac{1}{\omega C}\frac{V}{R}$$

and then

$$\frac{V_L}{V} = \frac{\omega L}{R} \text{ and } \frac{V_C}{V} = \frac{1}{\omega CR}$$

These ratios of the voltages across the reactances to the applied voltage are known as the circuit magnification or Q-factor. We notice from the forms of the ratios that the smaller the circuit resistance R, the greater the Q-factor.

Q-factor may also be defined as

$$\frac{\text{reactance of one kind}}{\text{total circuit resistance}}$$

Hence

$$Q = \frac{\omega L}{R} = \frac{1}{\omega CR}$$

(13)   By noting that at resonance, $\omega = \dfrac{1}{\sqrt{(LC)}}$

show that Q may be expressed as $\dfrac{1}{R}\sqrt{\dfrac{L}{C}}$

What factors would you take into consideration of you wanted to design a circuit having a very high Q-factor?

A coil of inductance L henrys and resistance R ohms, having no association with a resonant circuit would appear to have a Q-factor which could take any value, depending on ω. This is true if the coil is to be used over a wide range of frequencies, but in practice an inductance is designed only for a particular frequency range. As ω increases Q will increase provided R remains constant; but resistance depends on the cross-sectional area of the conductor through which the current flows and at high frequencies (those in excess of some 50 kHz) the current tends to flow in the area nearer the surface of the conductor. This is known as the skin effect, and it results in an effective reduction in the cross-sectional area. The relationship between resistance and frequency is not a simple one, but by the use of stranded wire and special methods of winding, the ratio $\omega L/R$ can be made to remain reasonably constant over the frequency range for which the coil is designed.

For inductances using laminated iron cores, such as low-frequency transformers, the Q-factor is quite low, generally of the order of 5 to 20. Coils employed at high radio frequencies have Q-factors of the order of 100 to 300, the higher values being brought about by the use of ferrite cores. Capacitors, unlike inductors, have very low effective series resistance and so on the whole exhibit much higher Q values than the average coil. In practice, therefore, where both L and C are used in resonant circuits, it is usual to treat the bulk of the circuit resistance as residing in the coil and the total effective Q-factor of the complete circuit approximates to that for the coil alone.

The advantage of high Qs in radio-frequency circuits is that such circuits then exhibit a sharp resonance curve (of I plotted against f) and are capable of discriminating between input signals which have frequencies relatively close together. Such circuits are said to be highly selective; see *Figure 5.12(a)*. Low Q circuits on the other hand have a flat response and are not selective in this way, see diagram *(b)*.

We will expand on this aspect of frequency discrimination in the next chapter.

At power frequencies of 50 Hz or so, a high Q circuit is very undesirable. Suppose a capacitor marked as being of 500 V working was used in a series circuit on 240 V mains supplies. This might appear to be adequately rated since the peak voltage of the supply will be 340 V. But at resonance, with a Q

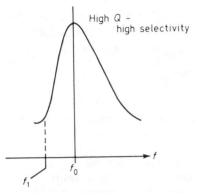

High Q –
high selectivity

Frequency f_1 is easily
distinguished from f_0

(a)

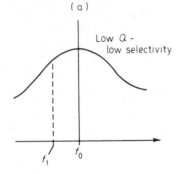

Low Q –
low selectivity

Frequency f_1 is not easily
distinguished from f_0

(b)

Figure 5.12

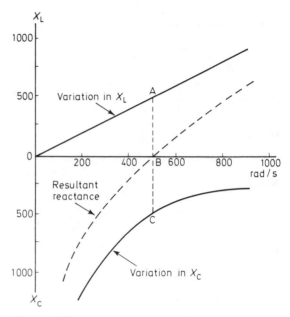

Figure 5.13

factor as low as 5, the voltage across the capacitor will reach some 1200 V and this will most probably lead to dielectric breakdown.

Example (14)   For the frequency range $\omega = 0$ to $\omega = 500$ rad/s, draw on the same pair of axes the reactance–frequency curves for (a) an inductance of 1 H, (b) a capacitance of 4 µF.

From your diagram find the frequency of resonance of a circuit made up of an inductance of 1 H in series with a capacitance of 4 µF.

Solution   Inductive reactance is given by $X_L = 2\pi fL = \omega L$. Hence X_L is directly proportional to ω and so the reactance–frequency curve will be a straight line passing through the origin. We require only two points to establish this line, the origin being one of them. For the other, take $\omega = 500$ rad/s, and then $X_L = 500 \times 1 = 500 \, \Omega$. The straight line is shown in the diagram of *Figure 5.13*.

Capacitive reactance is given by $X_C = 1/\omega C$, hence X_C is inversely proportional to ω and the curve this time will be a rectangular hyperbola. We require several points to obtain its shape reasonably accurately. Taking a range of values for ω we can draw up the following table:

ω	200	400	600	800	rad/s
X_C	1250	625	416	312	Ω

The curve is plotted on the axis of negative ohms as we conventionally take capacitive reactance as being negative.

Resonance occurs when the inductive and capacitive reactances are equal in magnitude and opposite in sign. From a study of the diagram this occurs at a frequency of 500 rad/s, or $500/2\pi = 80$ Hz, where on the perpendicular line ABC, AB = BC.

Parallel circuits

By this stage we should be thoroughly familiar with the phase relationships between voltage and current for the three basic circuit elements of resistance, inductance and capacitance, and there should be no difficulty in applying them to parallel combinations of these elements.

When resistance, inductance and capacitance are connected in various parallel combinations, problems

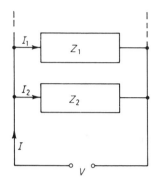

Figure 5.14

arising are solved in general in a manner similar to the methods used for parallel resistor circuits but bearing in mind that the branch currents are no longer necessarily in phase. Whereas in series circuits, the total circuit impedance is the phasor sum of the individual impedances, in parallel circuits the total circuit impedance is given by the phasor reciprocal sum

$$\frac{1}{Z} = \frac{1}{Z_1} + \frac{1}{Z_2} + \ldots$$

Z_1, Z_2 etc. being the individual branch impedances which, of course, must take into account the various phase relationships. *Figure 5.14* shows the general circuit details. There is one other basic difference between series and parallel circuit considerations and that lies in the choice of a reference phasor. In series circuits where, as we have seen, the current is common to all parts of the circuit, we have taken the current as our reference phasor. In parallel circuits the applied voltage V is common to all branches of the circuit; as a consequence we now choose V as our reference phasor and relate the various branch currents I_1, I_2 etc. to it in their respective magnitudes and phases. Further, the total circuit current I is the phasor sum of the individual branch currents, so that

$$I = I_1 + I_2 + \ldots$$

These branch currents are, or course, given by the ratio of the applied voltage V (common to all branches) and the branch impedances Z_1, Z_2 etc. Hence,

$$I = \frac{V}{Z} = \frac{V}{Z_1} + \frac{V}{Z_2} + \ldots$$

from which, by elimination of the common factor V, we arrive at the impedance formula given earlier.

To sum up: when drawing phasor diagrams for parallel circuits, we draw the applied voltage phasor as reference and set off from it the phasors for the various branch currents, scaled (if necessary) to their proper magnitudes and positioned to their proper phase angles relative to V, lagging or leading as the case may be. In an inductive circuit a lagging current will then appear in the fourth quadrant, i.e. ϕ will be negative. For a capacitive circuit a leading current will appear in the first quadrant so that ϕ will be positive.

The most simple parallel combinations are those of resistance in parallel with either a pure inductance or capacitance, and we shall begin our study with these elementary examples.

RESISTANCE AND INDUCTANCE

Figure 5.15(a) shows the circuit diagram with the branch impedances and currents indicated. The inductance branch is, of course, purely reactive in this instance. From the diagram, the current I_R in R is in phase with V, and the current I_L in L lags V by 90°. The phasor diagram is then drawn as shown in *Figure 5.15(b)*; since V is a common factor here, an impedance diagram can be deduced and is shown at *(c)*.

From diagram *(b)* the circuit current

$$I = \sqrt{(I_R^2 + I_L^2)}$$

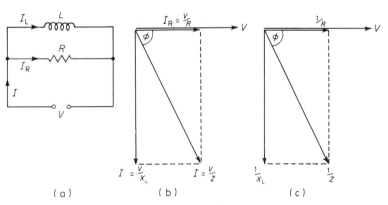

Figure 5.15

and the phase angle between the supply current and voltage is

$$\phi = \tan^{-1} \frac{I_L}{I_R}$$

And from diagram *(c)*

$$\frac{1}{Z^2} = \frac{1}{R^2} + \frac{1}{X_L^2}$$

so that

$$\frac{1}{Z} = \sqrt{\frac{1}{R^2} + \frac{1}{X_L^2}}$$

and phase angle $\phi = \tan^{-1} R/X_L$.

These are the basic relationships in terms of current and impedance of the parallel arrangement of resistance and inductance. As always, make no attempt to 'memorize' these relationships; draw the phasor diagram and apply Pythagoras and basic trigonometry. The next worked example will illustrate the method to adopt.

Example (15)   A coil of inductance 0.5 H is connected in parallel with a 200 Ω resistor across a 50 V 800 Hz supply. Determine (a) the current in each parallel branch, (b) the current drawn from the supply, (c) the phase angle between supply voltage and current.

Solution   *Figure 5.16(a)* illustrates the problem.

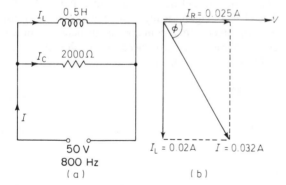

Figure 5.16

(a)   Voltage across both R and L is the applied voltage $V = 50$ V

Reactance of $L = 2\pi f L = 2\pi \times 800 \times 0.5$

$$= 2513 \ \Omega$$

∴   current $I_L = \dfrac{V}{X_L} = \dfrac{50}{2513}$ A $= 0.02$ A
(very nearly)

and the phase angle between I_L and V is 90° lagging.

Current $I_R = \dfrac{V}{R} = \dfrac{50}{2000}$ A $= 0.025$ A

and I_R and V are in phase.

(b)   From the phasor diagram in *Figure 5.16(b)* the supply current

$$I = \sqrt{(0.025^2 + 0.02^2)} = 0.032 \text{ A}$$

(c)   The phase angle between V and I, using the phasor diagram is

$$\phi = -\tan^{-1} \frac{0.02}{0.025} = -\tan^{-1} 0.8$$

$$= -38.6°$$

Now complete the following problems by yourself.

(16)   An inductive reactance of 32 Ω is connected in parallel with a resistance of value 24 Ω to a 240 V a.c. supply. Find (a) the supply current, (b) the phase angle between the supply current and voltage, (c) the circuit impedance.

(17)   The supply current from a 100 V 500 Hz source to a circuit comprising a 500 Ω resistor in parallel with a pure inductance is found to be 0.25 A. Calculate (a) the current in the inductance, (b) the reactance of the inductance, (c) the value of the inductance in millihenrys.

RESISTANCE AND CAPACITANCE

This parallel arrangement is treated in exactly the same way as the previous case. *Figure 5.17(a)* shows the circuit diagram, with the phasor diagram and impedance triangle, drawn at *(b)* and *(c)*, respectively.

Current I_R is in phase with supply voltage V, current I_c leads V by 90°. As before

$$I = \sqrt{(I_R^2 + I_c^2)}$$

and

$$\phi = \tan^{-1} \frac{I_C}{I_R} = \tan^{-1} \frac{R}{X}$$

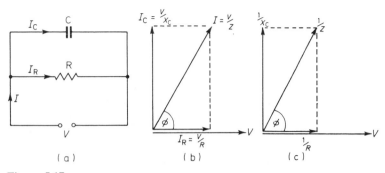

Figure 5.17

You should have no difficulty in following the next example.

Example (18)   When a 100 V a.c. supply is connected to a circuit made up of a resistor in parallel with a pure capacitor, the circuit current is 2 A at a phase angle of 30° leading the applied voltage. Draw the phasor diagram and calculate (a) the value of the resistance and the capacitive reactance, (b) the circuit impedance.

Solution   (a) The phasor diagram is drawn as shown in *Figure 5.18*, knowing that the resultant $I(=2 \text{ A})$ leads the voltage V by 30°. Completing the rectangle to provide the components of I as right-angled phasors now gives us the relative magnitudes of I_R and I_C:

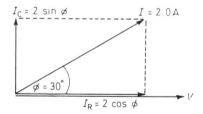

Figure 5.18

$$I_R = 2 \cos \phi = 2 \cos 30° = 1.732 \text{ A}$$

$$I_C = 2 \sin \phi = 2 \sin 30° = 1.0 \text{ A}$$

From this $R = \dfrac{V}{I_R} = \dfrac{100}{1.732} = 57.7 \, \Omega$

$$X_C = \dfrac{V}{I_C} = \dfrac{100}{1} = 100 \, \Omega$$

(b)   The circuit impedance is easily calculated:

$$Z = \frac{\text{applied voltage}}{\text{circuit current}} = \frac{100}{2} = 50 \, \Omega$$

(19)   A 1 μF capacitor is connected in parallel with a 1500 Ω resistor to a 40 V 50 Hz supply. Find the supply current and its phase angle relative to the supply voltage.

(20)   A 60 Ω resistor and a 50 μF capacitor are connected in parallel across a 50 Hz a.c. supply. The current through the resistor is 1.5 A. Calculate the current through the capacitor and the total circuit current.

You should now be ready to tackle parallel circuits where reactance is combined with reactance.

INDUCTANCE AND CAPACITANCE

Although a pure inductance (or a pure capacitance, for that matter) does not exist in practical form, we shall consider the case of a pure inductance L connected in parallel with a capacitor C as an introduction to the real-life circuit where the inductance does possess the bulk of the resistance.

Figure 5.19(a) shows the circuit, with the phasor diagram drawn at *(b)*. Reference phasor V is, as before, drawn horizontally; the current in the inductance, I_L, then lags on this voltage by 90° while the current in the capacitor, I_C, leads the voltage by 90°. Nothing out of the ordinary so far. Since there is no resistance in the circuit, the phasors representing I_L and I_C are in phase opposition, and the resultant current is the difference between I_L and I_C.

As the frequency is gradually increased from zero, I_L will decrease from infinity and I_C will increase from

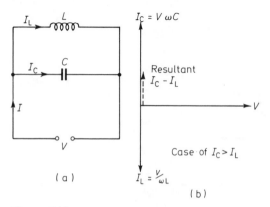

Figure 5.19

zero. So, depending on the magnitudes of I_L and I_C the resultant current may lead the voltage, lag the voltage or be zero. In this last condition we are clearly at some frequency where I_L and I_C are equal in magnitude, hence at this frequency the circuit will present zero reactance to the source of supply and become purely resistive. But by supposition the circuit is free from resistance; therefore, under the condition considered, $I_L = I_C$ and the circuit draws zero current. Hence we conclude that its impedance must be infinite. This is the condition of parallel resonance.

We can calculate the frequency of resonance by considering the circuit impedance. This will be the applied voltage divided by the net circuit current. So

$$Z = \frac{V}{I_L - I_C}$$

where

$$I = \frac{V}{\omega L} \text{ and } I_C = V\omega C$$

Hence

$$Z = \frac{V}{V/\omega L - V\omega C} = \frac{\omega L}{\omega^2 LC - 1}$$

But for this to be infinite, the denominator must be zero, therefore

$$\omega^2 LC = 1$$

$$\therefore \quad \omega^2 = \frac{1}{LC} \text{ or } f_o = \frac{1}{2\pi\sqrt{LC}}$$

using the symbol f_o for the resonant frequency.

This expression for the parallel resonant frequency is seen to be the same as the one previously calculated for series resonance, but in this case the impedance at resonance is infinite, unlike the series case where, at resonance, the circuit impedance reduces to

that of the circuit resistance, and this is normally very small. Also, although the current drawn from the supply is (theoretically) zero, the currents flowing in the inductive and capacitive branches may be very large.

These currents, confined to the parallel loop of the circuit are known as circulating currents. Electric charges flow backwards and forwards round the loop and there is a constant interchange of energy between the magnetic field of the inductance and the electric field of the capacitance.

If, as we have assumed, the circuit is free of resistance, there is no dissipation of energy and the oscillatory interchange in the closed loop proceeds indefinitely. This situation must, of course, be theoretical; in practice, some resistance is always present so the circulating currents cannot be maintained without some energy being drawn from the supply.

This practical case will now have to be considered. But try the following problem before going on.

---

(21)   A 0.02 μF capacitor is in parallel with a pure inductance of 0.25 H. What is the resonant frequency of the circuit? What will be the circuit impedance (a) 500 Hz below resonance, (b) 500 Hz above resonance?

TRUE PARALLEL RESONANCE

In the circuit of *Figure 5.20* let an inductance L henrys having a resistance $R \, \Omega$ be connected in parallel with a capacitor C farads. The resistance will now modify the arguments we used in the previous section where it was assumed that the circuit branches were pure reactances.

The current through C leads the applied voltage by 90° and is represented in *Figure 5.21(a)* by the phasor I_C. The current I_L through the inductive branch lags on the applied voltage by an angle ψ, say, but ψ is not 90° as it would be if there were no resistance in

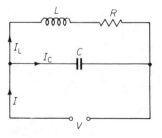

Figure 5.20

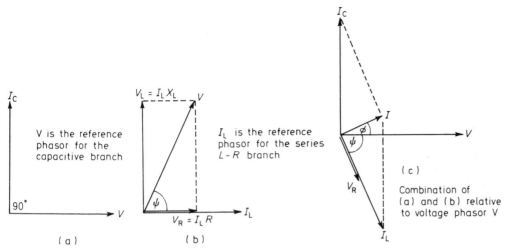

Figure 5.21

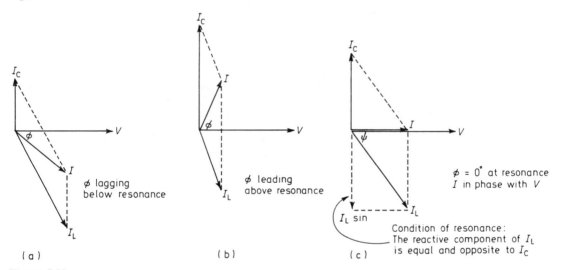

Figure 5.22

the inductive branch, but $\psi = \tan^{-1} X_L/R$, since we treat this branch as a series combination of L and R. V_R is in phase with I_L and the phasor diagram of this part of the circuit is shown in diagram *(b)*.

There is a common phasor V to both diagrams *(a)* and *(b)*, so we can combine them as at *(c)* to provide a diagram of the complete circuit. The supply current I is then the phasor sum of I_C and I_L and this makes a phase angle ϕ with the applied voltage.

The relative magnitudes of I_C and I_L will change with frequency, hence ϕ will, with one exception, either load or lag on the applied voltage. The one exception is when ϕ is zero and I and V are in phase. Analogous to the series circuit, this will be the condition of parallel resonance, this time with the resistance of the inductor taken into account. The diagrams of

Figures 5.22(a), *(b)* and *(c)*, respectively, are for frequencies below resonance, above resonance and at resonance. In this last situation it is seen that the reactive component of the inductor current, shown in broken line, is equal to the capacitor current, leaving only the real component of the inductor current as the total current which is in the phase with the applied voltage. The diagrams are not to scale; at resonance the I phasor is very small since ψ approximates closely to $90°$.

From the diagram at *(c)*, we have then

$$I_C = I_L \sin \psi$$

$$\frac{V}{X_C} = \frac{V}{\sqrt{(R^2 + X_L^2)}} \cdot \frac{X_L}{\sqrt{(R^2 + X_L^2)}}$$

$$\therefore \quad X_C X_L = R^2 + X_L^2$$

Let the resonant frequency be f_o, then

$$\frac{2\pi f_o L}{2\pi f_o C} = R^2 + (2\pi f_o L)^2$$

$$\therefore \quad \frac{L}{C} = R^2 + (2\pi f_o)^2 L^2$$

$$\therefore \quad f_o = \frac{1}{2\pi}\sqrt{\frac{1}{LC} - \frac{R^2}{L^2}}$$

We see from this result that the presence of resistance in the circuit does now affect the resonant frequency. However, if we set $R = 0$ (or assume that R is very small, the term R^2/L^2 will also be very small), then the expression reduces to

$$f_o = \frac{1}{2\pi\sqrt{LC}}$$

which is the result obtained for the series resonant circuit. When R is very small in a parallel circuit, the error introduced by using this simpler formula is negligible.

We have already noted that in a parallel circuit made up of pure reactances the impedance of a parallel circuit in which the inductive branch contains resistance will, while not exhibiting an infinite impedance, at least have an impedance which is very large. We can test this out by again referring to *Figure 5.22(c)*. For here

$$I_C = I \cdot \tan\phi, \text{ since } I_C = I_L \sin\phi \text{ and } I = I_L \cos\phi$$

Let the impedance at resonance be R_D, the dynamic resistance.

Then

$$R_D = \frac{V}{I} = \frac{V}{I_C} \cdot (\tan\phi)$$

$$= V\left[\frac{X_C}{V} \cdot \frac{X_L}{R}\right] \text{ since } I_C = \frac{V}{X_C} \text{ and } \tan\phi = \frac{X_L}{R}$$

$$\therefore \quad R_D = \frac{X_C \cdot X_L}{R} = \frac{\omega L}{\omega CR} = \frac{L}{CR}\,\Omega$$

As the current and voltage are in phase, this impedance is purely resistive. You should notice that the circuit resistance R affects R_D; as R becomes small, R_D becomes large, tending to infinity as R approaches zero. It can be proved that for R very small, R_D is a maximum at resonance. For this reason a parallel resonant circuit is known as a rejector circuit because it offers a very high opposition to the flow of current at the resonant frequency.

Figure 5.23 shows the variation of circuit imped-

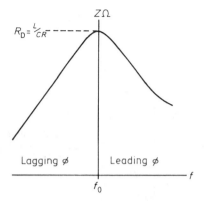

Figure 5.23

ance as the frequency is varied on either side of resonance. This is the response curve of the parallel tuned circuit.

Example (22)   An inductance $L = 0.05$ H having a resistance $R = 100\,\Omega$ is tuned to resonance by parallel capacitor $C = 0.01\,\mu$F. Calculate the frequency of resonance (a) taking account of R, (b) ignoring R.

Solution   (a) Taking account of R, we insert the appropriate figures into the expression for resonant frequency:

$$f_o = \frac{1}{2\pi}\sqrt{\left[\frac{1}{LC} - \frac{R^2}{L^2}\right]}$$

$$= \frac{1}{2\pi}\sqrt{\left[\frac{10^6}{0.05\times 0.01} - \left(\frac{100}{0.05}\right)^2\right]}$$

$$= \frac{1}{2\pi}\sqrt{\left[\frac{4\times 10^{10}}{25}\right]} = \frac{1}{5\pi}\times 10^5$$

$$= 7110 \text{ Hz}$$

(b)   Ignoring R: $f_o = \frac{1}{2\pi}\sqrt{\left[\frac{10^6}{0.05\times 0.01}\right]}$

$$= \frac{1}{2\pi}\sqrt{\left[\frac{10^{10}}{5}\right]}$$

$$= 7118 \text{ Hz}$$

This is a difference of about 0.001%.

Try this next problem for yourself. Make sure you work in the proper units.

(23)   A circuit comprising a 200 µH coil of resistance 30 Ω is in parallel with a 200 pF capacitor. Find the approximate frequency of resonance and the dynamic resistance.

CURRENT MAGNIFICATION

The currents I_C and I_L circulating in the closed loop of a parallel resonant circuit are, as we noted earlier, very large compared with the current I drawn from the supply. If the circuit resistance is small, so that angle ψ in *Figure 5.22(c)* is almost 90°, we can assume that $I_L = I_C$ since sin ψ is then almost unity. Then since $I_L = V/\omega L$, $I_C = V\omega C$ and $I = V/R_D$ we have

$$I_L = \frac{IR_D}{\omega L} \text{ and } I_C = IR_D\omega C$$

The ratio $I_L/I = I_C/I$ is the current magnification of the circuit, and hence represents the Q-factor of the circuit. So

$$Q = \frac{R_D}{\omega L} = \omega R_D C$$

whence

$$I_C = I_L = IQ$$

Hence, in a parallel resonant circuit, the currents circulating in the circuit branches may be many hundreds of times greater than the total feed current supplied from the source. You should compare this result with the voltage magnification factor obtained in the case of series resonance.

We can, of course, look at this result as saying that the impedance of a parallel circuit at resonance is equal to Q times the reactance of either branch.

The next worked example will illustrate some points about Q-factor.

Example (10)   A circuit consisting of an inductor of 0.05 H and resistance 5 Ω is in parallel with a capacitor of 0.1 µF. Calculate the frequency of resonance. Find for the circuit at this frequency (a) the impedance, (b) the Q-factor.

$$f_o = \frac{1}{2\pi} \sqrt{\left[\frac{1}{LC} - \frac{R^2}{L^2}\right]}$$

$$= \frac{1}{2\pi} \sqrt{\left[\frac{10^6}{0.05 \times 0.1} - \frac{25 \times 10^4}{25}\right]}$$

$$= \frac{1}{2\pi} \sqrt{\left[\frac{10^9}{5} - 10^4\right]}$$

Clearly the R^2/L^2 term can be neglected compared to $1/LC$; the resistance is only 5 Ω and the simpler expression for f_o could normally be used straightaway.

$$f_o = \frac{1}{2\pi} \sqrt{[2 \times 10^8]} = \frac{10^4}{\sqrt{2}\pi}$$

$$= 2250 \text{ Hz}$$

(a)   At resonance, the impedance $= R_D = \dfrac{L}{CR}$

$$\therefore \quad R_D = \frac{0.05 \times 10^6}{0.1 \times 5}$$

$$= 100 \text{ k}\Omega$$

(b) $Q = \dfrac{R_D}{\omega L}$

$$= \frac{100 \times 10^3}{2\pi \times 2250 \times 0.05} = 141$$

For this last part of the question we need not have used R_D in our Q calculation, for since the Q of the circuit is dominated by the Q of the coil (where the bulk of the resistance resides) we could, of course, have used the general expression for Q

$$Q = \frac{\omega L}{R} = \frac{2\pi \times 0.05 \times 2250}{5}$$

$$= 141, \text{ as before}$$

SUMMARY

- Inductive reactance $X = 2\pi fL$ where L is in henrys
- Capacitive reactance

$$X_C = \frac{1}{2\pi fC} \text{ where } C \text{ is in farads}$$

$$= \frac{10^6}{2\pi fC} \text{ where } C \text{ is in µF}$$

- L and R in series:

$$Z = \sqrt{R^2 + X_L^2}, \ \tan \phi = \frac{X_L}{R}$$

- C and R in series:

$$Z = \sqrt{R^2 + X_C^2}, \ \tan \phi = \frac{X_C}{R}$$

- At resonance:

series RLC $f_o = \dfrac{1}{2\pi\sqrt{LC}}$ Hz

parallel RLC $f_o = \dfrac{1}{2\pi}\sqrt{\left(\dfrac{1}{LC} - \dfrac{R^2}{L^2}\right)}$ Hz

- Impedance at resonance:

series circuit $Z_o = R$

parallel (dynamic) $Z_o = \dfrac{L}{CR}$

- Q-factor: $\dfrac{\omega L}{R} = \dfrac{1}{\omega CR} = \dfrac{1}{R}\sqrt{\dfrac{L}{C}}$

REVIEW QUESTIONS

1 Explain the difference between resistance, reactance and impedance.
2 Explain what you understand by the term 50° lagging; 50° leading.
3 Why is Q called the circuit magnification factor?
4 Upon what factors does the value of Q depend?
5 What units is Q measured in?
6 Sketch the impedance triangle for (a) a series R–L circuit, (b) a parallel R–C circuit.
7 How can the voltage across each series element in an RLC circuit be greater than the supply voltage?
8 Above ω_o is a series RLC circuit capacitive or inductive? Below ω_o?
9 What is the impedance of (a) a series circuit, (b) a parallel circuit, at resonance?
10 In a parallel resonant circuit, how can the element currents be greater than the current supplied from the source?

EXERCISES AND PROBLEMS

(25) Find the reactance of a 100 mH coil when it is connected to an a.c. supply of (a) 50 Hz, (b) 80 Hz, (c) 1 kHz.

(26) An inductor has a reactance of 377 Ω when connected to a 120 Hz supply. What is its inductance?

(27) At what frequency will a 80 mH inductance have a reactance of 302 Ω?

(28) Calculate the reactance of a 0.5 μF capacitor at frequencies of (a) 50 Hz, (b) 800 Hz, (c) 2 kHz.

(29) A capacitor has a reactance of a 36 Ω when connected to a 50 Hz supply. What is its capacitance?

(30) At what frequency will a capacitor of 1 nF have a reactance of 19 890 Ω?

(31) What is the impedance of a coil which has a resistance of 5 Ω and a reactance of 12 Ω?

(32) A circuit of impedance 12.5 Ω is made up of a coil wired in series with a resistance of 1.5 Ω. If the resistance of the coil itself is 6 Ω, calculate its reactance.

(33) A coil, when connected to a 12 V d.c. supply, draws a current of 3 A, but when connected to a 50 V 100 Hz supply draws 10 A. Find (a) the resistance, (b) the reactance, and (c) the inductance of the coil.

(34) Five 10 μF capacitors are connected (a) in parallel, (b) in series to a 100 V 50 Hz supply. What will be the circuit current in each case?

(35) A 1 μF capacitor is wired in series with a 300 Ω resistor across an a.c. supply of 15 V. If a current of 25 mA flows in the circuit, what is the supply frequency?

(36) A 100 V, 60 W lamp is to be operated on 250 V 50 Hz mains. Calculate the value of (a) a resistor, (b) an inductor (assumed to have negligible self-resistance) required to be connected in series with the lamp.

(37) When connected to a 220 V 50 Hz supply, an inductor draws a current of 5 A. When a resistance of 10 Ω is added in series with the

coil, the current falls to 4.4 A. What is the resistance and inductance of the coil, and what is the phase angle?

(38)   A voltage represented by $v = 282.8 \sin 3142t$ is applied to a circuit of impedance 50 Ω, and the current lags the supply voltage by 20°. Find (a) the frequency, (b) the peak current, (c) the component values, i.e. R and L, used in the circuit.

(39)   The Q-factor of a coil having an inductance of 1 mH is measured at a frequency of 1 kHz and found to be 75. What is the effective resistance of the coil?

(40)   An inductor has a Q-factor of 45 at a frequency of 600 kHz. What will be its Q at 1 MHz assuming that its effective resistance is 50 per cent greater at the higher frequency than at the lower?

(41)   A p.d. of 100 V is applied to an inductance of 0.1 H having a self-resistance of 100 Ω, in series with a 0.25 μF capacitor. Find the resonant frequency of this circuit, its impedance at resonance, and the voltage across the capacitor at resonance.

(42)   A circuit made up of a series arrangement of L and C resonates at a frequency f_o. When L is increased by 100 μH, the frequency decreases to $0.5f_o$. What is the value of L?

(43)   A circuit consists of a 200 μH coil of resistance 40 Ω and is tuned with a parallel capacitor of 200 pF. What is its resonant impedance?

(44)   A coil of inductance 160 μH and negligible resistance is tuned by a parallel capacitor which can be varied from 45 pF to 475 pF. What frequency range will the circuit cover? Would the range be any different if the components were joined in series?

(45)   Discuss briefly the variation of impedance of (a) a series circuit, (b) a parallel circuit, near to and at resonance.

(46)   An inductor is tuned to series resonance at 1500 kHz by a capacitor of 1600 pF. What is the new frequency of resonance if the capacitor is increased *by* 4025 pF?

6

Alternating quantities 3: power and selectivity

Power in steady state a.c. circuits
Active and reactive power
Power factor
The wattmeter
Loss angle
Selectivity and bandwidth
Acceptors and rejectors

The power developed in a d.c. circuit is expressed by the product of the voltage across, and the current flowing in, the circuit. In a.c. circuits the instantaneous power is similarly given by the product of the instantaneous values of voltage and current, so that

$$P = vi = i^2R = \frac{v^2}{R} \text{ watts}$$

These expressions for power are, however, of little value; we are concerned with what is called the 'true power' and this is defined as the average value of the power over a period of time. It might at first be thought that the product of the r.m.s. voltage and current would provide us with the true power in a circuit, but this is not so. In fact, such a product will give us what is known as the 'apparent power', and we shall see that this is not the same as the true power.

ACTIVE POWER

If an alternating voltage V volts is applied to a resistance R ohms and a current I amperes flows, then $V = IR$. The power dissipated as heat at any instant is i^2R watts, and a curve of power dissipation plotted against time will be as illustrated by the broken line in *Figure 6.1*. The curve is a sine-squared curve of peak amplitude $\hat{I}^2$ and its average height will be half of this; the average power consumption in the resistance is therefore

$$P = \frac{\hat{I}^2R}{2} \text{ W}$$

But $\hat{I} = \sqrt{2}I$, hence the power dissipated in terms of r.m.s. is I^2R watts, which is the same as the dissipation for d.c. This of course follows from our original definition of r.m.s. values. What we are simply saying, in other words, is that since power in a resistance is proportional to the square of the current (or the voltage) the power is always positive and energy is dissipated throughout the entire cycle. This is known as *active power*.

REACTIVE POWER

Unlike the case of a pure resistance, inductive and capacitive circuit elements store, but do not dissipate, energy. When current in an inductance is increasing, the energy, expressed by $\frac{1}{2}Li^2$, is transferred from the source of supply to the magnetic field, but when the current decreases this energy is returned. In the same way, when the voltage across a capacitor is increasing, the energy expressed by $\frac{1}{2}Cv^2$ is transferred from

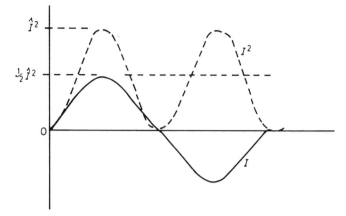

Figure 6.1

the source to the electric field established in the dielectric and power is positive. When the voltage decreases, this energy is returned to the source. We are, of course, assuming that our inductive and capacitive elements are pure examples, untainted by the power consuming properties of resistance; hence stored energy has nowhere to go but back to the source.

We conclude that in a reactive element there is no net energy transfer, so the average power is zero. There is, however, a periodic storage of energy and its return, and the magnitude of the power variation can be expressed as $P_x = I^2X$ or V^2/X. This is called the *reactive power* because of its similarity to active power I^2R.

We can illustrate these effects by simple power diagrams as *Figure 6.1* showed us for the case of resistance.

Pure reactance

Figure 6.2 shows the voltage and current curves for a purely inductive circuit. The current lags 90° behind the voltage, and the power delivered to the circuit over a complete cycle is found by multiplying together the ordinates of the current and voltage curves at every instant of time throughout the cycle.

The power curve is this time shown shaded in the figure. Notice that this power curve is a sine wave and that it is symmetrically disposed about the horizontal time axis, half of its area being above and half below the time axis. When this power curve is positive and above the axis, energy is being delivered to, and stored in, the magnetic field of the inductor; when the power curve is negative, the magnetic field is collapsing and the energy is being returned to the source of supply. You should make a note that the

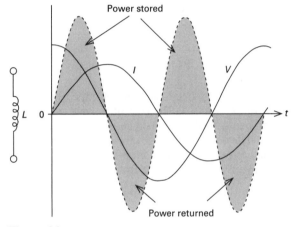

Figure 6.2

positive cycles of the power curve correspond with a rising current and the negative cycles with a falling current. The interchange therefore takes place during each half cycle: a quarter cycle of storage and a quarter cycle of energy return. Since the power curve is symmetrical about the zero axis (unlike the previous case of the resistive load where the curve was wholly above this axis) the total power delivered to the circuit over each complete cycle is zero.

In the same way, if an alternating voltage is applied to a pure capacitance, the power consumption of the circuit will again be zero. The energy involved in charging the capacitor over one quarter cycle is returned to the source during the next quarter cycle. *Figure 6.3* shows the relationship for this case.

In general, whenever the current in a circuit is 90° out of phase with the applied voltage, the power consumption is zero. We will now use this fact to find the power consumption of circuits where, as always, resistive elements are present and the phase angle

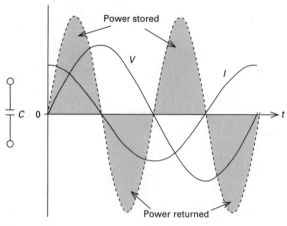

Figure 6.3

between currents and voltages is some angle φ different from 90°.

POWER FACTOR

Let V volts be applied to a circuit of impedance $|Z|$ as shown in *Figure 6.4*. The current flowing will be $V/|Z|$ amperes, leading or lagging on the voltage by some angle φ. Let the voltage phasor be V, then OI will be the current phasor representing $V/|Z|$. This current will have two mutually perpendicular components:

$I_R = I \cdot \cos \phi$ in phase with V

$I_x = I \cdot \sin \phi$ in quadrature with V

To find the true power dissipation in the circuit, we replace the real current $V/|Z|$ by these separate current components. We can then sketch a power phasor diagram, obtained by multiplying the phasors of *Figure 6.4* by V. This is shown in *Figure 6.5*.

Consider first the current represented by $I \cdot \sin \phi$. This is 90° out of phase with the applied voltage,

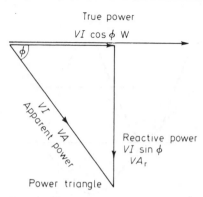

Figure 6.5

hence the quantity $VI \cdot \sin \phi$ represents power which simply supplies the magnetic or electrostatic fields of the reactive elements; power which surges backwards and forwards between the source and the load and so contributes nothing whatever to the true power input to the circuit. This is the reactive power and is measured not in watts but in *volt-amperes reactive*, VA_r.

The other current components represented by $I \cdot \cos \phi$ is in phase with the applied voltage, hence the power dissipated is now positive and expressed by the product of I and $\cos \phi$. Thus in the power triangle of *Figure 6.5*, the quantity $VI \cos \phi$ represents the actual or *true power* consumed in the circuit, and is measured in watts. So the power dissipated in an a.c. circuit depends not only on the values of the current and the voltage but also on the phase angle between them. Only if $\cos \phi = 1$ (which means that $\phi = 0$ and hence current and voltage will be in phase) is the true power actually given by the product VI. Most electrical machines, transformers and the like have their power ratings expressed in volt-amperes.

The quantity $\cos \phi$ is called the *power factor* of the circuit. Power factor, therefore, lies between zero and unity; it is that fraction of the *apparent* power which provides the true power.

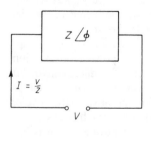

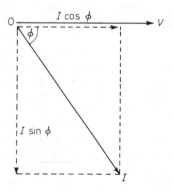

Figure 6.4

Hence

true power = apparent power × power factor

since the ratio true power/apparent power = $VI \cdot \cos \phi /
VI = \cos \phi$.

If the load impedance is $|Z| = \sqrt{(R^2 + X^2)}$ then

$$\cos \phi = \frac{R}{|Z|} = \frac{R}{\sqrt{R^2 + X^2}} = \frac{\text{resistance}}{\text{impedance}}$$

We may derive alternative forms of $VI \cdot \cos \phi$:
$I^2 |Z| \cdot \cos \phi$ and $(V^2 / |Z|) \cos \phi$. Try to deduce these
for yourself.

Follow the next examples carefully. They will illustrate some important principles.

Example (1)   A power line supplies 20 A to a welding transformer. If the line voltage is 240 V and a wattmeter indicates a power of 4 kW, what is the power factor of the equipment?

Solution   As we will see in a few pages, a wattmeter always indicates the *true* power. Hence the true power in this example is 4000 W.

The apparent power = $V \times I$ = 240 × 20 volt-amps

$$\therefore \quad \text{power factor} = \frac{\text{true power}}{\text{apparent power}}$$

$$= \frac{4000}{4800} = 0.83$$

Example (2)   A 50 Hz alternating current supplies a coil of inductance L and resistance R. The voltage across the coil is measured as 25 V r.m.s. and a wattmeter indicates a true power of 20 W delivered to the coil. Find values for L and R.

Solution   From $P = I^2 R$ we have $R = P/I^2 = 20/2^2 = 5 \ \Omega$.

Also $\cos \phi = \dfrac{P}{VI} = \dfrac{20}{2 \times 25} = 0.4$

Then $\phi = 66.4°$

Now $P_x = VI \sin \phi = 25 \times 2 \sin 66.4° = 45.8 \ VA$

Then $X_L = \dfrac{P_x}{I^2} = \dfrac{45.8}{2^2} = 11.45 \ \Omega$

$\therefore \ L = \dfrac{X_L}{2\pi f} = \dfrac{11.45}{2\pi \cdot 50} = 0.036$ H or 36 mH

Example (3)   A 240 V electric motor delivers an output power equivalent to 200 W at 80 per

cent efficiency. If the power factor of the circuit is 0.85, what current is taken from the supply? If the same motor was operated at unity power factor, what would be the reduction in current for the same output power?

Solution   Output power = 2000 W, input power

$$= \frac{2000}{0.8} = 2500 \ \text{W}$$

Power = $VI \cos \phi$

$$\therefore \quad I = \frac{P}{V \cos \phi} = \frac{2500}{240 \times 0.85} = 12.25 \ \text{A}$$

For $\cos \phi = 1$, current I would be $\dfrac{2500}{240} =$

10.4 A

This last example illustrates that at unity power factor there is a reduction in the current taken by the motor for the same power output. This would apply equally to any other piece of electrical equipment. When machines operate at low values of power factor, they draw more current than is necessary. This means that if the power factor of a circuit is low due to a preponderance of (most usually) inductive or capacitive reactance, that is, if ϕ is large, the source of supply has to turn out heavy currents without a corresponding useful power in the load, the power losses in the supply cables are higher than they need be, and the efficiency of the load equipment is reduced because the losses due to the resistance of the conductors in the load are higher than necessary.

The power factor of a machine is a feature of its design, and since most machines consist largely of coils wound on iron pole pieces and armatures they are inevitably highly inductive. Hence the phase angle is large so $\cos \phi$ is small and lagging. To overcome these problems, power factor correction is applied to many pieces of apparatus by the addition of reactance of such a kind that the phase angle is brought towards 0°.

Power factor correction

Parallel resonance can be applied to the problems associated with power factor correction or improvement, so making the generating equipment at mains supply frequency more efficient. So for reasons of economy particularly, it is necessary to make the loading placed on generating equipment as resistive as possible. For inductive loads this can be achieved by

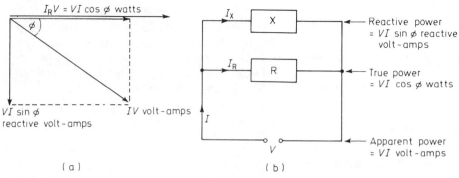

Figure 6.6

placing a suitable capacitor in parallel with the load. If the systems can be made to approach resonance in this way, the resistive condition brings the power factor close to unity.

Calculations on power factor correction are usually made with the various parallel loads described in terms of watts, volt-amperes or reactive volt-amperes. Suppose an inductive load has a lagging phase angle ϕ, see *Figure 6.6(a)*. Each power phasor is obtained by taking the current phasors for the circuit branches and multiplying them by the common voltage V. Diagram *(b)* shows how these power components are distributed in the actual circuit elements. If now a capacitor is placed in parallel with the load, there will be an additional reactive power leading the voltage by 90°. This will have no effect on the power supplied but it will act in opposition to the existing reactive power component $VI \sin \phi$ which simply represents power which surges between the generator and the load without doing any useful work. Phase angle ϕ will be reduced to zero when the reactive powers are equal; the true power and the apparent power phasors will then coincide. It is important at this point that you do not confuse these reactive power phasors with real power phasors; power is *not* a vector quantity and you can only add real powers arithmetically.

This next example will illustrate the foregoing.

Example (4) At a full load power of 500 W an induction motor running on 240 V 50 Hz mains supplies, has a lagging power factor of 0.7. Calculate the value of capacitance required in parallel with the motor to bring the power factor to unity.

Solution   We find first the motor current I:

$$VI \cos \phi = 500 = 240 \times I \times 0.7$$

$$\therefore \quad I = 2.98 \text{ A}$$

This current lags the applied voltage by angle ϕ. Looking at *Figure 6.7* we see that the in-phase component of the current is $I . \cos \phi = 2.98 \times 0.7 = 2.086$ A, and the reactive component $I . \sin \phi = I\sqrt{(1 - \cos^2\phi)} = 2.98\sqrt{(1 - 0.7^2)} = 2.98 \times 0.714 = 2.13$ A.

To bring the power factor to unity, the added capacitor must draw a current of 2.13 A *ahead* of the voltage, that is

$$X_c \text{ must equal } \frac{240}{2.13} = 112.6 \ \Omega$$

from which $C = \dfrac{10^6}{100\pi} . \dfrac{1}{112.6} = 28.3 \ \mu\text{F}$

The effect of bringing the power factor to unity has been to reduce the total current to simply that of the in-phase component, that is, 2.086 A.

Figures 6.7(a) and *(b)* respectively, show the conditions before and after correction.

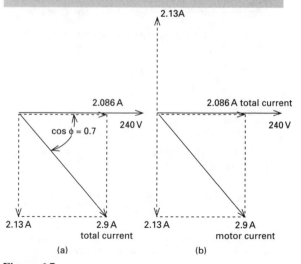

Figure 6.7

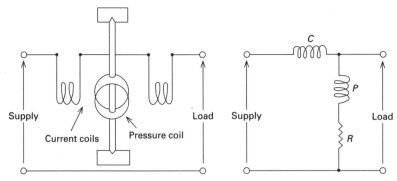

Figure 6.8

THE WATTMETER

We have seen that we are concerned in practice with the average power taken over a period of time, and this is the power indicated by a wattmeter connected into a circuit. The dynamometer wattmeter is possibly the most common of power measuring instruments although an induction type is also used. Such instruments are designed only for use at low frequencies.

The dynamometer wattmeter which is illustrated in simplified form in *Figure 6.8*, consists of two fixed current coils, wound with heavy gauge wire and connected in series with the load, and a second coil, known as the voltage (or pressure) coil, which is mounted on jewelled bearings and able to rotate within the influence of the magnetic field established by the current coils. The movement of the pressure coil is controlled (as it is in ordinary moving-coil instruments) by two mutually opposing spiral hair springs of phosphor bronze, which also serve as connecting leads into and out of the coil. Attached to the pivot upon which the coil rotates is a pointer which moves over a power-graduated scale. These features have been omitted from the diagram for the sake of clarity.

When both fixed and moving coils carry current, there is a force of repulsion between the fields created, and a torque is exerted on the pressure coil which is proportional to the product of the currents in the two coil systems. The moving coil accordingly rotates until this torque is balanced by the restoring action of the springs, and the pointer then indicates a definite position on the scale. The power indicated is the average or true power.

The series coils carry the whole of the load current; the moving coil is arranged to have a very high self-resistance or has a high resistance R wired in series with it as a multiplier device, and is wound to have a low self-inductance, so that the current through it will be very nearly in phase with the voltage across

it. It therefore carries a current proportional to the supply voltage. Now the phase difference between the magnetic fields will be the same as that between the currents they are carrying, hence the deflecting torque on the pressure coil at any instant is proportional to the *instantaneous* product of the voltage and the current, that is, to the instantaneous power. As the inertia of the moving coil system prevents the pointer from following the variations in torque throughout a cycle, the pointer indicates a direction resulting from the average torque throughout the cycle, that is, it indicates the true power.

For heavy currents or high voltages, the coils are not wired directly into the circuit; the fixed coils are supplied from a current transformer and the moving coil is supplied from a voltage transformer. It is impossible to avoid inductive reactance appearing in the coils of the dynamometer wattmeter however carefully designed, and so the instrument is restricted to low frequency use where the variation of impedance with frequency is negligibly small.

(5) A 100 V 50 Hz supply is connected to a 100 mH inductor which has a resistance of 10 Ω. Determine (a) the circuit current, (b) the total power absorbed, (c) the power factor.

(6) A 500 V, 50 Hz motor takes a full load current of 40 A at a power factor of 0.85 lagging. If a capacitor of 80 μF is connected across the motor terminals, show that the power factor will rise to 0.97 and the full load current fall to about 35 A.

Power measurements

We have just looked at the dynamometer wattmeter for the measurement of power in an a.c. circuit, but there are other methods which involve the readings

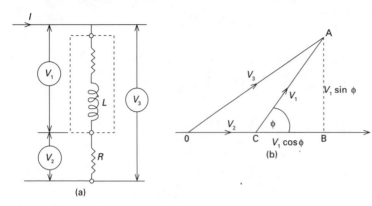

Figure 6.9

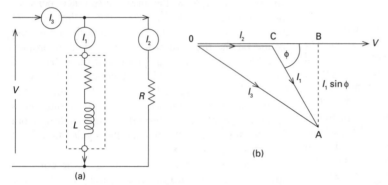

Figure 6.10

obtained from three voltmeters or three ammeters. Although three meters are mentioned for each method, only one instrument is actually used to make three distinct measurements.

1   In the three voltmeter method, assuming we are working on an inductive circuit, the system is arranged as shown in *Figure 6.9(a)*. One measurement is taken across the inductive load (V_1); one measurement across a non-inductive resistor R which is wired in series with the load (V_2); and one other measurement across the entire series combination of load and R (V_3).

The phasor diagram for these voltages is as shown in diagram *(b)*:

V_1 leads the current by phase angle ϕ;
V_2 is in phase with I, hence $I = V_2/R$;
V_3 is the supply voltage, the phasor sum of V_1 and V_2

The load power = $V_1 I \cos \phi = \dfrac{V_1 V_2}{R} \cos \phi$, since

$I = \dfrac{V_2}{R}$

Now CB = $V_1 \cos \phi$, AB = $V_1 \sin \phi$, OA = V_3
Also OA^2 = OB^2 + AB^2 and substituting

$$V_3^2 = (V_2 + V_1 \cos \phi)^2 + (V_1 \sin \phi)^2$$
$$= V_2^2 + 2V_1V_2 \cos \phi + V_1^2 \cos^2 \phi + V_1^2 \sin^2 \phi$$
$$= V_2^2 + 2V_1V_2 \cos \phi + V_1^2 (\cos^2 \phi + \sin^2 \phi)$$

But $\cos^2 \phi + \sin^2 \phi = 1$

$\therefore$ $V_1V_2 \cos \phi = \frac{1}{2}(V_3^2 - V_1^2 - V_2^2)$

Load power = $\dfrac{V_3^2 - V_1^2 - V_2^2}{2R}$ watts

For this method to give reasonable accuracy ϕ must be large, that is, when V_1 is of the same sort of magnitude as V_2. The voltmeter readings must also be taken with considerable care.

The three ammeter method which follows has the advantage that the load is connected directly to the supply terminals, the resistor R now being in parallel.

2   The three ammeter method is set up as shown in *Figure 6.10(a)*, where R is the non-inductive

resistor. Here one measurement is taken of the current in the load (I_1); one measurement of the current in R (I_2); and one measurement of the total supply current (I_3).

The phasor diagram for these currents is given in diagram (b):

I_1 lags on the supply voltage by phase angle ϕ;
I_2 is in phase with the voltage;
I_3 is the supply current, the phasor sum of I_1 and I_2.
The load power = $VI_1 \cos \phi = I^2R \times I_1 \cos \phi = I_1I_2R \cos \phi$

The procedure now follows that used for the three voltmeter method and you should try this for yourself. You should get

$$I_1I_2 \cos \phi = \tfrac{1}{2}(I_3^2 - I_2^2 - I_1^2)$$

$$\therefore \quad \text{Load power} = \frac{R(I_3^2 - I_2^2 - I_1^2)}{2} \text{ watts}$$

Notice that the factor $R/2$ appears this time, not $1/2R$ as it was for the three voltmeter method.

For the best accuracy ϕ should be large, so that I_1 is approximately equal to I_2. Because the power dissipated in R may be comparable with the load power, this method is suitable for relatively low power measurements.

Example (7)   An inductive load connected to a 240 V a.c. supply draws a current of 1.6 A; a non-inductive resistor wired in parallel with this load draws 1.2 A. If the total supply current is 2.3 A, find (a) the load impedance, (b) the power absorbed by the load, (c) the load power factor.

Solution

(a)   Load impedance = $\dfrac{250}{1.6}$ = 156.25 Ω

(b) $R = \dfrac{250}{12} = 208.3 \ \Omega$

Power absorbed by the load

$= \dfrac{R}{2}(I_3^2 - I_2^2 - I_1^2)$

$= 104.15(2.3^2 - 1.6^2 - 1.2^2) = 134.4$ W

(c)   Power factor of load = $\dfrac{134.4}{250 \times 1.6}$ = 0.336

lagging

Example (8)   In *Figure 6.9*, assume the supply to be 50 V a.c. and $R = 8 \ \Omega$. A voltmeter con-

---

nected across the load indicates 24 V and when connected across R indicates 32 V. Calculate (a) the power absorbed by the load, (b) the power absorbed by R, (c) the total circuit power dissipation, (d) the power factor of the load and of the whole circuit

Solution   Here $V_1 = 24$ V, $V_2 = 32$ V, $V_3 = 50$ V; and the circuit current $I = \dfrac{50}{8}$ = 6.25 A.

(a)   Power absorbed in the load

$= \dfrac{1}{2R}(V_3^2 - V_2^2 - V_1^2)$

$= \dfrac{1}{16}(50^2 - 32^2 - 24^2) = 56.25$ W

(b)   Power absorbed by $R = V_2^2R = \dfrac{32^2}{8}$

$= 128$ W

(c)   Total power absorbed = 56.25 + 128 = 184.25 W

(d)   Power factor of load = $\dfrac{56.25}{24 \times 6.25}$ = 0.375

lagging

Power factor of circuit = $\dfrac{184.25}{50 \times 8}$ = 0.46

Power in parallel circuits

You will probably have realized by now that what has been said about power in series circuits applies equally well to parallel arrangements. Power will be dissipated only in the resistive elements and the true power absorbed will be expressed by $VI \cos \phi$.

The next two worked examples will illustrate cases of parallel circuit power problems.

Example (9)   A 240 V induction motor is connected in parallel with a 100 Ω non-reactive resistor. The motor draws a current of 3 A and the total circuit current is 4 A. Calculate the power and the power factor of (a) the motor, (b) the complete circuit.

Solution   *Figure 6.11* shows the circuit and the relevant phasor diagram. The current in the resistor $I_R = 240/100 = 2.4$ A and this current is in phase with V.

The current in the motor branch, I_m, lags V by an angle θ (which is not 90°) and the resultant

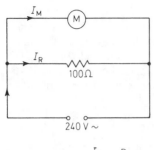

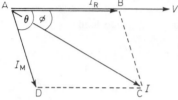

A phasor diagram with labels A, I_R, B, V, θ, ϕ, I_M, D, C, I

Figure 6.11

circuit current I is the diagonal of the parallelogram ABCD. We require to find the phase angle ϕ between V and I. A scaled diagram could be used here but the cosine rule from trigonometry is best for this purpose; we are not dealing with a right-angled triangle. Then from triangle ABC:

$$I_m^2 = I^2 + I_R^2 - 2II_R \cos \phi$$

or

$$3^2 = 4^2 + 2.4^2 - (2 \times 4 \times 2.4) \cdot \cos \phi$$

and from this we find $\cos \phi = 0.665$

This is the power factor of the complete circuit. Now the power taken by the circuit $= VI \cos \phi = 240 \times 4 \times 0.665 = 638.4$ W.

The power dissipated in the resistor $= I_R^2 R = 2.4^2 \times 100 = 576$ W

Hence the power taken by the motor $= 638.4 - 576 = 62.4$ W

Now for the motor

$$\cos \theta = \frac{\text{motor power}}{\text{volts} \times \text{motor current}}$$

$$= \frac{62.4}{240 \times 3} = 0.087 \text{ lagging}$$

This very small value for the power factor gives $\theta = 85°$; this means that the motor is highly inductive.

Example (10)   Two components X and Y are connected in parallel across a 100 V 400 Hz supply. Component X draws a current of 1.5 A at unity power factor, and Y draws a current of 2 A at 0.7660 lagging. Identify and evaluate the two component pieces.

Solution   For unity power factor, component X must be purely resistive; for the lagging power factor component Y will be inductive. The circuit therefore comprises a resistor R in parallel with an inductor L having some self-resistance R_L. *Figure 6.12* shows the circuit and the phasor diagram.

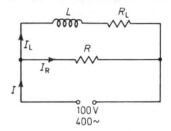

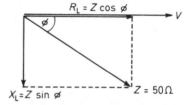

Figure 6.12

Resistance of branch X: $R = \dfrac{V}{I_R} = \dfrac{100}{1.5} = 66.7 \ \Omega$

Impedance of branch Y: $Z = \dfrac{V}{I_L} = \dfrac{100}{2} = 50 \ \Omega$

This impedance is made up of X_L and R_L in series, having a phase angle $\phi = \cos^{-1} 0.7660$; hence from the diagram we have

$$X_L = Z \cdot \sin \phi = 50 \cdot \sin 40°$$

$$= 50 \times 0.6428 = 32.14 \ \Omega$$

Hence

$$L = \frac{X}{2\pi f} = \frac{32.14}{2\pi \times 400} \text{ H}$$

$$= 0.0128 \text{ H or } 12.8 \text{ mH}$$

You should now have no difficulty in calculating that $R = 38.3 \ \Omega$.

LOSS ANGLE

Losses occur in both inductive and capacitive elements, the bulk of such loss being confined in practical circuits to the resistive part of the inductor. A capacitor, on the other hand, has very small overall losses, both as regards resistance and those losses associated with the dielectric. It is usual in the case of capacitors to express such losses either in the form of a very small resistance placed in series with an otherwise ideal component, or as a very large resistance placed in parallel with an ideal component. For any particular capacitor the values of such loss resistances can be found which give models identical in phase angle and impedance for either case.

In the model of *Figure 6.13(a)*, let the total equivalent series resistance be R. Because of the presence of this resistance the current will not lead the voltage by 90° but by some angle ψ which is less than 90°. *Figure 6.13(b)* shows the impedance triangle for this circuit; we define ψ as the loss angle of a capacitor, where $\psi = (90° - \phi)$ and $\cos \phi$ is the power factor. In general, the losses in a capacitor being small, phase angle ϕ will be close enough to 90° to make $\cos \phi$ very small and for this reason it is usual to express the loss angle ψ in a circular (radian) measure. A loss angle of 10^{-5} radian would be typical for a good quality capacitor at low frequencies.

Now from the impedance triangle $R/Z = \cos \phi = R/(\sqrt{R^2 + X^2})$ and when R is very small compared with X, R^2 in the denominator may be neglected, giving

power factor $= \cos \phi = \omega CR$

If the equivalent model is considered to have the loss resistance in parallel with the ideal capacitor the sides of the impedance triangle become $1/Z$, $1/R$ and $1/X$, respectively, and the power factor is then Z/R or $\cos \phi$

$= 1/\omega CR$. The following examples will illustrate the above.

Example (11)   A 0.001 µF capacitor has a loss angle of 0.5°. Find the equivalent series loss resistance and the power dissipated in this resistance when 100 V at a frequency for which $\omega = 10^6$ radians per second is applied.

Solution   For $\omega = 10^6 X = \dfrac{1}{\omega C} = \dfrac{10^6}{0.001 \times 10^6}$

$= 1000 \, \Omega$

Since the phase difference is small, we can write

Power factor $= \omega CR$

Now 0.5° in radian measure $= \dfrac{0.5 \times \pi}{180} = 8.7 \times 10^{-3}$ rad

$R = 8.7 \times 10^{-3} \times X = 8.7 \, \Omega$

Notice how the use is made of circular measure; we *cannot* substitute angles expressed in degrees directly into the equations.

Since R is small compared with X we can say $Z = X = 1000 \, \Omega$, therefore the circuit current $I = 100/1000 = 0.1 \, A$

$\therefore$ power dissipated $= I^2 R = 0.1^2 \times 8.7$
$= 0.087$ W or 87 mW

Alternately, we could say power $= VI \cos \phi$ and this will lead to the same result. $\phi = 89.5°$ remember!

Example (12)   An alternating voltage of 10 V at a frequency for which $\omega = 10^4$ rad/sec is applied across a 10 nF capacitor. What current flows in the circuit? If the power dissipated is 100 µW, calculate (a) the loss angle, (b) the equivalent series resistance, (c) the equivalent parallel resistance.

Solution   Here we have $\omega = 10^4, C = 10 \times 10^{-9}$ F. Then

$$X = \dfrac{1}{\omega C} = \dfrac{10^9}{10^4 \times 10} = 10^4 \, \Omega$$

capacitor current

$$I = \dfrac{V}{X} = \dfrac{10}{10^4} = 10^{-3} \, A \text{ (1 mA)}$$

(a)   Power factor $= \cos \phi = \dfrac{\text{true power}}{\text{apparent power}}$

$$= \dfrac{100 \times 10^{-6}}{10^{-3}} = 10^{-1}$$

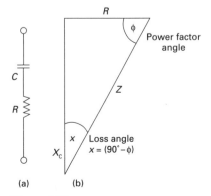

(a)   (b)

Figure 6.13

But loss angle $\psi = 90° - \phi$ so that $\cos \phi = \sin \psi$ $= 10^{-1}$ since $\sin \psi = \psi$ for small angles.

$$\therefore \quad \text{loss angle} = 10^{-1} \text{ radian}$$

As 10^{-1} radian is about 5.7°, this capacitor would not be considered particularly good. Let us have a look at the loss resistance.

(b)   Let the equivalent series resistance be R_s. Then since the power dissipated in it is 100 μW at a current of 10^{-3} A we get

$$100 \times 10^{-6} = (10^{-3})^2 \times R_s$$

$$R_s = 10^2 \text{ or } 100 \, \Omega$$

(c)   Let the equivalent parallel resistance be R_ρ. Then since 100 μW of power is dissipated at 10 V we get

$$100 \times 10^{-6} = \frac{10^2}{R_\rho}$$

$$R_\rho = 10^6 \, \Omega \text{ or } 1 \text{ M}\Omega$$

The two models are illustrated in *Figure 6.14*; these are identical and represent an ideal capacitor with either series or parallel loss resistances.

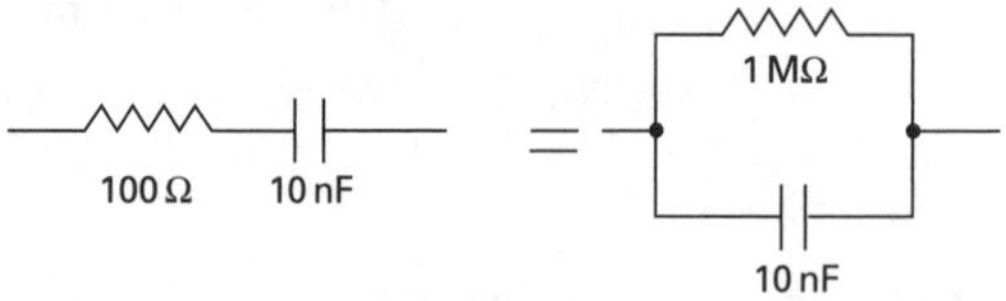

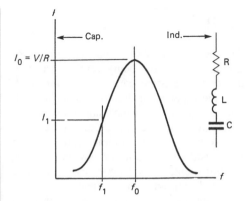

Figure 6.15

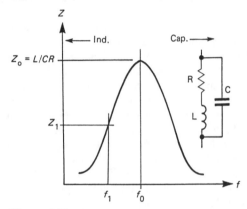

Figure 6.16

Figure 6.14

(13)   If the product of the equivalent series resistance and the capacitance of a capacitor at a certain frequency is 20×10^{-10} and the power factor is 0.001, at what frequency was the measurement made?

SELECTIVITY AND BANDWIDTH

The ability of a resonant circuit, series or parallel, to discriminate between input signals of different frequencies defines its selectivity. If a constant voltage is applied to a series circuit and the frequency is varied, the current will become a maximum when the impedance of the circuit is a minimum, that is, when the voltage applied is at the circuit's resonant frequency. We recall that the impedance is then purely resistive

and equal to the circuit R. At frequencies above and below resonance, the impedance increases, becoming capacitive below resonance and inductive above, and the circuit current falls from its maximum value of I_o ($=V/R$) in the manner indicated in *Figure 6.15*. Clearly, the 'peakiness' of this response curve is an indication of the ability of the circuit to pick out one particular frequency (f_o) from some other frequency relatively close by (such as f_1), the circuit currents flowing at f_o and f_1 being I_o and I_1, respectively.

The response curve obtained for a parallel resonant circuit is similar, except that this time the impedance is a maximum at resonance and the current drawn a minimum. By using impedance Z, therefore, in place of current I on the vertical axis of *Figure 6.15*, the diagram can be used to illustrate the response of a parallel circuit, and this has been done in *Figure 6.16*. This time the impedance (which is resistive at resonance and equal to L/CR, remember) becomes inductive below resonance and capacitive above. The peakiness of the curve is again an indication of its selectivity.

Ideally we want a resonant circuit to pick out a precise band of frequencies centred about the resonant

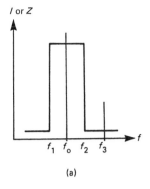

(a)

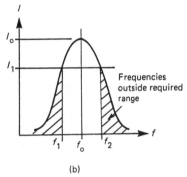

Frequencies
outside required
range

(b)

Figure 6.17

frequency; all other frequencies must be rejected. Such an ideal response curve would be rectangular in shape, with vertical sides and a flat top – rather as shown in *Figure 6.17(a)*. Only those frequencies lying between the limits of f_1 and f_2 would be of interest to us. When we tune into a radio signal, we select first the frequency on which that particular station is transmitting, what is called the carrier frequency. The actual voice or music content of the signal then has to be received on a band of frequencies which extends for a certain 'distance' on either side of the carrier frequency. For medium-wave reception, the distance f_1 to f_2 is about 9 kHz. With our ideal response curve, any interference on a frequency such as f_3 outside the limits of our wanted range would not disturb our enjoyment of the programme.

The response curve we actually have, of course, is not rectangular and unless it has steep sides, some frequencies outside the wanted limits of f_1 and f_2 will influence reception, see *Figure 6.17(b)*. The frequency range f_1 to f_2 is known as the bandwidth of the system and for high discrimination between what we want and what we don't want the ratio

$$\frac{f_0}{f_2 - f_1}$$

must clearly be large.

Now it might appear at first consideration that the bandwidth of a particular circuit might be anything, being dependent on the level selected for the current I_1 flowing at frequency f_1 or f_2. And so it would be if the response level I_1 was chosen arbitrarily. But I_1 must obviously always be stated as a definite proportion of the maximum I_0 if a meaningful measure of the circuit selectivity is to be obtained.

Let the level I_1 be chosen such that at those points on the response curve, the circuit reactance is equal to its resistance. That is, let $R = X$, then

$$Z = \sqrt{(R^2 + R^2)}$$
$$= R\sqrt{2}$$

So,

$$I_1 = \frac{V}{Z} = \frac{V}{R\sqrt{2}} = \frac{I_0}{\sqrt{2}} = 0.707 I_0$$

If, then, we measure frequencies f_1 and f_2 at points where the response has fallen to 0.707 of its peak value at resonance, we obtain a bandwidth figure which, in proportion to the resonant frequency f_0, is a measure of selectivity. We take note of two important points:

(a) A ratio I_0 to $0.707I_0$ corresponds to a power ratio of -3 dB. The frequencies f_1 and f_2 are known as the 'half-power points' and the bandwidth of a resonant circuit is always stated with reference to the half-power points.

(b) The phase angle at f_1 and f_2 is 45°, this follows from the fact that $\tan^{-1} R/X = \tan^{-1} 1$, since $R = X$. Whether the current leads or lags on the voltage by this angle depends on which side of resonance we consider, i.e. on whether the circuit is inductive or capacitive.

In *Figure 6.17(b)* the capacitance is predominant at f_1;

$$\therefore \quad \frac{1}{\omega_1 C - \omega_1 L} = R$$

and so

$$\frac{1}{C} - \omega_1^2 L = \omega_1 R \quad \text{(i)}$$

At f_2 the inductive reactance is predominant, and so

$$\omega_2^2 L - \frac{1}{C} = \omega_2 R \quad \text{(ii)}$$

Adding (i) and (ii):

$$\omega_2^2 L - \omega_1^2 L = \omega_1 R + \omega_2 R$$

and

$$L(\omega_2 - \omega_1) = R$$

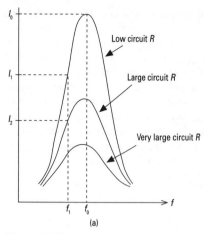

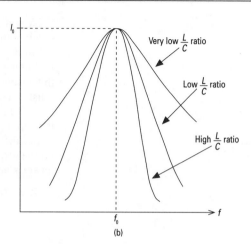

Figure 6.18

Dividing across by ω_o $(=2\pi f_o)$ we have finally

$$\frac{\omega_2 - \omega_1}{\omega_o} = \frac{R}{\omega_o L}$$

from which

$$\frac{f_o}{f_2 - f_1} = \frac{\omega_o L}{R} = Q$$

Thus by choosing this particular response level, we have shown that Q is a direct measure of selectivity.

Factors affecting response

We return now to the series resonant circuit to look at the actual factors which affect the peakiness of the response curve.

1 The circuit resistance. The height of the response curve is inversely proportional to the circuit resistance, the current at resonance I_o being V/R. The greater R, therefore, the lower the peak current, and *Figure 6.18(a)* shows the effect of different circuit resistances. So selectivity varies with $1/R$, for with the low resistance curve, the current at frequency f_1 (which, for example, is 10 per cent below the resonant frequency f_o) is plainly much greater than current I_2 produced by the high resistance curve at that point.

2 The ratio of inductance L to capacitance C. The response *at* resonance depends on the circuit resistance only; the response *off* resonance depends on resistance *and* reactance. That is, the rate at which the impedance of the circuit varies with frequency above and below the resonant frequency depends on the value of R and the ratio of L to C. It is found that with a low L/C ratio the reactance

increases only slowly as the frequency departs from resonance, so that the circuit impedance remains low, giving a flat, unselective response curve. *Figure 6.18(b)* shows the effect of different L/C ratios in a series circuit. By comparing the two diagrams we can see that the effect of decreasing the circuit resistance is the same as that of increasing the L/C ratio. In practice, the coil contributing the inductance in a series circuit must possess the bulk of the resistance, so that it is not possible to improve the L/C ratio by increasing L without at the same time altering the resistance.

ACCEPTORS AND REJECTORS

A resonant circuit made up of inductance and capacitance in series is known as an acceptor circuit. A resonant circuit made up of inductance and capacitance in parallel is known as a rejector circuit. The essential characteristics are that the acceptor circuit provides an easy path for current at the resonant frequency and a more difficult one for all others, whereas the rejector makes a difficult path for current at the resonant frequency and an easier path for all others.

In radio reception, the required frequency is selected from all others by the use of a rejector circuit connected between the aerial system and earth, see *Figure 6.19*. The circuit is resonated by varying the capacitance and, at the frequency of resonance, when the impedance is a maximum, the greatest voltage is developed across the circuit and applied to the receiver input.

Acceptor circuits have their uses but they are not so often used as rejectors. One use of a rejector is to

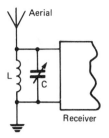

Figure 6.19

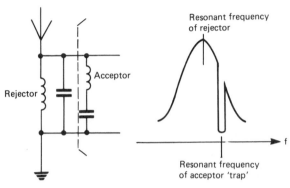

Figure 6.20

bypass frequencies which are very close to the wanted frequency and which need to be suppressed. An acceptor can then be connected across a rejector as shown in *Figure 6.20*, where it behaves as a very low resistance to the unwanted signal. It is then known as a 'trap', and the diagram illustrates its behaviour in removing part of the frequency spectrum. Its presence does not affect the functioning of the rejector; at the frequency to which the rejector is tuned, the acceptor looks like a capacitance, hence only modifies the actual tuning capacitance of the rejector. Such systems as this are found on fixed tuned circuits, particularly in television receivers.

SUMMARY

For sinusoidal quantities:

- Power (average or true) $P = VI \cos \phi$ W
- Reactive power $P_x = VI \sin \phi$ volt-amps reactive
- Apparent power $P_a = VI = I^2 Z = \dfrac{V^2}{Z}$ volt-amps
- Power factor $= \cos \phi = \dfrac{\text{true power}}{\text{apparent power}}$

Power measurement:

- Three voltmeter method: load power
  $$= \frac{V_1^2 - V_2^2 - V_3^2}{2R} \text{ W}$$
- Three ammeter method: load power
  $$= \frac{I_1^2 - I_2^2 - I_3^2}{R} \text{ W}$$

For low loss capacitor:

- Power factor $= \cos \phi = \omega CR$
- Loss angle $\psi = (90° - \phi)$
- The frequency selectivity of a resonant circuit depends on the bandwidth BW where $BW = f_2 - f_1$

  Also $\dfrac{f_o}{\text{BW}} = \dfrac{f_o}{f_2 - f_1} = Q$

- Selectivity varies as $1/R$ and as the ratio L/C. Bandwidth is measured at the points f_1 and f_2 where the response has fallen to 0.71 of its peak resonant value. These are the half power or -3 dB points.

REVIEW QUESTIONS

1 Explain what is meant by 'active power' and 'reactive power'.
2 What is the difference between the terms 'volt-amperes' and watts?
3 Power factor $= \cos \phi$. Explain what this expression actually means.
4 An industrial user with a low power factor pays a higher charge than a consumer with a high power factor. Why?
5 Explain the basic operation of a dynamometer wattmeter.
6 What is meant by the loss angle of a capacitor?
7 What is actually happening when you turn the tuning scale of a radio receiver?
8 Is a series resonant circuit capacitive or inductive above resonance? Below resonance?
9 What does the term 'half-power points' indicate?
10 What are acceptors, rejectors and traps? Give an example of each and where they might be used.

EXERCISES AND PROBLEMS

(14)   Complete the following statements:

(a)   A resistive circuit has . . . power factor.
(b)   A leading power factor indicates a (an) . . . circuit.
(c)   The object of power factor correction is to bring the power factor close to. . . .

(15)   A circuit absorbing 850 W of power is fed from a 200 V supply. If the current drawn is 4.5 A, find the phase angle.

(16)   The current taken by an induction motor is 1.2 A when the applied voltage is 250 V. A wattmeter indicates 135 W. What is the power factor and phase angle?

(17)   A coil of inductance 3 H and resistance 500 Ω is connected to a 500 V 50 Hz supply. What is the power factor of the coil at this frequency?

(18)   An inductor of 0.318 H has an impedance of 200 Ω. It is wired in parallel with a resistance of 200 Ω to a 100 V 50 Hz supply. Find (a) the resistance of the inductor, (b) the total power dissipated in the circuit.

(19)   An alternating voltage of angular frequency $\omega = 10^4$ rad/sec is connected to the circuit of *Figure 6.21*. The voltage measured across the 750 Ω resistor is found to be 3 V. Calculate (a) the supply voltage, (b) the total power supplied, (c) the power factor. It is desired to increase the power factor to unity. What value of capacitor would be required for this and how would it be connected?

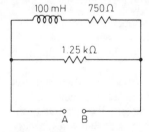

100 mH   750 Ω

1.25 kΩ

A   B

Figure 6.21

(20)   A transmitting capacitor has a capacitance of 0.025 μF. It has a power factor of 5×10^{-4}

and is carrying a current of 100 A at a frequency of 25 kHz. What power is being dissipated in the capacitor?

(21)   The parallel circuit of *Figure 6.20* is connected to a 100 V supply having an angular frequency $\omega = 1000$ rad/s. Calculate the total circuit current and the branch currents. Illustrate your answers by means of a scaled phasor diagram.

(22)   Power is supplied to a workshop at 440 V, and the power lines feeding the workshop have a resistance of 0.2 Ω. If the workshop load is 25 kW at a power factor of 0.7, what power is wasted as heat in the power lines? What would this loss reduce to if the load was adjusted to have unity power factor?

(23)   A circuit is made up of a coil of inductance 180 μH and resistance 10 Ω in series with a capacitor of 300 pF. Calculate (a) the circuit Q-factor, (b) the bandwidth at the resonant frequency.

(24)   A tuned circuit, resonant at a frequency f_o, has two frequencies f_1 and f_2, respectively, below and above f_o, at which the current in the circuit is one-half the value measured at f_o. Show that

$$f_o^2 = f_1 \times f_2 \text{ and that } f_2 - f_1 = \frac{\sqrt{3} \cdot R}{2\pi L}$$

(25)   Show that if $Q \geq 10$, then

$$f_1 = f_o - \frac{f_o}{2Q} \text{ and } f_2 = f_o + \frac{f_o}{2Q}$$

What do these results demonstrate?

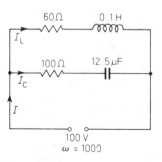

60 Ω   0.1 H

I_L

100 Ω   12.5 μF

I_C

I

100 V
$\omega = 1000$

Figure 6.22

7

Transformer principles

A transformer transfers electrical energy from one circuit to another by means of a magnetic field which is common to both circuits. Its behaviour can be explained in terms of a magnetic circuit excited by an alternating supply. One of the most important advantages of alternating supplies over continuous supplies is the ease with which a transformation from a low to a high voltage, or conversely, may be accomplished. Such transformations are one of the main jobs of the appropriately named power transformer.

In its most common form, the transformer consists of two (often more) coils wound on the same magnetic core but insulated from it. A changing voltage applied to the input or *primary* coil sets up an alternating magnetic flux in the core which links with the turns of the output or *secondary* coil and induces a voltage in it. This is illustrated in *Figure 7.1*. Here an alternating voltage V_1 produces a current I_1 in the primary coil having N_1 turns of wire. This establishes an alternating flux Φ which links with the adjacent coil having N_2 turns of wire and induces an e.m.f. e_2

in this coil. Since the e.m.f. generated in the second coil is due entirely to the common flux produced by the current in the first coil, the circuits are coupled by mutual induction. This is the fundamental principle of all transformer action.

Since there is no electrical connection between the input and output coils, the transformer insulates one circuit from another while permitting an exchange of energy between them. Further, since only alternating currents are transformed, an output circuit can be completely isolated from direct-current components of the input.

Transformers take a variety of forms. There are those which are designed to operate at relatively high frequencies (radio frequencies) covering the frequency spectrum from about 50 kHz to possibly as high as 100 MHz. Others are designed to cover the lower range of audio frequencies from, perhaps, some 100 Hz or so to about 20 kHz, while a third group are made exclusively for use over the power frequency range of 40 to 60 Hz. The physical appearance of each of these types of transformer varies considerably, particularly between those used at radio frequencies and those used at power frequencies. We will consider briefly the general constructional features of these transformer types.

HIGH FREQUENCY TRANSFORMERS

In general these consist of two coils wound on a bakelite, paxolin or polystyrene tube, the coil ends

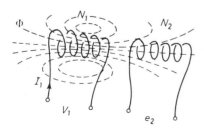

Figure 7.1

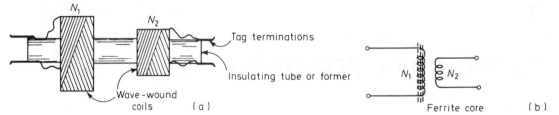

Figure 7.2

being brought out to solder tags rivetted either to the tube itself or to an attached ring of insulating material firmly glued to the tube. In other cases, the wires are taken directly to the appropriate points on the circuit board. Unless they consist of only a few turns of wire, in which case they are generally wound as single layer coils along the length of the tube, they are usually 'wave-wound'. This method of winding is not done solely for the sake of neatness; the open type of winding which results reduces the self-capacitance of the coil while at the same time leading to compactness and consistency of inductance. A typical appearance is illustrated in *Figure 7.2*.

The separation of the coils is relatively large, often a number of centimetres. The coils are then said to be *loosely* coupled. On the other hand, *tightly* coupled coils are found where the coils are closely side by side or, in some cases, wound one above the other. The coils do not necessarily have the same number of turns or the same gauge of wire.

In transformers of the loosely coupled type it is clear that most of the magnetic flux produced by the current in the primary coil does not link with the turns of the secondary coil. As the secondary coil is located further away from the primary, so the flux linking with it is reduced, and hence the secondary induced e.m.f. and the mutual inductance decrease. The proportion of the total flux that links with the secondary is stated as the *coefficient* of coupling, symbol k. For reasons which do not concern us here, the value of k is often as small as 0.01 or less.

Radio-frequency transformers often have an iron-dust or ferrite core threaded into the supporting tube which can be used to adjust the coil inductance or modify the coefficient of coupling. In other cases the coils may be wound directly on to a length of ferrite material where the system then performs as an aerial for such things as radio receivers. In nearly all cases, one of the coils will be tuned with a parallel capacitor to be resonant at one particular frequency or adjustable over a range of frequencies. As we have noted above, transformers of this kind are found in circuits which operate within the frequency range of some

50 kHz to 100 MHz. *Figure 7.2* shows the circuit symbol for a radio-frequency transformer; if a ferrite core is employed, this is indicated as a series of broken lines.

LOW FREQUENCY TRANSFORMERS

The general construction forms of transformers designed to cover the audio-frequency range and those to be used on the fixed power frequency of 50 Hz are very similar. The detailed design variations follow from the fact that in the audio case a range of frequencies has to be catered for at relatively low power requirements, while at 50 Hz, though the frequency is fixed, considerable power outputs are often the order of the day. In both applications the object is to get as much of the primary flux as possible linking with the secondary coil; in other words, very tight coupling is called for. This is a quite different situation from that found in radio-frequency transformers, and it is not usual to work on audio and power transformers in terms of mutual inductance or coupling coefficients.

Transformers for these frequencies (in their most basic form) consist essentially of two windings wound upon a closed magnetic circuit, the assembly and form of which will be discussed later. As before, one of the windings is referred to as the primary and the other as the secondary coil, and the most basic arrangement is shown in *Figure 7.3(a)*. Here again, N_1 and N_2 represent the number of turns on the primary and secondary coils, respectively.

The closed iron core is used to obtain a high flux and to ensure that the coupling between the coils is as tight as possible. In practice, both coils are usually wound on a centre limb of the core and not on the side limbs as the diagram illustrates for purposes of clarity, but this does not invalidate the principle of operation in any way. The circuit symbol for an iron-cored transformer is shown in *Figure 7.3(b)*; here the iron core is depicted by the solid lines between the coils.

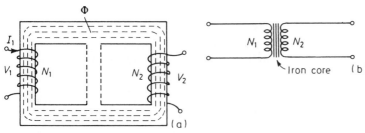

Figure 7.3

TRANSFORMATION RATIO

Referring to *Figure 7.3(a)* again, if an alternating voltage V_1 is applied to the primary coil, both the resulting current I_1 and the core flux Φ will be alternating. The changing flux will link with both windings and there will be induced e.m.f.s in both windings in accordance with Faraday's laws of induction.

Flux Φ linking with turns N_1 will induce an e.m.f. E_1 proportional to ΦN_1: this is an e.m.f. of self-induction. The same flux linking with turns N_2 will induce an e.m.f. E_2 proportional to ΦN_2; this is an e.m.f. of mutual induction. Assuming that all the flux links with both coils, then

$$\frac{E_1}{E_2} = \frac{\Phi N_1}{\Phi N_2} = \frac{N_1}{N_2}$$

So the flux linkages are in the ratio of the number of turns and so also are the induced voltages due to the flux changes. Recalling our earlier work on magnetic induction, we know that the primary induced e.m.f. E_1 is equal and opposite to the input voltage V_1 (back-e.m.f., remember?), and clearly the secondary voltage V_2 is equal to E_2. Hence we can write

$$V_2 = \frac{N_2}{N_1} \cdot V_1$$

The ratio N_2/N_1 is called the *transformation ratio* of the transformer.

When the secondary has fewer turns than the primary, the ratio is less than unity and $V_2 < V_1$. The transformer is then known as a 'step-down' transformer. When the secondary has more turns than the primary, the ratio is greater than unity and $V_2 > V_1$. The transformer is now a 'step-up' transformer. Note that it is to the voltage change that these terms apply.

On occasion, a transformer may have an equal number of turns on both primary and secondary. In this case $V_2 = V_1$ and the transformer is known as a 'one-to-one' or 'isolating' transformer, its function being to isolate electrically (from a d.c. point of view)

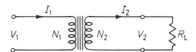

Figure 7.4

one circuit from another without altering the a.c. conditions. All transformers are isolators in this sense.

In everything we have said above, we have assumed that the transformers are ideal, that is, 100 per cent efficient. In practical transformers this cannot be true and there is always some power loss in the device. The expressions for transformation of voltage are therefore only approximations, although most power transformers have efficiencies above 95 per cent.

We shall next consider the operation of the iron-cored transformer in more detail and return to the differences between fixed frequency power transformers and audio-frequency range transformers later on.

CURRENT TRANSFORMATION

In *Figure 7.4* a load resistance R_L has been connected across the secondary terminals of the transformer. The secondary voltage V_2 will now cause a current I_2 to flow through R_L and power will be dissipated in R_L. This power must, of course, originate at the generator connected to the primary terminals. Now since the flux is the same for each winding, the ampere-turns must also be equal, so that $I_1N_1 = I_2N_2$ or

$$\frac{N_2}{N_1} = \frac{I_1}{I_2}$$

but

$$\frac{N_2}{N_1} = \frac{V_2}{V_1}$$

so that

$$\frac{I_1}{I_2} = \frac{V_2}{V_1} \tag{7.1}$$

Hence, an increase or step-up in voltage from primary to secondary must be accompanied by a corresponding decrease or step-down in current from primary to secondary, and conversely. The ratio of primary to secondary current is therefore inversely proportional to the turns ratio, that is

$$I_2 = \left(\frac{N_1}{N_2}\right) I_1 \qquad (7.2)$$

We can, of course, look at equation (7.1) above in terms of power. I_1V_1 is the power input to the primary, and I_2V_2 is the power output at the secondary into load resistor R_L. So $I_1V_1 = I_2V_2$, an obvious relationship since (assuming our transformer is an ideal component) the power input to the primary must be equal to the power delivered at the secondary. Follow the next worked example carefully.

Example (1)   An ideal transformer has 1000 primary turns and 3500 secondary turns. If 250 V is applied to the primary terminals, what voltage will appear at the secondary terminals? If a resistor of 7000 Ω is connected to the secondary, what will be the power dissipated in the resistor, and what will then be the primary current?

Solution   Transformation ratio

$$\frac{N_2}{N_1} = \frac{3500}{1000} = 3.5$$

$$V_2 = 3.5V_1 = 3.5 \times 250 = 875 \text{ V}$$

When this voltage is applied to the 7000 ohm resistor, the power dissipated will be

$$\frac{V_2^2}{R} = \frac{875^2}{7000} = 109.4 \text{ W}$$

This must also be the power into the primary, V_1I_1

$$\therefore \quad I_1 = \frac{\text{power}}{V_1} = \frac{109.4}{250} = 0.44 \text{ A}$$

You should perhaps not need reminding that the voltages and currents in all transformer problems and notes are alternating quantities, stated in their r.m.s. values.

Can you think of an alternative way of solving the above problem?

(2)   An ideal transformer having a primary winding of 1500 turns has a 25 Ω resistor connected across its secondary terminals. A 100 V alternating input produces 75 V across the resistor. Calculate (a) the number of turns on the secondary winding, (b) the current in each winding, (c) the power dissipated in the load.

THE UNLOADED IDEAL TRANSFORMER

We now examine the operation of a transformer in rather more detail, although we shall still assume that we have an ideal, loss-free component. Let the transformer of *Figure 7.5* be such a transformer and suppose further that the secondary terminals are left disconnected, so that the transformer is unloaded. Then as far as the input is concerned, the primary winding will appear as a very large inductance and a small current i_m lagging 90° on the applied voltage V_1 will flow when V_1 is connected. This current is known as the magnetizing current. It will produce an alternating flux Φ in the core of the transformer which is in phase with the current and hence also lagging by 90° on the applied voltage.

We can therefore construct a phasor diagram to illustrate this: the relationship between the applied voltage, the magnetizing current and the core flux is shown in *Figure 7.6(a)*. Notice particularly that we have taken the flux Φ as our reference phasor, this being the only quantity which is common to both primary and secondary circuits.

The alternating flux produces in the primary winding an alternating back-e.m.f. which, in a purely inductive circuit, is at all times equal and opposite to the applied voltage. Since the back-e.m.f. must depend on the magnitude of the alternating flux, it follows that this magnitude is determined solely by the magnitude of the applied voltage. This is a condition of fundamental importance in the whole of transformer theory: If the applied voltage V_1 is constant, then the

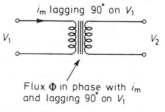

Flux Φ in phase with i_m
and lagging 90° on V_1

Figure 7.5

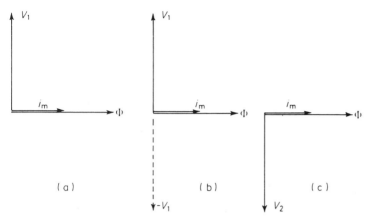

Figure 7.6

flux Φ is constant. This statement is true irrespective of any other currents which may be caused to flow in either or both of the transformer windings because of secondary loading or any other reason. *Figure 7.6(b)* now shows us the phasor diagram with the addition of the back-e.m.f. equal to $-V_1$.

The alternating flux which threads the primary winding and induces in it the back-e.m.f. $-V_1$ links also with the secondary winding, consequently there is induced in the secondary an e.m.f. which is in phase with the primary back-e.m.f. and *Figure 7.6(c)* shows the phasor representation of this. The magnitude of this secondary e.m.f. is easily calculated, for V_2 is produced by the flux linking with the secondary turns, and as the flux is common to both primary and secondary windings the ratio of the primary induced e.m.f. to that of the secondary must be the same as the ratio of primary to secondary turns. But the primary induced e.m.f. is the back-e.m.f. in the primary and this is equal in magnitude but opposite in phase to V_1. Hence the secondary voltage V_2 is such that

$$\frac{V_2}{V_1} = \frac{N_2}{N_1} \text{ or } V_2 = \frac{N_2}{N_1} \cdot V_1$$

a relationship already derived by an alternative approach in the previous section.

The unloaded real transformer

So far we have assumed the transformer to be an ideal device. Although the iron-cored power transformer is indeed a highly efficient unit, losses are inevitable in practice, and the operating condition of real transformers will now be considered.

When the secondary winding of a transformer is on open circuit, the primary winding behaves as a very large inductive impedance. The applied voltage V_1 has to perform two functions: (a) cause a current to flow against the effect of the coil resistance, and (b) balance the induced e.m.f. (the back-e.m.f.) in the winding due to its self-inductance. For a coil of many turns, wound on an iron core, the inductance will be very large relative to the resistance, hence the voltage drop due to the inductive reactance will closely match that of the applied voltage. Hence a small current will flow in the winding (I_o) lagging behind the applied V_1 by an angle ϕ_o which is nearly 90°. This current produces an alternating flux Φ in the core which in turn induces both primary and secondary e.m.f.s E_1 and E_2 lagging the flux by 90°. The actual values of E_1 and E_2 will be proportional to the number of turns on the primary and secondary windings, respectively, since practically all the flux set up by the primary current will link with the secondary.

Since the difference between the applied voltage V_1 and the primary induced e.m.f. E_1 will be very small (in an ideal inductor the back-e.m.f. would exactly equal the applied voltage), the induced e.m.f. is practically antiphase to V_1. This situation is shown in the phasor diagram of *Figure 7.7* where it has been as-

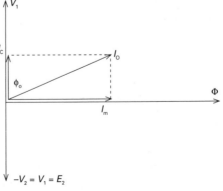

Figure 7.7

sumed for simplicity that the primary and secondary turns are equal, so making E_2 equal to E_1.

The no-load current I_o can be split into two components: (a) a component $I_o \sin \Phi_o$ in phase with the flux Φ, a wattless or reactive component just sufficient to produce the flux Φ which sets up the e.m.f. of self-induction E_1 in the primary coil almost antiphase to V_1 and nearly equal to it in magnitude; this component is the true magnetizing current I_m; (b) a component $I_c = I_o \cos \phi_o$ in phase with the applied voltage V_1 which is a power or active component supplying the transformer losses. These are losses in the core and the small but finite heat loss due to the resistance of the primary winding.

Hence, neglecting the resistive loss

Iron loss $= V_1 I_c$ watts

From the phasor diagram we see that

No-load current $I_o = \sqrt{(I_c^2 + I_m^2)}$

No-load power factor $= \cos \phi_o = \dfrac{I_c}{I_o}$

This latter is practically zero.

Example (3)   A transformer has its primary terminals connected to a 240 V, 50 Hz supply and the primary current is then measured as 0.05 A. If the secondary is on open circuit and the power absorbed is 5.5 W, find the iron loss current and the magnetizing current.

Solution   The power component

$= V_1 I_o \cos \phi_o = 5.5$ W

∴   The iron loss current

$I_c = I_o \cos \phi_o = 5.5/240 = 0.023$ A

The magnetizing current is then

$I_m = \sqrt{(I_o^2 - I_c^2)}$

$= \sqrt{0.05^2 - 0.023^2} = 0.044$ A

The loaded condition

To simplify the analysis of a loaded transformer, we will assume that the transformer losses are only those associated with the iron circuit and neglect the resistive losses of the windings themselves. With this assumption it follows that the secondary terminal voltage V_2 will be identical to the secondary induced e.m.f. E_2, and that the primary applied voltage V_1 is equal in magnitude and opposite in phase to the self-induced

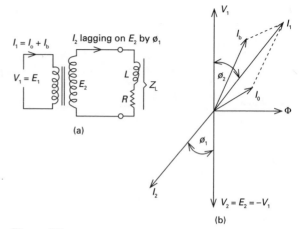

Figure 7.8

E_1 in the primary coil. Further, if we take the primary and secondary to have equal numbers of turns, $N_1 = N_2$ and $E_1 = E_2$.

Let the secondary terminals of this transformer now be connected to a load which, by way of example, is assumed to be a positive impedance, that is, an inductive load with self-resistance, as shown in *Figure 7.8*. The induced secondary e.m.f. E_2 ($=V_2$) will at once cause a current I_2 to flow through the load impedance and the secondary winding. This current will lag by some angle ϕ_1 on the e.m.f. and will attempt to create a magnetic flux of its own in the transformer core. It is here that the constancy of the flux Φ must be taken into account; the core flux depends *only* on the applied primary voltage V_1 and *cannot* change unless V_1 changes. This condition is true irrespective of any other currents which may be caused to flow in either or both of the transformer windings. Some action must therefore take place to nullify the effect of the secondary current on the core flux; what happens is that an additional primary current appears of such a magnitude and phase that the effect of the secondary current is exactly neutralized. The demagnetizing ampere-turns of the secondary are, in fact, neutralized by the increase in the primary ampere-turns.

This new primary current which is known as the 'balancing current' I_b must therefore be 180° out of phase with I_2 and of such a magnitude that the total effective resultant flux introduced by the two additional currents into the core is zero.

We can now draw a phasor diagram for the loaded transformer as has been done in *Figure 7.8*. The total primary current I_1 is the phasor sum of the no-load current I_o and the balancing current I_b, lagging on the primary voltage by ϕ_2. In the diagram the I_o phasor has, for clarity, been drawn far greater relative to the other currents than it would be in an actual component. In practice, I_b is very much greater than I_o, and both I_1 and I_b may be considered as practically equal in magnitude. The angle ϕ_1 by which the secondary current lags the secondary e.m.f. is then nearly the same as that by which the primary balancing current lags behind the primary voltage, that is, $\phi_1 \simeq \phi_2$. This can be expressed in another way by saying that the power factor is nearly the same for both primary and secondary coils.

Hence, since $\cos \phi_1 \simeq \cos \phi_2$

$$I_1 E_1 \cos \phi \simeq E_2 I_2 \cos \phi_1$$

and

input power $\simeq$ output power

The transformer thus transforms a high-voltage, low-current into a low-voltage, high-current relationship (or conversely) without altering to any significant degree the phase angle between them.

Example (4)   A transformer having 1000 turns on the primary and 200 turns on the secondary coil has a load connected across the secondary terminals consisting of an inductance of 0.2 H and self-resistance 25 Ω. Find the primary current and the phase angle between it and the applied voltage when the primary is connected to a 240 V 50 Hz supply.

Solution   The turns ratio here is 200/1000 or 1/5, a step-down component. The secondary load impedance

$$Z_L = \sqrt{(R^2 + X_L^2)}$$
$$= \sqrt{[25^2 + (20\pi)^2]} = 67.6 \ \Omega$$

The secondary e.m.f. $E_2 = 240 \times \dfrac{1}{5} = 48$ V

∴   secondary current $I = \dfrac{48}{67.6} = 0.71$ A

Hence primary current $I_1 = 0.71 \times \dfrac{1}{5}$

$$= 0.142 \ A$$

Angle of lag $\phi = \tan^{-1} \dfrac{X_L}{R} = \tan^{-1} \dfrac{20\pi}{25}$

$$= 68°$$

Here are a couple of examples to work out for yourselves:

(5)   Draw the phasor diagram of a loaded transformer but having a purely resistive secondary load. Assume a one-to-one turns ratio.

(6)   An ideal transformer has 500 primary turns and 7000 secondary turns. If 240 V is applied to the primary, what voltage will appear across the secondary? If a secondary load resistance is connected of such a value that a secondary current of 0.1 A flows, what will be the primary current?

INPUT RESISTANCE

When the secondary winding of a transformer is unloaded, the primary impedance is very high and the current which flows is small. When a load is connected to the secondary, the primary current increases and the primary impedance falls. Now the impedance seen at the primary terminals is given by the ratio V_1/I_1 and since I_1 will depend on the secondary load, so too will the input impedance. The relationship between the magnitude of the secondary load and the effective primary impedance is a very important one in transformer theory and we now investigate it. To simplify things, we will assume that the secondary load is purely resistive, given by R_L.

In *Figure 7.9* let R_L be connected to the secondary terminals and let the primary input be a constant V_1 volts. We have already shown that

$$\frac{V_2}{V_1} = \frac{N_2}{N_1} \text{ and } \frac{I_1}{I_2} = \frac{N_2}{N_1}$$

Then by multiplication of these

$$\frac{V_2}{V_1} \times \frac{I_1}{I_2} = \left[\frac{N_2}{N_1}\right]^2$$

so that

$$\frac{V_2}{I_2} \div \frac{V_1}{I_1} = \left[\frac{N_2}{N_1}\right]^2$$

But V_2/I_2 = load resistance R_L and V_1/I_1 = the input resistance R_{in}

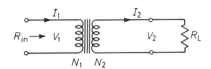

Figure 7.9

$$\therefore \quad \frac{R_L}{R_{in}} = \left[\frac{N_2}{N_1}\right]^2$$

or

$$R_{in} = R_L \left[\frac{N_1}{N_2}\right]^2 \qquad (7.3)$$

Now N_1/N_2 which is the inverse of the turns ratio will be greater than unity if $N_1 > N_2$, that is, for a step-down transformer, and less than unity if $N_1 < N_2$, that is, for a step-up transformer. The input resistance of the transformer can therefore be made to suit any required value for a given output load by the proper choice of the turns ratio.

This is an important property of a transformer because it can be used as a resistance transforming device, distinct from its more obvious properties of voltage and current transformation. Stated in words, we say: when a circuit includes a transformer, the resistance of the circuit as seen at the primary terminals is increased or decreased by a factor $[N_1/N_2]^2$. Notice particularly that the resistance ratio depends on the square of N_1/N_2 and not directly on this ratio as current and voltage ratios do. It helps to keep in mind that the resistance value appears to increase when referred to a winding with a greater number of turns, and to decrease when referred to a winding with a fewer number of turns.

The following worked examples should make things clear for you.

Example (7)   A step-down transformer has a turns ratio of 10:1. If a resistance of 1 kΩ is connected to the secondary terminals, what effective resistance does the transformer offer at the primary terminals?

Solution   In this transformer $N_1 = 10$ and $N_2 = 1$ (strictly in terms of the turns ratio),

$$\therefore \quad \frac{N_1}{N_2} = 10$$

The secondary load $R_L = 1000\ \Omega$

$\therefore$ primary resistance

$$R_{in} = \left[\frac{N_1}{N_2}\right]^2 R_2 = 10^2 \times 1000$$

$$= 100\ k\Omega$$

This result is illustrated in *Figure 7.10*. As far as the circuit connected to the primary is concerned, the transformer and its 1 kΩ load ap-

Figure 7.10

pears as a single 100 kΩ resistor. The secondary load has been 'transferred' or 'reflected' to the primary circuit without in any way modifying the characteristics or behaviour of the circuit; that is, the two circuits shown in the diagram are completely identical as far as the a.c. conditions are concerned.

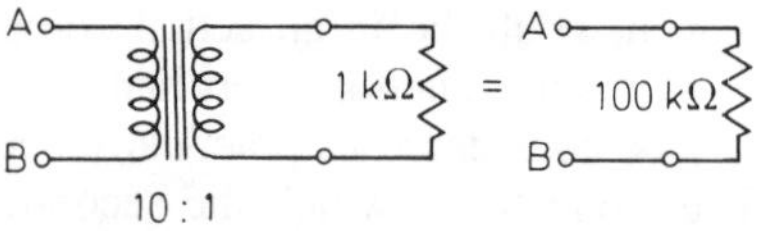

Example (8)   The resistance of a loudspeaker is 8 Ω and it is to be connected to an output transistor in an amplifier which requires a load resistance of 100 Ω. What should be the turns ratio of a transformer used to connect the transistor to the loudspeaker?

Solution   This arrangement is a common use for a transformer: as an output transformer connecting an amplifier to a loudspeaker. *Figure 7.11* shows a simplified circuit of this kind.

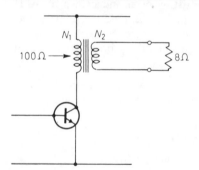

Figure 7.11

Using our transformation formula

$$R_{in} = \left[\frac{N_1}{N_2}\right]^2 R_L$$

we have

$$100 = \left[\frac{N_1}{N_2}\right]^2 \times 8$$

$$\therefore \quad \left[\frac{N_1}{N_2}\right]^2 = \frac{100}{8} = 12.5$$

$$\frac{N_1}{N_2} = 3.54$$

Hence we require a step-down transformer with a turns ratio of 3.54:1.

MAXIMUM POWER TRANSFER

The previous example raises a very important subject in electrical and electronic principles, i.e. the transfer of electrical power from a generator to a load. In *Figure 7.12*, a generator of e.m.f. E volts and internal resistance r Ω is connected to a load resistor R_L. Current I will flow in the circuit and power will be delivered to the load resistor (where it is required) and to the internal resistance of the generator (where it is not required and is consequently wasted).

What is the relationship between r and R_L for the greatest power to be delivered to the load? We have already proved that this will occur when $r = R_L$ and under this condition it is not difficult to see that the maximum possible power delivered to the load is one-half of the total power supplied by the generator.

Having recalled this important fact, it is now necessary to make the best practical use of it. In most circuits, the internal resistance of the generator and the actual resistance of the load are rarely identical. In the problem above, for example, the load resistance was the 8 Ω of the loudspeaker unit, while the resistance offered by the generator (the transistor amplifier) was 100 Ω. If the loudspeaker is connected directly to the transistor output, only a small part of the available power will be developed in it. By interposing the transformer, the effective resistance of the loudspeaker is increased to 100 Ω and the maximum power is then transferred to it.

Such a process is known as resistance (or impedance) matching, so a transformer with a suitable turns ratio affords us a convenient way of reflecting resistance from one circuit to another to maximize the power transfer from a source to load.

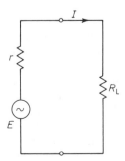

Figure 7.12

Example (9)   A generator of e.m.f. 10 V and internal resistance 50 Ω is connected directly to a load resistance of 150 Ω. What power is dissipated in the load? A transformer is now employed to match the load to the generator. Find the necessary turns ratio and calculate the power then dissipated in the load.

Solution   Figure 7.13(a) shows the direct connection. The circuit current $I = 10/200 = 0.05$ A and hence the power dissipated in the load

$$P_L = I^2R = (0.05)^2 \times 150$$

$$= 0.375 \text{ W}$$

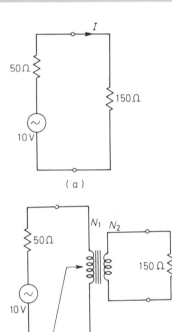

Figure 7.13

Figure 7.13(b) shows the transformer connected between generator and load. For maximum power in the load, the primary of the transformer must present a resistance of 50 Ω to the generator, that is, a value equal to its own internal resistance. Hence, since

$$R_{in} = \left[\frac{N_1}{N_2}\right]^2 R_L$$

$$\left[\frac{N_1}{N_2}\right]^2 = \frac{R_{in}}{R_L} = \frac{50}{150} = \frac{1}{3}$$

Inverting

$$\left[\frac{N_2}{N_1}\right]^2 = 3 \quad \text{or} \quad \left[\frac{N_2}{N_1}\right] = 1.73$$

The transformer has to be a step-up transformer with turns ratio 1:1.73. The power in the load will now be

$$P = I^2 R_L = \left[\frac{10}{100}\right]^2 \times 50 = 0.5 \text{ W}$$

This is the maximum possible power.

Try the next problem for yourself. It will illustrate in graphical form how maximum load power occurs when the resistances of the load and generator are equal.

(10)   Assume in the previous problem that R_L can be varied between 100 Ω and 200 Ω in 10 Ω steps. For the direct connection as indicated in *Figure 7.13(a)*, calculate the circuit current I for each step in the value of R_L and find the corresponding power developed in R_L.

Plot a graph of load P_L (vertically) against R_L (horizontally) and verify that the maximum power is 0.5 W when the value of R_L is equal to 50 Ω.

TRANSFORMER LOSSES AND MATERIALS

When iron cores are used in transformers, two fundamental power losses make their appearance. These losses are:

(a)   eddy current losses;
(b)   hysteresis losses.

Eddy current loss results because energy is dissipated as heat in the metal of the core when an alternating current flows in the surrounding coils. The alternating flux, as well as inducing voltages in the coils, produces voltages in the core itself, causing small local currents to circulate in the iron circuit. This loss is reduced by using insulated laminations for the core, so breaking up direct current paths through the material and presenting instead a high resistance to the

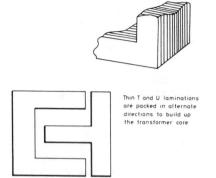

Thin T and U laminations are packed in alternate directions to build up the transformer core

Figure 7.14

circulation of such currents. *Figure 7.14* shows the general form of iron core construction for low-frequency and power transformers. For transformers designed for the higher audio frequencies, the thickness of the laminations is reduced to ensure that the induced voltages do not lead to increased eddy current magnitudes.

At very high frequencies, when it becomes impracticable to make laminations any thinner, core losses are reduced by using iron-dust or granulated iron cores. Such cores are used at radio frequencies, generally for tuning purposes, and are available for frequencies up to many tens of megahertz. The losses are very small, the binding material between the iron granulations acting as an efficient insulator which breaks up the eddy current paths to a negligible size.

Hysteresis is that property of a magnetic material by which its magnetization depends not only on the applied magnetizing force but also on its previous magnetic state. In a magnetic cycle, while the primary current goes through one complete cycle of alternation, the flux density lags behind the magnetizing force; this represents a power loss which appears in the form of heat. You will recall that the area of the hysteresis loop for a particular iron material is proportional to the loss.

Hysteresis loss is reduced by the proper choice of core material. Silicon iron or 'Stalloy' is generally used for low-frequency transformer cores. Iron losses increase as the frequency is raised, but not equally. Hysteresis loss is proportional to the frequency but eddy current loss is proportional to the square of the frequency. The power wasted in iron losses is all supplied by the magnetizing current and consequently this current does not lag by exactly 90° on the applied primary voltage as we previously assumed. When transformers are further examined in later parts of the course, factors such as this have to be taken into account.

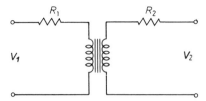

Figure 7.15

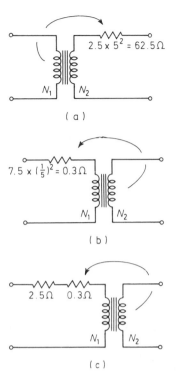

Figure 7.16

There is one other fundamental reason for power loss in a transformer, one that is common to all other components as well. This is loss due to the resistance of the windings, referred to as the copper loss to distinguish it from the iron losses in the core. Resistance losses may be viewed as internal resistances in much the same way as the internal resistance of a cell or generator is considered; as small series resistances incorporated into the otherwise loss-free windings of the transformer. *Figure 7.15* shows the representation of copper losses. Once these are considered, of course, the voltage actually effective on the primary coil is no longer V_1 as we have so far assumed, but less than V_1 by an amount equal to the voltage drop across the series loss resistance R_1. In the same way, the actual voltage appearing at the secondary terminals is less than V_2 by the amount of voltage drop across the series loss resistance R_1. Clearly, since the power loss is expressed by I^2R, the loss increases with the load or the square of the load current.

We can use our resistance-transformation property of a transformer effectively to transfer the copper losses to one winding only, and this often simplifies problems where the copper losses have to be taken into account. The next example will illustrate this.

Example (11)   A transformer has a turns ratio of 1:5. The resistance of the primary is 2.5 Ω and of the secondary is 7.5 Ω. Find the equivalent resistance of the primary in terms of the secondary, the secondary in terms of the primary, and the total resistance in terms of the primary.

Solution   In this problem the ratio

$$\frac{N_1}{N_2} = \frac{1}{5}$$

From *Figure 7.16(a)*, the equivalent resistance of the primary in terms of the secondary = 2.5 × 5² = 62.5 Ω. The equivalent resistance of the secondary in terms of the primary = 7.5 × [1/5]² = 0.3 Ω. This is illustrated in *Figure 7.16(b)*.

The total resistance in terms of the primary

= primary resistance + transferred secondary resistance

= 2.5 + 0.3 = 2.8 Ω

This is shown in *Figure 7.16(c)*.

The above example shows us that we can use our transformation relationships in either direction through the transformer (for what is a step-up ratio in one direction becomes a step-down ratio in the other), and also that all the resistances associated with transformer windings can be transferred entirely to one side of the transformer. As far as the a.c. conditions are concerned, all three diagrams of *Figure 7.16* are identical.

A transformer model

We can make up a model of a transformer by assuming an ideal unit and degrading this by the addition of hypothetical components representing the various losses we have mentioned above. We have already noted that the copper losses may be considered as small resistances placed in series with the primary and

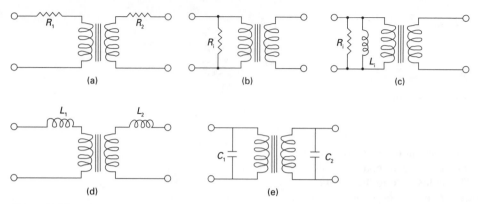

Figure 7.17

secondary windings and this is represented in *Figure 7.17(a)*. In addition to this, the hysteresis and eddy current losses may be simulated by a small resistance added in parallel to the primary coil as illustrated in diagram *(b)*. The power wasted in iron losses is all supplied by the magnetizing current and consequently this current does not lag by exactly 90° on the applied voltage. This effect can be represented by the addition of an inductance in parallel with the primary coil and the iron loss resistance already shown, diagram *(c)*.

We have not so far mentioned flux leakage or self-capacity, both of which must be added to a model representation to complete the picture as it were. The flux threading the secondary coil is always less than the primary flux because of leakage from the iron circuit into the surroundings. The effect of this leakage can be simulated by the addition of a small series inductance in both primary and secondary windings as shown in diagram *(d)*.

The windings of a transformer have capacitance between each turn and the next and the self-capacity of the total windings may be quite appreciable. At power frequencies, however, the self-capacity is not important, but it becomes so at the higher audio frequencies. Such capacitance can be represented as parallel capacitors wired to the primary and secondary terminals, see diagram *(e)*.

A complete equivalent model for an iron-cored transformer is shown in *Figure 7.18*, all the second-

ary loss, leakage and capacitance being transferred to the primary circuit in the way we have earlier investigated. In this representation:

R_c = the coil resistances referred to the primary

$$= R_1 + \left[\frac{N_1}{N_2}\right]^2 R_2$$

L_1 = the leakage inductance referred to the primary

$$= L_1 + \left[\frac{N_1}{N_2}\right]^2 L_2$$

C_w = the winding capacitances referred to the primary

$$= C_1 + \left[\frac{N_2}{N_1}\right]^2 C_2$$

L_p = the effective inductance of the primary with the secondary on open-circuit such that $\dfrac{V}{\omega L_p}$ gives the magnetizing current

R_s = the primary shunt resistance such that $\dfrac{V}{R_s}$ gives the eddy current and hysteresis losses

The equivalent circuit can often be simplified by neglecting any losses which are normally small under the conditions in which the transformer is used. Thus in the case of transformers used between transistor stages in audio amplifiers, the winding resistances are usually small as also are the iron losses. Here, however, the winding capacitances become important. In the figure, therefore, R_c and R_s may be omitted but C_w must be maintained. For transformers used on power frequency, these capacitances are rarely of any significance.

What is actually omitted and what is retained, however, is not invariable over a *range* of frequencies.

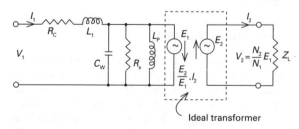

Figure 7.18

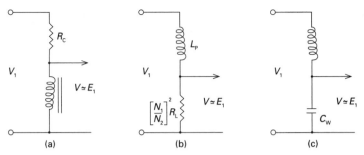

Figure 7.19

The equivalent circuit is often quite different at the high-frequency end of the range from what it is at the low-frequency end. If a reasonably uniform response is required over, say, the audio range of frequencies such transformers, which include inter-stage types, input and output impedance matching types, and so on, call for careful design considerations. At very low frequencies the primary circuit will approximate to that shown in *Figure 7.19(a)*. Here the series inductance L_1 is neglected together with the shunt resistance and the winding capacitance C_w. Hence the ratio of the terminal voltage V_1 and the effective voltage V on the primary will be small, so V_1/V_2 will be small. At some mid-frequency, C_w and L_p will resonate so that the circuit will approximate to diagram *(b)*. The transferred load resistance $(N_1/N_2)^2 R_L$ will be very much greater than the reactance of L_1 and hence the ratio V_2/V_1 will approach that of the turns ratio. At very high frequencies the parallel effect of capacitance becomes of importance, having a very low reactance in comparison with the series L_1 which will have a relatively large inductance. Again, the ratio of the source voltage V_1 to the effective primary voltage V_p will become small, so that V_2/V_1 will also be small. Diagram *(c)* shows the condition this time.

The full transformation ratio of voltage is consequently never attained and the frequency vs. gain characteristic shows a fall-off in response at low and high frequencies with a resonant 'hump' appearing somewhere in the range. By careful design this hump can be arranged to occur towards the upper frequency parts of the range and so extend what would otherwise be a region of falling gain. By sectionalizing the windings, the capacitances can be reduced, and the low frequency response can be helped by maintaining a high primary inductance, often by keeping d.c. out of the winding. The proper choice of the core material characteristics also has a large influence on transformer performance. A typical response curve is shown in *Figure 7.20*.

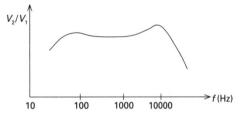

Figure 7.20

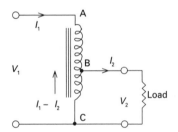

Figure 7.21

The auto-transformer

This is not a device which is used in cars, as one of my students once envisaged. The auto-transformer is a transformer with a single winding where the primary is part of the secondary or vice versa. *Figure 7.21* shows a step-down form of this transformer, the whole winding between terminals A and C forming the primary, and that part of the winding between B and C forming the secondary.

The transformation ratio is, as for the double-wound device, the ratio of the number of turns on the secondary or output winding BC to the number of turns on the primary or input winding AC, and the same relationships for current and voltage ratios still apply, that is

$$\frac{V_1}{V_2} = \frac{\text{turns AC}}{\text{turns BC}} = \frac{I_2}{I_1}$$

The chief advantage that the auto-transformer has lies in the fact that the secondary section of the winding carries (for a step-down unit) both the primary and secondary currents, but these are in phase opposition and the net current in BC $(I_1 - I_2)$ is obviously less than the full primary current I_1. The presence of the secondary current thus gives a reduction in the primary copper loss and there is no secondary copper loss at all. For the same output and voltage ratio, therefore, the auto-transformer requires less copper than a double-wound transformer. Though it must be added that the saving is only appreciable when the transformation ratio is close to unity, for then I_1 and I_2 become almost equal and the losses become very small.

A disadvantage of the auto-transformer is that the two windings are not electrically separate, hence where it is used on mains supplies a dangerous shock may be given on the secondary side in case of breakdown or other fault. The transformer may be used as a step-up device by changing over the primary and secondary connections.

By making the point B adjustable over the length of the winding (which in this case is wound on a toroidal core), the device becomes what is generally known as a 'Variac' providing a wide variation in the output voltage V_2. The action is similar to a potential divider with the important difference that in the case of the latter the voltage can only be stepped down. With Variacs in general, by using the unit in step-up mode, an output from zero to 270 V is available from a nominal 240 V mains input.

REGULATION AND EFFICIENCY

In most power transformers, the expenditure of energy due to the iron and copper losses is not very great. The efficiency of a transformer is expressed as a percentage:

$$\frac{\text{output power}}{\text{input power}} \times 100\%$$

The input, however, is equal to the output plus the losses, so the efficiency may be expressed as

$$\frac{\text{output power}}{\text{output power} + \text{copper losses} + \text{iron losses}} \times 100\%$$

$$\frac{\text{power delivered by secondary}}{\text{total power supplied by the primary}} \times 100\%$$

Iron losses are reasonably constant, but copper losses vary as the square of the currents flowing. It can be proved that the efficiency is greatest when the copper

losses are equal to the iron losses. To measure efficiency it is convenient to put the transformer on a pure resistive load and measure the power output by means of an ammeter and voltmeter. The input power may be measured by a wattmeter.

As more current is drawn from the secondary of a transformer, the terminal voltage falls because of the increased copper loss. The difference between the secondary p.d. at no load and the secondary p.d. at full load is expressed as a percentage and is known as the regulation of the transformer.

Regulation

$$= \frac{\text{No load voltage} - \text{full load voltage}}{\text{Full load voltage}} \times 100\%$$

Regulations of some 1 to 2 per cent are general in well-designed transformers. This is as far as we need to go into transformer theory for the present.

SUMMARY

- For an ideal transformer $\dfrac{V_2}{V_1} = \dfrac{N_2}{N_1} = \dfrac{I_1}{I_2}$

- The total primary current $I_1 = I_2 \left[\dfrac{N_2}{N_1} \right]$

- The equivalent effective a.c. resistance referred to the primary

$$Z_1 = Z_2 \left[\frac{N_1}{N_2} \right]^2 = Z_2 \left[\frac{V_1}{V_2} \right]^2$$

- The equivalent effective a.c. resistance referred to the secondary

$$Z_2 = Z_1 \left[\frac{N_2}{N_1} \right]^2 = Z_1 \left[\frac{V_2}{V_1} \right]^2$$

- Efficiency of a transformer η

$$= \frac{\text{output power}}{\text{output power} + \text{losses}} \times 100\%$$

$$= \frac{\text{power delivered by secondary}}{\text{total power consumed by primary}} \times 100\%$$

- Regulation of a transformer

$$\frac{\text{no load voltage} - \text{full load voltage}}{\text{full load voltage}} \times 100\%$$

- Efficiency is greatest when iron losses = copper losses

REVIEW QUESTIONS

1   Why is a closed iron core preferable to an open iron core?
2   When is a transformer said to be ideal?
3   What is meant by iron losses? What are the two principle iron losses?
4   How can eddy current loss be minimized?
5   Why is the induced e.m.f. different from the terminal voltage?
6   Give three examples of the usage of transformers.
7   Why does the gain response of an audio transformer fall off at both low and high frequencies?
8   What is meant by the regulation of a transformer? The efficiency?
9   State the advantages and disadvantages of an auto-transformer.
10 Sketch a phasor diagram (after *Figure 7.8*) illustrating conditions for a capacitive (leading current) load.

EXERCISES AND PROBLEMS

(12)   A power transformer draws a primary current of 250 mA from a 250 V supply when the secondary is open-circuited. If 22 W of power is absorbed, find the iron loss current and the magnetizing current.

(13)   A transformer has a turns ratio of 5:1. A 240 V supply is connected to the primary terminals and a purely resistive load of 150 Ω is connected across the secondary. Calculate (a) the secondary voltage, (b) the secondary current, (c) the primary current, (d) the power dissipated in the load.

(14)   A 240/6 V transformer supplies a number of 6 V 0.5 W filament bulbs connected in parallel. Calculate (a) the greatest number of such lamps this transformer can supply without being overloaded, (b) the primary current under full load. The transformer is rated at 100 W.

(15)   A transformer with a primary coil of 1750 turns has a 50 Ω non-inductive resistor connected across its secondary terminals. A 24 V supply is connected to the primary and 45 V is developed across the resistor. Calculate (a) the transformation ratio, (b) the number of secondary turns, (c) the current in each winding, (d) the power dissipated in the resistor.

(16)   A step-down transformer has a turns ratio of 7.5:2. If a resistance of 2 kΩ is connected to the secondary, what effective resistance does the transformer offer at the primary?

(17)   A loudspeaker of resistance 16 Ω is to be connected to the output terminals of an amplifier which requires a load resistance of 200 Ω. What should be the transformation of the transformer? If an exact ratio was not available, do you think it would be preferable to use a transformer with (a) a lower ratio or (b) a higher ratio? Give reasons for your choice.

(18)   The primary winding of a 550/110 V transformer has a resistance of 1.5 Ω and a secondary resistance of 0.075 Ω. Calculate the equivalent resistance of this transformer referred to (a) the primary, (b) the secondary.

(19)   A generator of e.m.f. 25 V and internal resistance 100 Ω is connected directly to a load resistor of 50 Ω. What power is dissipated in the load? A transformer is now used to match the load to the generator. Find the required turns ratio and calculate the power now dissipated in the load.

(20)   A transformer has primary and secondary winding resistances of 5 Ω and 8 Ω, respectively. What will be the secondary terminal voltage (a) on open-circuit, (b) with a load resistor of 500 Ω, given that the primary input is 100 V and the transformation ratio is 1:3?

(21)   A 10 μF capacitor of negligible loss is connected to the secondary terminals of a step-down transformer of turns ratio 2:1. What is the equivalent input capacity of the transformer?

(22)   A 9:1 step-down transformer has a primary resistance of 3.5 Ω and a secondary resistance of 0.02 Ω. When the transformer is delivering a full load current of 50 A to a resistive load, the secondary terminal voltage is 80 V. If the iron loss of the transformer is 60 W on no load, find (a) the full load copper loss, (b) the regulation, (c) the efficiency.

(23)   A transformer with a turns ratio of 1:2 is fed from a 250 V supply having an internal

resistance of 20 Ω. If the transformer losses are neglected, calculate the power dissipated in a secondary inductive load where $X = 800$ Ω and $R = 520$ Ω.

(24)   A tape recorder head has an internal resistance of 10 Ω and generates an e.m.f. of 1 mV. It is connected by way of a transformer of turns ratio 1:N to an amplifier which has an input resistance of 100 Ω. Show that the power delivered to the amplifier is given by

$$P = \left[N + \frac{10}{N} \right]^{-2} \mu W$$

and hence find the value of N for which the transfer of power is a maximum. What is this maximum power?

(25)   An auto-transformer is used to transform from 500 to 440 V into a purely resistive load

equivalent to a power consumption of 2 kW. Neglecting losses, find the current in the windings AB and BC. (Use *Figure 7.21*.)

(26)   An auto-transformer steps up voltage in the ratio 240 V to 440 V. If the total number of turns on the core is 1360, estimate the number of primary and secondary turns. This transformer supplies a load drawing 9.6 A; what current is carried by the common section of the winding?

(27)   The iron losses of a transformer having a step-down ratio of 6 to 1 are 55 W which may be assumed constant. If the primary resistance is 2.5 Ω and the secondary resistance 0.15 Ω, and if the secondary p.d. is 100 V when delivering a full-load current of 10 A to a resistive load, find (a) the full load copper loss, (b) the efficiency.

8

Three-phase systems

Limitations of single-phase systems
Two and three-phase generation
Advantages of polyphase circuits
Star and delta connections
Power in a balanced system
Three-phase transformers

Up to present, in our study of steady state alternating systems, we have dealt only with the aspects of single-phase working. There is an objection to single-phase supplies, however, which as we have seen earlier, concerns problems of power. Any power absorbed varies continuously throughout the cycle and unless the power factor is unity when current and voltage are exactly in phase, can become negative (or reversed) during part of the time. *Figure 8.1* demonstrates this effect; here there are two periods of negative power during which the flow of power is from the load to the source. This reverse power comes from the field energy of the inductance which, in this example, creates the lag of the current. During these periods of reversed power, the induced e.m.f. across the inductance overcomes the voltage of the source and a reverse current flows. If we consider a capacitive circuit where the current is leading, the reverse power arises

from the electric field of the capacitor where the discharge of the stored energy overcomes the e.m.f. of the source.

We have already noted that the average power over a period of time is expressed by $VI \cdot \cos\phi$ and correction for the reverse power can up to a point be made by power factor adjustment. Notice from a study of the figure that alterations in the phase angle have the effect of altering the position of the power curve with respect to the horizontal axis X-Y. Only when the circuit is purely resistive does the power curve 'sit' upon this axis and eliminate the regions of reverse power.

POLYPHASE SYSTEMS

This chapter will deal with circuits which are supplied by certain special combinations and arrangements of sinusoidal sources; all of these circuits will contain several sources. Certain connections of alternating supplies (and their respective loads) are advantageous from an economical viewpoint, especially where considerable power transmissions have to be interconnected over large regions as for the national grid system. We shall see that by using such polyphase systems as these methods are called, the periods of negative power which are an unwanted feature of single-phase working may be eliminated and a larger ratio of power output to equipment weight and size can be obtained.

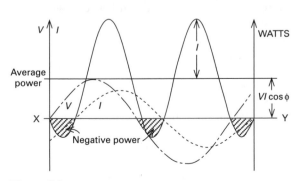

Figure 8.1

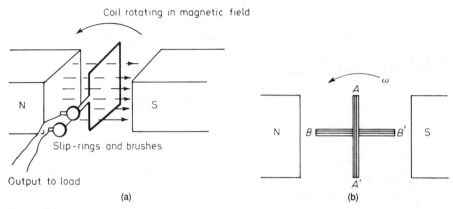

Figure 8.2

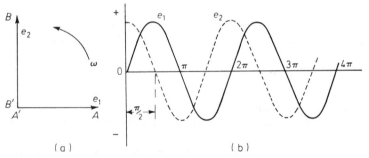

Figure 8.3

All the sinusoidal sources to be discussed will be taken to operate at a single frequency, and this assumption is true of the national grid which operates at a frequency of 50 hertz.

POLYPHASE GENERATION

Although the single-phase a.c. generator sketched in *Figure 8.2(a)* is theoretically sound for demonstrating a sinusoidal output wave, a number of practical limitations prevent its direct application to electricity production. The pole pieces, for example, need proper shaping and it is much better to have the field rotate while the coil remains stationary. This overcomes the problem of bringing the generated e.m.f. and the current associated with it out to an external load by way of slip rings or any other kind of sliding contacts. This is of importance where heavy currents and high voltages are concerned, and particularly so where more than a single coil is involved.

Suppose that instead of a single coil, two identically wound coils are mounted mutually at right angles to each other and rotated in a magnetic field as shown in *Figure 8.2(b)*. The e.m.f.s generated in each of these coils will be sinusoidal in form and will be of the same frequency and peak amplitudes but will differ in phase by 90°.

Hence if

$$e_1 = \hat{E} \sin \omega t$$

then

$$e_2 = \hat{E} \sin \left(\omega t + \frac{\pi}{2} \right)$$

The phasor diagram and the graphs of the two output e.m.f.s are shown in *Figure 8.3*. This arrangement provides the basic form of a two-phase a.c. generator and the supply is known as a two-phase supply. If the currents supplied are equal and lag (or lead) by the same angle ϕ, they will be, for the lagging case, *Figure 8.4(a)*:

$$i_1 = \hat{I} \sin (\omega t - \phi)$$

$$i_2 = \hat{I} \sin \left(\omega t + \frac{\pi}{2} - \phi \right)$$

The instantaneous value of the power supplied is:

$$e_1 i_1 + e_2 i_2 = \hat{E}\hat{I} \left[\sin \omega t \cdot \sin (\omega t - \phi) \right.$$
$$\left. + \sin \left(\omega t + \frac{\pi}{2} \right) \cdot \sin \left(\omega t + \frac{\pi}{2} - \phi \right) \right]$$

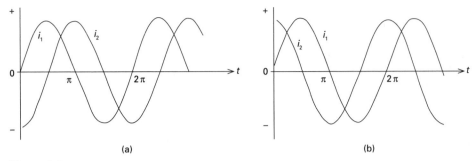

(a)   (b)

Figure 8.4

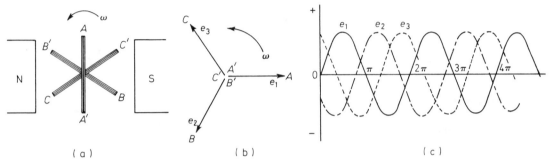

(a)   (b)   (c)

Figure 8.5

$$= \hat{E}\hat{I} \, [\sin \omega t \, . \, \sin (\omega t - \phi)$$
$$+ \cos \omega t \, . \, \cos (\omega t - \phi)]$$

This line is the trigonometric identity $\sin A \, . \, \sin B + \cos A \, . \, \cos B = \cos (A - B)$ where $A = \omega t$ and $B = \omega t - \phi$.

$\therefore \quad \cos (A - B) = \cos \omega t - (\omega t - \phi) = \cos \phi$ and the instantaneous power is given by $\hat{E}\hat{I} \cos \phi = 2EI \cos \phi$

This is a constant, hence the power is constant throughout the cycle.

The situation for leading currents is shown in *Figure 8.4(b)* where the identical products for $e_1 i_1 + e_2 i_2$ are obtained.

Suppose now that three identical coils, fixed 120° apart, are rotated in the magnetic field of the generator. The e.m.f.s generated will again be sinusoidal in form and have the same frequency and peak amplitudes but will differ in phase from each other by 120°. This time we have a three-phase supply; hence, if

$$e_1 = \hat{E} \sin \omega t$$

then

$$e_2 = \hat{E} \sin \left(\omega t - \frac{2\pi}{3} \right)$$

and

$$e_3 = \hat{E} \sin \left(\omega t - \frac{4\pi}{3} \right)$$

and the phasor diagram and the graphs of the three voltages will now be as shown in *Figure 8.5*.

In the same way it is possible to generate four-, six- or twelve-phase alternating systems, all having a uniformity of power output throughout the cycle. Three-phase generation is the most common practical method.

As we are going to be involved with three voltages, it will be necessary for us to be able to distinguish one from the others at any particular instant of time. The sequence in which the voltages are generated depends on the direction of rotation of the coils within the magnetic field. In *Figure 8.5*, the voltage in coil A (e_1) is instantaneously zero but about to begin its cycle. The other two voltages, e_2 and e_3, are part way through a cycle and are instantaneously of negative and positive polarity, respectively. This sequence is evident from the phasor diagram where for an anti-clockwise rotation the phasors would pass a fixed point in the order $e_1 - e_2 - e_3 - e_1 - e_2 \dots$ This can also be seen from the graphical plot where the maximum values occur in the same order. If the coils were rotated in the opposite direction, the sequence would become $e_3 - e_2 - e_1 - e_3 \dots$

In order to ensure that the three-phase voltages are

in their proper positions relative to each other it is necessary to mark the corresponding ends of the coils so that the same direction or sense of the voltages is obtained in each winding. This has been done in the diagrams of *Figures 8.5(a)* and *(b)* where A', B', and C' denote the respective finishes. The generated voltages will then be correctly displaced by 120° when the three starts (or the three finishes) are displaced by the same angle. In practical systems, the phases are distinguished by colour coding the wires conventionally as RED, YELLOW and BLUE phases. The three voltages are known as the phase e.m.f.s and the order in which they attain their peak amplitudes is the phase sequence.

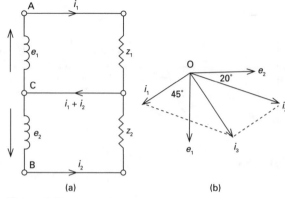

(a)   (b)

Figure 8.6

CONNECTION OF PHASES

Two-phase, three-wire

To provide an introduction to the methods of interconnection of the phases generated by three-phase machines, we will look briefly at a two-phase, three-wire system of distribution. It must be pointed out that this particular system is not of much practical significance and is rarely used, but it will illustrate the principles that we will need to understand for the practical three-phase systems which follow.

The arrangement is shown in *Figure 8.6(a)* where two sources are represented by the two coils generating e.m.f.s e_1 and e_2, respectively. The generators are connected to load impedances Z_1 and Z_2 by means of three wires or lines: the line wires A and B and the common return wire C. The two sources are always so arranged that when the alternating voltage wave is positive and is plotted above the axis as in *Figure 8.3* earlier, then the line wires A and B, respectively, are positive to the common return of line C.

Thus A will be positive to C during the first and second quarter cycles shown in the figure, and B will be positive to C during the first and fourth quarter cycles. Current from e_1 flows out along wire A and back along wire C, while current from e_2 flows along wire B and similarly returns along wire C. The current in C will therefore be the phasor sum of these two currents. This situation is illustrated in *Figure 8.6(b)*. Here e_1 lags e_2 by 90°; the currents i_1 and i_2 depend on the impedances presented by the loads Z_1 and Z_2, so each current will lag or lead on the voltage by the phase angle of the load characteristic and will be equal to e_1/Z_1 and e_2/Z_2. In the figure, the current i_1 is represented by phasor Oi_1 lagging e_1 by 45°; similarly, current i_2 is represented by phasor Oi_2 lag-

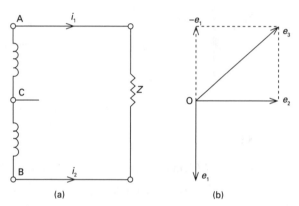

(a)   (b)

Figure 8.7

ging e_2 by 20°. These figures are for illustration only. For other impedance magnitudes and phase angles, of course, the currents would take up correspondingly different positions. The resultant current i_3 in the return wire is obtained by completion of the parallelogram Oi_1 and Oi_2. Now a polyphase system is said to be balanced when the e.m.f.s and phase angles are the same for each phase, that is, the load impedances are then balanced. The currents are then equal in magnitude and make equal angles with Oe_1 and Oe_2, respectively. In this case it should afford you no difficulty to deduce that i_3 is $\sqrt{2}$ times the current in either of the line wires, since $i_1 = i_2$ are separated by 90°.

With the three-wire arrangement, loads could be connected between the line wires A and B as before at Z in *Figure 8.7(a)*, the return wire not being connected. This time the voltage available between A and B, i.e. the amount by which A is positive to B, will be the difference between the amount by which A is positive to point C and the amount by which B is positive to point C. This follows from the fact that for this two-wire arrangement the two sources are so connected that when e_1 and e_2 are positive, the wires

A and B, respectively, are positive to C. If one voltage is negative, the voltage between A and B will be arithmetically the sum of the two instantaneous voltages, but this, however, is automatically allowed for by taking the algebraic difference and making due allowance for the signs. The phasor diagram of *Figure 8.7(b)* shows how this difference is found, by reversing the voltage e_1 and adding, by completion of the rectangle Oe_2 Oe_3 Oe_1. This voltage is plainly $\sqrt{2}e_1$, so that the voltage between A and B is $\sqrt{2}$ times the voltage of each source, and in phase is 45° ahead of the source e_2.

Three-phase, three-wire

Fundamentally each of the three outputs from a three-phase generator or alternator may be independently connected to its own particular load and three completely separate circuits are then obtained. As *Figure 8.8* shows, this arrangement required six wires (or lines) between alternator and loads. If the loads are completely identical, the system is said to be balanced and the e.m.f.s, currents and phase angles will be equal for each output. However, a considerable

saving is possible if the phases are interconnected in such a way that six lines are unnecessary.

Referring to *Figure 8.5(b)* it is not difficult to see that the resultant of the three voltage phasors there shown is zero, since the resultant of any two of them will be equal and opposite to the third. Now these phasors may represent currents as well as voltages, hence we may say that the instantaneous sum of the e.m.f.s or currents in a balanced three-phase system is always zero:

$$e_1 + e_2 + e_3 = 0, \ i_1 + i_2 + i_3 = 0$$

It is this property which allows interconnection between the three outputs from the alternator so that only three or four lines are needed.

The two methods of interconnection are known as (a) star or Y, (b) delta or mesh. We will investigate these in turn.

STAR CONNECTION

By connecting together the starts (or the finishes) of the windings on the alternator as a common point, a star-connected circuit is obtained. This is shown in *Figure 8.9*. The common point of the three windings is called the neutral or star point, and the line joining the common point of the star-connected loads to this point is the neutral line.

Let e_1, e_2 and e_3 be the phase voltages and let i_1, i_2 and i_3 be the respective phase currents. The voltage between any pair of lines is equal to the phasor difference between two of the phase voltages. Let the phase voltages be represented by the 120° spaced phasors V_p shown in *Figure 8.10*. Then the line voltages are represented by the sides of the equilateral triangle and the neutral point is at the centre of the triangle. The line-to-neutral voltage has a horizontal projection, $V_p \cdot \cos 30°$ or $V_p \frac{\sqrt{3}}{2}$ volts. Since the base length is the sum of two such projections it follows that

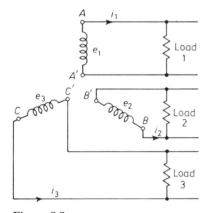

Figure 8.8

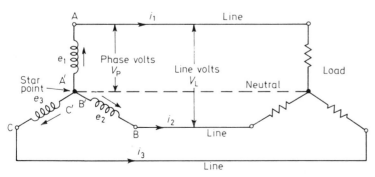

Figure 8.9

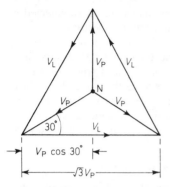

Figure 8.10

$$V_L = 2V_p \left[\frac{\sqrt{3}}{2} \right] = \sqrt{3} V_p$$

So the line voltages are equal to $\sqrt{3}$ times the phase voltage. Also, the line current equals the phase current.

We notice the following important points from the phasor diagram of *Figure 8.10*.

(a) the line voltages are 120° apart,
(b) the line voltages are 30° ahead of the phase voltages,
(c) the angle between the line currents and the corresponding line voltages is $(30° \pm \theta)$, lagging for $+\theta$, leading for $-\theta$.

Example (1)   What is the line voltage on a system where the phase voltage is 240 V ?

Solution

$$V_L = \sqrt{3} \cdot V_p$$
$$= 1.732 \times 240 = 415 \text{ V}$$

It may have occurred to you that the neutral line is unnecessary if the loads are exactly balanced. For since the line current is equal to the phase current, the three currents meeting at the star point of the load add together to give the resultant current in the neutral; but for equal currents spaced 120° apart the result is zero. Hence there is no neutral current and the neutral line can be removed without in any way upsetting the circuit conditions.

In a real life situation it is not possible that the loads on the system, made up as they are of domestic and industrial demands, can be balanced, hence the four-wire system is necessary for the distribution of a.c. supplies. Domestic requirements, heating, cooking and lighting can then be supplied at single-phase voltage (usually 240 V) line to neutral, while workshops and industrial plant are supplied at line voltage, $\sqrt{3} \times 240 = 415$ V. This leads to a more efficient use of equipment such as motors and transformers which, by being designed for three-phase operation, constitute balanced loads and give improved performances at higher voltage.

DELTA CONNECTION

Connection of coil terminations A to B′, B to C′ and C to A′ results in the alternative form of three-phase connection shown in *Figure 8.11*. This is the delta or mesh connection. Here again the load is assumed to be balanced and this time there is no neutral point or neutral line. The voltage between any pair of lines V_L is this time equal to the voltage across a phase of the alternator, V_p. Hence $V_L = V_p$.

The line currents are each made up of two components, one flowing to the load and one flowing from the load. So, for a situation that is very similar to that

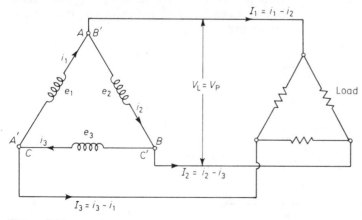

Figure 8.11

for line voltages in star connection, the line current equals the phasor difference between two of the phase currents.

Then since

$I_1 = I_2 = I_3 = I_L$, the line current

$I_1 = i_1 - i_2 = 2I_p \cos 30°$

$I_1 = 2I_p \left[\dfrac{\sqrt{3}}{2} \right] = \sqrt{3} \cdot I_p$

and similarly for I_2 and I_3.

So line currents are each equal to $\sqrt{3}$ times the phase current. As for the star connection, you should make a note of the following points:

(a)   The line currents are 120° apart.
(b)   The line currents are 30° lagging on the phase currents.
(c)   The angle between line current and the corresponding line voltage is $(30° \pm \theta)$.

---

(2)   Sketch a phasor diagram showing line and phase currents for the delta circuit of *Figure 8.11*.

Example (3)   Three non-inductive resistors, each of 100 Ω, are connected (a) in star, (b) in delta to a 440 V three-phase supply. Calculate in each case (i) the phase voltage and phase current, (ii) the line current.

Solution   It is necessary in three-phase problems to start with one phase. The 440 V given in the question is the line voltage. This is always implied unless the contrary is expressly stated. So $V_L = 440$ V

For star connection: $V_p = \dfrac{440}{\sqrt{3}} = 254$ V

Since the load resistance is 100 Ω, phase current $= \dfrac{254}{100} = 2.54$ A

But for star connection $I_L = I_p$

$\therefore \quad I_L = 2.54$ A

For delta connection: $V_p = V_L = 440$ V

For a 100 Ω load $I_p = \dfrac{440}{100} = 4.4$ A

But for delta connection $I_L = \sqrt{3} \cdot I_p$

$\therefore \quad I_L = 7.62$ A

---

(4)   Each phase of a three-phase alternator develops an e.m.f. of 250 V. Find the line voltage for (a) star connection, (b) delta connection. If one of the alternator coils was reversed, what would be the line voltage in star connection?

POWER IN A BALANCED SYSTEM

We consider now the problem of power calculations in balanced star and delta load systems, and to make this a general study we let the load elements be impedances $(Z_1, Z_2$ and $Z_3)$. Since these impedances are equal in balanced loads, they will carry equal currents, hence the phase power will be one-third of the total power.

The star-connected impedances of *Figure 8.12(a)* carry the line currents and the voltage across each is the phase voltage. For a phase angle θ between voltage and current in the impedance, the phase power will be

$P_p = V_p I_L \cos \theta$

and the total power will be three times this:

$P_T = 3 V_p I_L \cos \theta$ (8.1)

but

$V_L = \sqrt{3} \cdot V_p$

$\therefore \quad P_T = \sqrt{3} \, V_L I_L \cos \theta$ (8.2)

Turning to the delta-connected load of *Figure 8.12(b)*, the voltage across each impedance is the line voltage and the current in each is the phase current. Again for a phase angle θ we get

$P_p = V_L I_p \cos \theta$

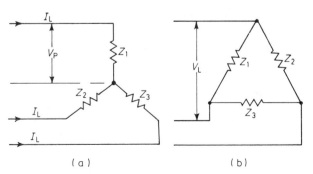

(a)   (b)

Figure 8.12

and total power

$$P_T = 3V_L I_p \cos \theta \qquad (8.3)$$

but

$$I_L = \sqrt{3} I_p$$

$$\therefore \quad P_T = \sqrt{3} V_L I_L \cos \theta \qquad (8.4)$$

Since equations (8.2) and (8.4) are identical, the total power in any balanced three-phase load is given by $\sqrt{3} V_L I_L \cos \phi$.

For purely resistive loads, $\phi = 0$ and $\cos \phi = 1$, hence $P_T = \sqrt{3} V_L I_L$.

Example (5) A balanced three-phase load is connected to a 440 V three-wire system. The input line current is 40 A and the total power input is 15 kW. Calculate the load power factor.

Solution The power factor is, of course, $\cos \phi$

Since $P_T = \sqrt{3} V_L I_L \cos \theta$

$$15 \times 10^3 = \sqrt{3} \times 440 \times 40 \times \cos \theta$$

$$\therefore \quad \cos \phi = \frac{15 \times 10^3}{3 \times 440 \times 40}$$

$$= 0.492$$

Example (6) A 415 V three-phase system supplies a balanced load, each arm of which has an impedance of 50 Ω and a phase angle of 36°. Find the line current for star and delta connection, and the total power dissipated.

Solution Working on one phase, in star connection $V = 415$ V

$$V_p = \frac{415}{\sqrt{3}} = 240 \text{ V}$$

For a load impedance of 50 Ω

$$I_p = \frac{240}{50} = 4.8 \text{ A}$$

but for star connection $I_L = I_p$

$$\therefore \quad I_L = 4.8 \text{ A}$$

For $\phi = 36°$, $\cos \phi = \cos 36° = 0.809$

$$\therefore \quad P_T = \sqrt{3} V_L V_L \cos \theta$$

$$= \sqrt{3} \times 415 \times 4.8 \times 0.809$$

$$= 2790 \text{ W}$$

For delta connection $V_p = V_L = 415$ V

$$I_p = \frac{415}{50} = 8.3 \text{ A}$$

but $I_L = \sqrt{3} I_p$

$$\therefore \quad I_L = 14.38 \text{ A}$$

$$P_T = \sqrt{3} V_L I_L \cos \phi$$

$$= \sqrt{3} \times 415 \times 14.38 \times 0.809$$

$$= 8370 \text{ W}$$

This example shows that although the expressions for total power are the same for both star and delta loads, the loads in delta take three times the power of the same loads in star.

Example (7) Three coils each of inductance 0.03 H and resistance 15 Ω are connected (a) in star, (b) in delta, to a three-phase supply, the line voltage being 440 V and the frequency 50 Hz. Calculate the line current for each case and find the total power consumed.

Solution (a) Star connected, see *Figure 8.13(a)*:

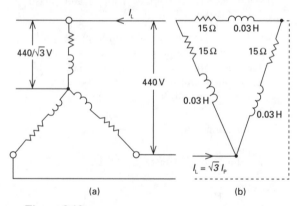

(a) (b)

Figure 8.13

Phase voltage $V_p = \dfrac{440}{\sqrt{3}} = 254$ V

$$\omega L = 2\pi \times 50 \times 0.03 = 9.43 \ \Omega$$

$$Z = \sqrt{(15^2 + 9.43^2)} = \sqrt{314} = 17.72 \ \Omega$$

$$\therefore \quad I_L = \frac{254}{17.72} = 14.33 \text{ A}$$

Also $\cos \phi = \dfrac{15}{17.72} = 0.85$

$\therefore$ Power $= \sqrt{3}V_L I_L$

$\qquad = \sqrt{3} \times 440 \times 14.33 \times 0.85$

$\qquad = 9283\ \text{W}$

(b)   Delta connected, see *Figure 8.13(b)*:

Phase current $I_p = \dfrac{440}{17.72} = 24.8\ \text{A}$

Line current $I_L = \sqrt{3} \times 24.8 = 42.9\ \text{A}$

$\therefore$ Power $= \sqrt{3} \times 440 \times 42.9 \times 0.85$

$\qquad = 27790\ \text{W}$

Notice again the relationship between power in star and delta, respectively.

(8)   Power expressions (8.2) and (8.4) give the true power dissipated in the load. Write down the expressions representing (a) the apparent power, (b) the reactive power in the load, stating the appropriate units for these expressions.

MEASUREMENT OF POWER

True power is measured by a wattmeter. As we have seen a wattmeter is an instrument with a moving voltage (or pressure) coil and a fixed current coil so arranged in relation to each other that the deflection obtained on the scale of the instrument is proportional to $VI \cdot \cos \phi$. In use, the wattmeter is wired into circuit so that the circuit current passes through the current coil and the circuit voltage is impressed across the voltage coil. *Figure 8.14* shows the general arrangement.

Power in three-phase circuits can be measured by using either two or three wattmeters, depending on whether the system is three- of four-line. The two wattmeter method is illustrated in *Figure 8.15*, the current coils of the meters being placed in series with any two of the three available lines. For a star-connected load

Voltage across $W_1 = e_1 - e_3$

Voltage across $W_2 = e_2 - e_3$

Total power $= e_1 i_1 + e_2 i_2 + e_3 i_3$

Since current i_3 is not passing through a meter, we can eliminate it from these equations. In any balanced system

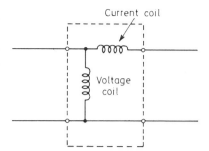

Figure 8.14

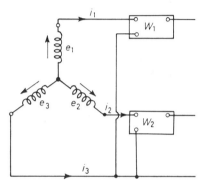

Figure 8.15

$i_1 + i_2 + i_3 = 0$

$i_3 = - i_1 - i_2$

Substituting,

total power $= e_1 i_1 = e_2 i_2 = e_3(-i_1 - i_2)$

$\qquad = i_1(e_1 - e_3) + i_2(e_2 - e_3)$

$\qquad = i_1$ (voltage across W_1) $+ i_2$ (voltage across W_2)

But each of these terms represents the wattmeter power readings, hence

$W = W_1$ reading $+ W_2$ reading

This result is equally true for a delta-connected load.

A four-wire star-connected load using three wattmeters is shown in *Figure 8.16*. One meter is placed in each of the lines. Each of these meters will indicate the respective phase power and by a simple analysis similar to that above, the total power is given by

$W = W_1$ reading $+ W_2$ reading $+ W_3$ reading

It is important to note that the readings obtained are not necessarily identical. For the two wattmeter method and a balanced load W_1 reading is always greater than W_2 unless the power factor is unity when the readings will be equal.

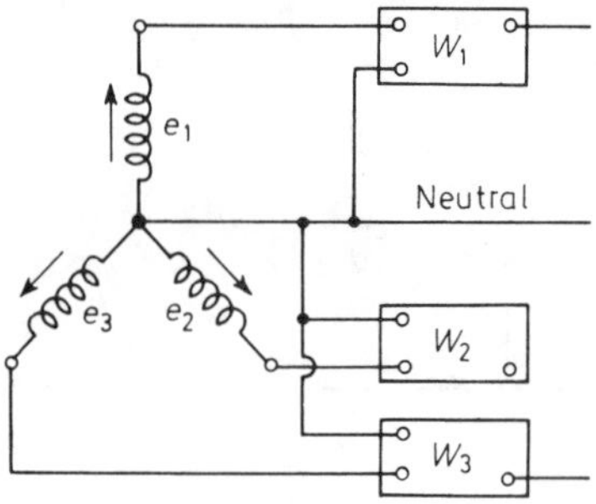

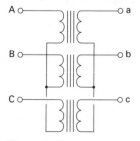

Figure 8.16

Power factor: two wattmeter method

For a balanced load the power factor can be calculated from the two wattmeter readings obtained from the above method of measurement.

For W_1 in the lagging line, it can be proved that

$$W_1 - W_2 = (W_1 + W_2) \frac{1}{\sqrt{3}} \tan \phi$$

$$\therefore \quad \tan \phi = \sqrt{3} \frac{W_1 - W_2}{W_1 + W_2} \quad (8.5)$$

From this equation ϕ, and hence $\cos \phi$, may be found.

For W_1 leading, W_1 and W_2 must be interchanged in (8.5) above.

Example (9)   A 240 V motor has a full load output of 7875 W. If the power factor is 0.85 and the full load efficiency 90 per cent, find the reading on each of two wattmeters connected to measure the input power.

Solution   Input power $= \dfrac{7875}{0.9} = 8750$ W

$\therefore \qquad W_1 + W_2 = 8750$ W

But $\cos \phi = 0.85$

$\therefore \qquad \phi = \tan^{-1} 0.85 = 31.8°$

Hence $\tan \phi = 0.62$

$\therefore \qquad 0.62 = \sqrt{3} \dfrac{W_1 - W_2}{W_1 + W_2} = \sqrt{3} \dfrac{W_1 - W_2}{8750}$

$\qquad W_1 - W_2 = \dfrac{0.62 \times 8750}{\sqrt{3}} = 3132$ W

Then $W_1 - W_2 = 3132$ W

$\qquad W_1 + W_2 = 8750$ W

from which, by adding $2W_1 = 11\,882$ W

Hence $W_1 = 5941$ W and $W_2 = 2809$ W

THREE-PHASE TRANSFORMERS

For three-phase working it is possible to use either three single-phase transformers of the type we have already described in Chapter 7, or a single three-phase unit. Employing three single-phase units has an advantage that a failure in the system involves the replacement of one spare single-phase transformer instead of a complete three-phase unit. This latter type of transformer has its own advantage in that it takes up less room than the three single units and overall is a cheaper option.

In general, it is not possible to obtain a polyphase output by direct transformer action, but a three-phase supply may be transformed into a similar three-phase system at a different voltage or converted into a six-phase supply; under certain circumstances a three-phase to single-phase transformation may be accomplished.

Technically, the difference between the single three-phase unit and the three single-phase units lies in the fact that there is direct magnetic coupling between the phases in the former but not in the latter.

Because of the large number of possible interconnections in both primary and secondary windings, there are many ways in which power can be supplied to the primary and taken from the secondary. We cannot discuss all of these possibilities in this elementary survey but the commonly used arrangements of star–star and delta–star transformers will be covered briefly.

Three Single-Phase Units

Figure 8.17 shows the primaries and secondaries of a bank of three single-phase transformers both connected in star. The three transformers act independently of each other, the load on each secondary phase being reflected in the manner we have already seen into the corresponding primary phase.

If the primaries are connected to a three-phase, four-wire system each primary receives a constant voltage. This arrangement is quite practical but it has a disadvantage in that the primary currents show no tendency to balance when the secondary loads are unbalanced.

Figure 8.17

If the primaries are supplied from a three-wire system, then the primary voltages are again constant but the primary phase voltages may be unbalanced, the star point potential being unfixed. The worst case turns up when one secondary phase is loaded. The transformer with the load will show an input impedance which will be very small relative to the two unloaded transformers; these will act simply as high inductive impedances. The loaded phase voltage will consequently fall to a low value, while the voltage across the other two phases will rise closely to that of the line voltage. If there is any possibility of severely unbalanced loading, this style of connection is unsuitable.

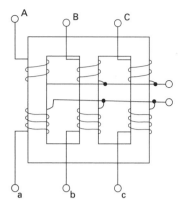

Figure 8.18

Three-limb single transformer

Figure 8.18 is a diagrammatic sketch of a three-limb core type of three-phase transformer, with both primaries and secondaries star connected. With the exception that a single core accommodates all the windings, this transformer gives results that are similar to the three single-phase units arrangement discussed above.

As the figure indicates, the primary and secondary coils of each phase enclose the same limb of the transformer core, the magnetic circuit for the flux produced in any one of the limbs being completed through the other two. The phase relationship of the flux produced in each limb is the same as that of the respective voltages; at any instant in time the phasor sum of the 'upwards' and 'downwards' flux is zero. Compare this with the text associated with *Figure 8.17* earlier. What this means is that the flux in any of the limbs is that which would be established if each of the phases were acted on by separate transformers. *Figure 8.19* shows the state of affairs at two specific instants of time for phase e_1 (from *Figure 8.8*).

Figure 8.19(a) shows the instant of maximum flux in limb A, this flux being equal to the sum of the other two fluxes established at the same moment in the other two limbs (or phases). Diagram *(b)* shows the instant of minimum flux in phase e_1, and this is simultaneously the moment at which the flux due to phases e_2 and e_3 produced mutual cancellation in limb A.

The representation of the star–star primary and secondary connection of *Figure 8.18* can be simplified to the skeleton form of the diagram depicted in *Figure 8.20(a)*. The star–star connection is an economical one for high voltage operation and finds applications in transformers used for supplying relatively light loads. Both neutral lines are available for earthing or for permitting a balanced four-wire supply to be used.

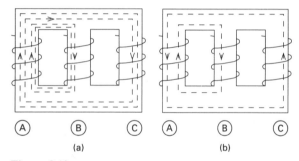

(a)   (b)

Figure 8.19

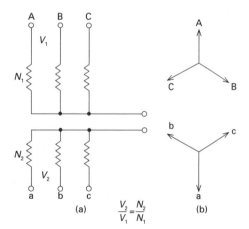

(a) $\dfrac{V_2}{V_1} = \dfrac{N_2}{N_1}$ (b)

Figure 8.20

The phasor diagram for the connection is shown in diagram *(b)* on the right of the figure. This is derived from the fact that the e.m.f.s induced in all windings on the same limb are in phase and in direct ratio to the number of turns in the windings.

Delta–star transformer

Figure 8.21 is a skeleton representation of the connections for delta-connected primaries and star-connected

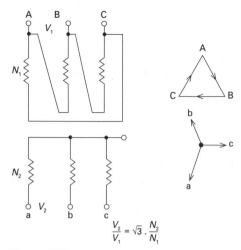

$$\frac{V_2}{V_1} = \sqrt{3} \cdot \frac{N_2}{N_1}$$

Figure 8.21

secondaries in a transformer. This arrangement is widely used for stepping down to supply a four-wire load which may be balanced or unbalanced, and for stepping up to supply a high tension transmission line. In cases where the secondary load is automatically balanced, the secondary neutral point may be earthed. As the figure suggests, a large unbalanced load does not disturb magnitude equilibrium since the primary current will flow in the corresponding winding only and the primary and secondary ampere-turns will be balanced in each limb.

This time the voltage transformation ratio is proportional to $\sqrt{3}$ times the turns ratio.

Example (10) A three-phase 3300/415 delta–star transformer supplies a line current of 85 A to a balanced three-phase load. Neglecting losses, calculate (a) the primary phase current, (b) the primary line current.

Solution As there are no losses we can say

$$\frac{\text{primary phase voltage}}{\text{secondary phase voltage}}$$

$$= \frac{\text{secondary phase current}}{\text{primary phase current}}$$

As the primary is delta connected the primary phase voltage = 3300 V.

For the star connected secondary, the secondary phase voltage = $415\sqrt{3}$ = 240 V, and the secondary phase current is 85 A.

Hence $\quad \dfrac{3300}{240} = \dfrac{85}{\text{primary}} I_p$

$\therefore$ primary I_p

$$85 \times \frac{240}{3300} = \frac{20\,400}{3300} = 6.18 \text{ A}$$

And primary $I_L = \sqrt{3} \times 6.18 = 10.7$ A

(11) Try the previous example by an alternative method, equating the input and output powers and taking the primary power factor ϕ_1 = secondary power factor ϕ_2.

(12) Estimate the turns ratio of a three-phase, delta–star transformer which has a primary line voltage of 6000 V and a secondary no-load voltage of 240 V. Ignore losses.

SUMMARY

- Assuming three-phase balanced loads are sinusoidal quantities:
- Star-connected load $V_L = \sqrt{3} \cdot V_p$

$$I_L = I_p$$

- Delta-connected load $V_L = V_p$

$$I_L = \sqrt{3} \cdot I_p$$

- Star- or delta-connected load

Apparent power $= \sqrt{3} \cdot V_L I_L = 3V_p I_p$

Total power $W = \sqrt{3} \cdot V_L I_L \cos \phi$

$$= 3V_p I_p \cos \phi$$

- Balanced or unbalanced loads, using 2 wattmeters:

Total power $W = W_1$ reading $+ W_2$ reading

- Balanced load:

$$\tan \phi = \sqrt{3}\, \frac{W_1 - W_2}{W_1 + W_2} \text{ for } W_1 \text{ lagging}$$

$$\tan \phi = \sqrt{3}\, \frac{W_2 - W_1}{W_2 + W_1} \text{ for } W_1 \text{ leading}$$

$$\cos \phi = \frac{1}{\sqrt{(1 + \tan^2 \phi)}}$$

REVIEW QUESTIONS

1   What is the pulsation frequency of single phase 50 Hz supplies?
2   What is meant by 'negative power' delivered to a reactive circuit? How can this be overcome?
3   Is power a scalar or a vector quantity?
4   Differentiate between 'line voltage' and 'phase voltage' in a three-phase system.
5   Can there be such a differentiation in single-phase systems?
6   When line and phase voltages in a load are the same, how is the load connected, star or delta?
7   Why are two wattmeters sufficient to find the power in a three-wire system?
8   Express the transformation ratio V_2/V_1 in terms of the winding ratios for a three-phase unit connected in (a) star–star, (b) delta–star.

EXERCISES AND PROBLEMS

(13)   A balanced three-phase load takes a line current of 30 A at a line voltage of 415 V. Calculate the apparent power supplied by the system.

(14)   A three-phase supply having a line voltage of 420 V provides a line current of 70 A into a balanced load. What is the load power?

(15)   A balanced delta-connected load carries phase currents of 26 A and the voltage across each phase of the load is 430 V. What is the line current and the line voltage?

(16)   A three-phase balanced load dissipates 25 kW at a power factor of 0.8 leading. If the line voltage is 440 V, calculate the line current.

(17)   Three non-inductive resistors, each of 100 Ω, are connected (a) in star, (b) in delta, to a 500 V three-phase supply. Calculate in each case (a) the phase voltages and phase currents, (b) the line currents.

(18)   In the previous example, calculate the total power dissipated in the load for each of the connections.

(19)   A three-phase system supplies 20 kW at a power factor of 0.85, the line voltage being 240 V. Calculate the line current I_L and the phase current for loads (a) star connected, (b) delta connected.

(20)   Three capacitors, each of 300 μF, are connected in star to a 500 V 50 Hz supply. Calculate the line current.

(21)   Three equal resistors form a load in star connection fed from a three-phase supply. Prove that the power is reduced by one-half if any one of the resistors is disconnected.

(22)   What is the turns ratio of a three-phase, delta–star transformer having a primary line voltage of 11 kV and a secondary line voltage of 550 V on no load.

(23)   Two single-phase wattmeters are connected in circuit for a test on a partially loaded three-phase motor. Wattmeter A reads 12.5 kW and wattmeter B reads 2.5 kW. If the efficiency of the motor is 60 per cent, calculate (a) the output power of the motor, (b) the power factor.

(24)   Design a three-phase heater for operation on a 220 V line by calculating the values of the three identical resistive elements and their power rating when they are connected in delta.

(25)   A three-phase transformer has a step-down ratio of 10:1. With an 11 kW line to line supply, find the secondary line voltage for each of the connections: (a) star–star, (b) delta–star, (c) star–delta, (d) delta–delta.

9

Diode applications

Diode non-linearity
Load Line Methods
Rectification and filtering
Ripple factor
Waveshaping circuits

The semiconductor diode is a two-terminal device which has a low resistance to the flow of current in one direction and a high resistance to the flow of current in the opposite direction. An ideal diode would have zero resistance in the 'forward' direction and infinite resistance in the 'reverse' direction; this implies that there would be no voltage drop across an ideal diode when current passed in the forward direction and no current would pass through the diode when it was connected in the reverse direction. Hence no power would be dissipated in the ideal element. Real diodes, however, exhibit a forward voltage drop of about 0.25 V for germanium and 0.7 V for silicon devices, and the reverse resistance, though very high, is finite.

The symbol for the diode is shown in *Figure 9.1(a)* with a characteristic curve relating the voltage across the diode to the current through it in diagram *(b)*. The arrowhead of the symbol (the *anode* terminal) indi-

cates the forward or low resistance direction of flow; clearly this shows the flow of conventional current when the anode is made positive. The true electronic flow is opposite to this, that is, from *cathode* to anode of the diode. In practical diodes the *marked* end is the cathode.

NON-LINEARITY

As *Figure 9.1(b)* indicates, the diode is a non-linear element. This means that it does not obey Ohm's law, a point we noted in the Chapter 1. The effective resistance of a diode is *not* independent of V and I, although any two related points along the characteristic curve will always satisfy Ohm's law. Most electronic components are non-linear: the transistor, field-effect transistor, voltage dependent resistors (VDRs), thermistors and electric lamps to name just a few. In making an analysis of such elements, therefore, we imagine the element to be ideal but 'degraded' to its real status by the addition of linear elements such as resistors and voltage sources; we then get a *model* which can be used to predict the behaviour of the real thing.

Example (1) Assuming ideal diodes, what currents will flow in each of the three branches of the circuit of *Figure 9.2* with (a) the battery connected as shown, (b) the battery reversed?

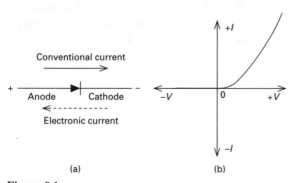

(a)   (b)

Figure 9.1

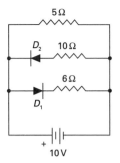

Figure 9.2

Solution (a) With the battery connected as shown, only diode D_1 will be conducting, its anode going to the positive terminal of the battery. Hence the current in the 6 Ω branch will be 10/6 = 1.67 A, in the 10 Ω branch zero, and in the 5 Ω branch 10/5 = 2 A.
(b) With the battery reversed, only diode D_2 will be conducting. Hence the current in the 6 Ω branch will be zero, in the 10 Ω branch 10/10 = 1 A, and in the 5 Ω branch unaffected at 2 A.

In this example we have, in fact, treated the diodes simply as perfect switches which are either opened or closed according to the direction of the applied voltage.

MODELS

A silicon diode does not conduct appreciably until the applied forward voltage exceeds about 0.7 V. It then passes a current but exhibits a small forward resistance R_f. Our model for such a diode can therefore be as shown in *Figure 9.3(a)* where a hypothetical 0.7 V battery is considered to oppose the applied voltage V and a small series resistor R_f represents the forward resistance. When reverse biased, the model

becomes that of diagram *(b)*; a very small reverse current is assumed to flow through parallel resistor R_r while the ideal diode element prevents any further flow through R_f. In modern silicon diodes R_r is large enough to be ignored in most practical cases.

Example (2)   What will be the branch currents in the circuit of *Figure 9.2* if diodes are used having the idealized characteristic of *Figure 9.4*?

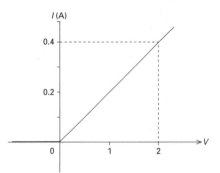

Figure 9.4

Solution The idealized characteristic, being a straight line, follows a linear (Ohm's law) relationship between V and I; hence we are treating the diodes as having a constant resistance given by the reciprocal gradient of the line or V/I. This is seen to be, taking any two convenient points, 2/0.4 = 5 Ω. Therefore, replacing the diodes by their models, we get the circuit of *Figure 9.5*.

With the battery connected as shown, D_1 only will conduct (as previously) and the current in the (5 + 6) = 11 Ω branch will be 10/11 = 0.91 A. With the battery reversed, only D_2 will conduct and the current in the (10 + 5) = 15 Ω branch will be 10/15 = 0.67 A. The current in the 5 Ω branch is unaffected and remains at 2 A for both orientations.

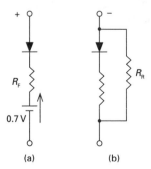

(a)   (b)

Figure 9.3

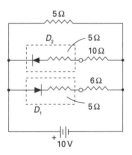

Figure 9.5

LOAD LINES

It is often quite legitimate to idealize diode character-istics in the way illustrated above since the curve approximates quite closely to a straight line when the device is fully conducting; see *Figure 9.1(b)* again. We cannot, however, always dismiss the non-linear conditions in a circuit by making such an approxima-tion. In the diode circuit of *Figure 9.6*, for instance, if we know the input voltage V, an approximation to the p.d. across a load resistor R_L can be obtained from Ohm's law as $V = I \cdot R_L$. The problem with a diode, or indeed with any other kind of non-linear element in the circuit, is to find the value of I which takes account of the voltage drop across the non-linear element as well as that across R_L.

From the figure, the supply voltage V is the sum of the drops across the diode, V_d and the load R_L:

$$V = V_d + I \cdot R_L$$

from which

$$I = \frac{V}{R_L} - \left(\frac{1}{R_L}\right) V_d$$

If this expression is plotted on the I/V_d characteristic axes of the diode or non-linear element, we get a straight line, since the variables are I and V_d, and V/R_L and $1/R_L$ are constants. The actual position of the line depends on V, for when $V = V_d$, $I = 0$, and when $V_d = 0$, $I = V/R_L$. The gradient of the line, which is known as a *load line*, is $-1/R_L$. The point P (see *Figure 9.7*) where the load line and the non-linear char-acteristic meet gives us the coordinates of the current flowing in the circuit and the volts drop across the diode for any particular value of V. Note that if the supply voltage V varies during operation, the load line moves parallel to itself.

When there is a single non-linear element in an otherwise linear circuit, construction of a load line provides a simple graphical method of getting a solution.

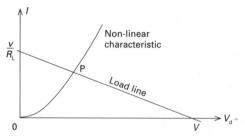

Figure 9.7

A direct mathematical method can be used if the equation of the non-linear device is known as our next example illustrates.

Example (3)   The equation for the I/V charac-teristic of a certain non-linear element is $I = 0.05V^2 + 0.1V$ A. Estimate the current in and the p.d. across the element when it is connected in series with a $20\,\Omega$ resistor and a battery of e.m.f. 10 V. What power is dissipated in the element?

Solution   *Figure 9.8* shows us the circuit where the element is assumed to be a diode. In reality, the characteristic of a diode is exponential and not square law, but the example illustrates the method of dealing with any non-linear device.

Let the p.d. across the diode be V_d, then the circuit current I will be, from the figure:

$$20I = 10 - V_d \text{ or } I = \frac{10 - V_d}{20} \text{ A}$$

But from the characteristic equation, $I = 0.05V_d^2 + 0.1V_d$ A

we have $\dfrac{10 - V_d}{20} = 0.05V_d^2 + 0.1V_d$

or, rearranging $V_d^2 + 3V_d - 10 = 0$

This is a quadratic which will factorize, and taking the positive root we find $V_d = 2$ V.

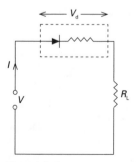

Figure 9.6

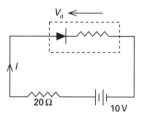

Figure 9.8

Hence, $I = 0.05(2)^2 + 0.1(2) = 0.4$ A

The power dissipated in the diode $= V_d I = 2 \times 0.4 = 0.8$ W.

(4)   Plot the I/V characteristic of a diode from the following table :

I (mA)	0	0.75	4.9	12	19.2	28	37
V (volts)	0	1	2	3	4	5	6

For a supply voltage of 6.5 V and a load resistance of 300 Ω use a load line graphical method to obtain an estimation of (a) the load current, (b) the p.d. across the diode, (c) the p.d. across the load.

RECTIFICATION

Rectifier is a name used generally when a diode is employed as a device by which a direct but pulsating current can be obtained from an alternating voltage source.

Half-wave rectifier

The circuit of *Figure 9.9(a)* shows a half-wave rectifier system. Current can only flow through the diode when the anode terminal is positive relative to the cathode, that is, when end X of the transformer winding is going through the positive half-cycle. Therefore current flows conventionally through the load resistor R_L during one half of each complete input cycle, hence the name of the circuit. Notice particu-

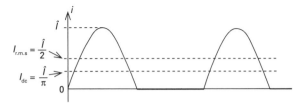

Figure 9.10

larly that the *cathode* terminal of the diode is the *positive* output level of the voltage across R_L. The input alternating voltage and the output unidirectional but pulsating current waves are shown in diagram *(b)*. If the load is purely resistive, the output voltage variations will be in phase with the current. For most practical work, the small forward voltage drop across the diode can be ignored.

For an input $v = \hat{V} . \sin \omega t$, the load current $i = v/R_L = \hat{V} . \sin \omega t/R_L$ for $0 < \omega t < \pi$ and $i = 0$ for $\pi < \omega t < 2\pi$. The d.c. component of the output, that is, the average or mean value is, as explained earlier, equal to $2/\pi = 0.637$ of the peak value over each half cycle; for a waveform of *alternate* half cycles this mean value falls to $1/\pi = 0.318$ of peak value. Hence $I_{dc} = \hat{I}/\pi$ and $\hat{I} = \hat{V}/R_L$.

The r.m.s. current is important in rectifier systems as its value determines the rating of the transformer. It can be proved that $I(\text{r.m.s.}) = \hat{I}/2$ for the half-wave rectifier. Both of these results are important and are illustrated in *Figure 9.10*.

Example (5)   The diode of *Figure 9.11(a)* is used as a half-wave rectifier with an a.c. input voltage $v = 20 \sin \omega t$ V. If the load resistor is 500 Ω and the average R_f is 50 Ω, find the load current.

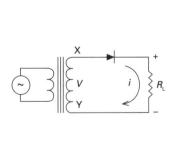

(a)

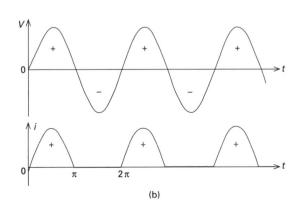

(b)

Figure 9.9

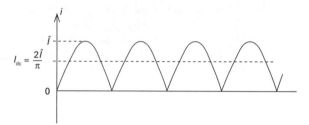

Figure 9.11

Figure 9.12

Solution   Replacing the diode with its model, diagram *(b)*, we have, for $0 < \omega t < \pi$ that v is positive and the diode is forward biased. Then

$$i = \frac{V}{R_L + R_f} = \frac{20 \sin \omega t}{550} = 0.036 \sin \omega t \text{ A}$$

For $\pi < \omega t < 2\pi$ v is negative and the diode is reverse biased, hence $i = 0$. Summarizing:

$i = 36 \sin \omega t$ mA for positive half cycles of v
$i = 0$ for negative half cycles of v

as sketched in *Figure 9.11(c)*.

Example (6)   An ideal diode is used as a half-wave rectifier with an input of 240 V r.m.s. at 50 Hz. For a load $R_L = 2500 \ \Omega$ estimate I_{dc}, V_{dc} and the load power dissipation.

Solution   For 240 V r.m.s. the peak value is $240\sqrt{2} = 340$ V. The load current at the peak voltage $= \hat{V}/R_L = 340/2500 = 0.136$ A. Hence the mean load current $I_{dc} = 0.318 \times 0.136 = \underline{0.043 \text{ A}}$.

Then $V_{dc} = I_{dc} \times R_L = 0.043 \times 2500 = \underline{107.5 \text{ V}}$
Power in the load $= I(\text{r.m.s.})^2 R_L$, but $I(\text{r.m.s.}) = \frac{\hat{I}}{2} = \frac{0.136}{2} = 0.068$ A.

$P = (0.068)^2 \times 2500 = \underline{11.56 \text{ W}}$

The frequency of the supply does not enter into the calculation.

(7)   If a standard moving-coil type d.c. meter was wired across the load resistor of the previous example, what would it read? Suppose the meter is now switched to its equivalent a.c. range. Would the indication change? If so, to what, and why?

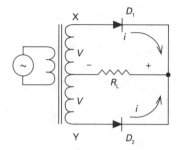

Figure 9.13

Full-wave rectification

By using either two diodes in a *biphase* circuit, or four diodes in a *bridge* circuit, full-wave rectification can be obtained. This means that the gaps in the half-wave output are 'filled in'.

The biphase circuit is shown in *Figure 9.12*, where the supply is derived from a centre-tapped transformer. Each half winding in conjunction with its own particular diode form a half-wave circuit similar to that shown in *Figure 9.9(a)* earlier, and these then have a common load R_L. It is not difficult to see that when terminal X is positive, diode D_1 is conducting and the p.d. across R_L will have the polarity indicated. The voltage across D_2 is in the reverse direction and so D_2 contributes no current to the load. During the following half cycle of input when terminal Y is positive, diode D_2 conducts and current again flows through R_L in the same direction as before, D_1 this time being reversed. Each of the two circuits therefore provides the current during alternate half cycles so that the whole of the input sine wave is used; the load waveform (for current or voltage) is shown in *Figure 9.13*. Take note from this that the d.c. or mean component of the full-wave output is *twice* that of the half wave, that is $I_{dc} = 2\hat{I}/\pi = 0.637\hat{I}$. The r.m.s. current is, as we should expect, $I = \hat{I}/\sqrt{2}$.

The bridge rectifier shown in *Figure 9.14* uses four diodes but does not require a centre-tapped transformer. Here we have, though it may not appear so at first glance, two half-wave rectifiers in series. Again

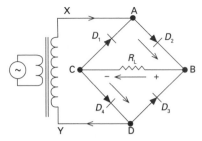

Figure 9.14

assuming terminal X to be positive, the current will now follow the path XABCDY, and the load current will have the polarity indicated. When the next input half cycle makes terminal Y positive, the current path becomes YDBCAX and the load current is still directed from B to C. The load waveform is the same as that already shown for the bi phase circuit. Other things being equal, the bi phase is the more efficient of the two circuits.

Example (8)   Write down for the two full-wave rectifiers discussed above, expressions for the mean current in the load taking the diodes to have forward resistance R_F.

For the bi-phase, the total circuit resistance (neglecting the transformer winding) is $(R_L + R_F)$ for either diode; then since $\hat{I} = \hat{V}/(R_L + R_F)$ the average current will be

$$I_{dc} = \frac{2\hat{V}}{\pi(R_L + R_F)}$$

For the bridge circuit, there are two diodes in series for each half cycle of rectification. Hence

$$I_{dc} = \frac{2\hat{V}}{\pi(R_L + 2R_F)}$$

(9)   Repeat the previous example for the r.m.s. current in the load for each of the circuits.

In the next example we look at a method of drawing a characteristic for a complete diode circuit.

Example (10)   A diode has the following forward characteristic and is wired to an adjustable d.c. source by way of a 40 Ω load resistor, R_L.

V_d (V)	0	0.5	1.0	1.5	2.0	2.5	3.0
I (mA)	0	1.0	3.5	8.5	15.5	25.0	36.0

Draw the diode characteristic and on the same axes the voltage current characteristic for the complete circuit

Solution   It is no problem to plot the diode characteristic and this is shown in *Figure 9.15*. To obtain the characteristic of the whole circuit requires us to estimate the setting of the applied voltage V which, with the series load resistor of 40 Ω, produces the stated diode voltage at the stated level of diode current. To do this, take a pair of related values from the diode curve, say, when $V_d = 2.0$ V, $I = 15.5$ mA. The voltage across the resistor $V_R = IR_L = 15.5 \times 10^{-3} \times 40 = 0.62$ V. Hence the applied voltage $V = 0.62 + 2.0 = 2.62$ V.

This gives us point Q which will lie on the circuit characteristic. Calculating the other points in the same way, we get

I (mA)	0	1.0	3.5	8.5	15.5	25.0	36.0
V	0	1.04	1.14	1.84	2.62	3.5	4.44

The broken line curve shows the complete circuit characteristic. From this, the current can be found for a given applied voltage, e.g. when $V = 2.5$, $I_d = 14$ mA.

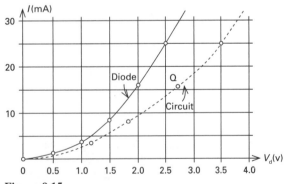

Figure 9.15

RIPPLE FACTOR

The rectifier circuits already discussed produce pulsating direct currents, the 'roughest' being that from the half-wave circuit. The output from either a bi-phase or a bridge rectifier is an improvement in so far as the average d.c. output is twice that of the half wave, but in neither example of a rectifier system is there any pretence at what we might call a 'smoothness' of the waveform. What we want from the rectifier system is an output which approximates very closely to that obtained from direct current sources such as batteries. We can then use these rectifier circuits to replace battery supplies, so making the equipment which is energized by such sources fully operational from mains supplies.

A measure of the smoothness (or otherwise) of the output current from a rectifier circuit is expressed by what is known as the ripple factor F_r where

$$F_r = \frac{\text{r.m.s. value of the alternating component}}{\text{average value of the d.c. component}}$$

$$= \frac{I_{ac}}{I_{dc}} = \frac{V_{ac}}{V_{dc}}$$

We have already seen that (a) for the half-wave circuit $I = \hat{I}/2$, $I_{dc} = \hat{I}/\pi$ and (b) for the full-wave circuit $I = \hat{I}/\sqrt{2}$, $I_{dc} = 2\hat{I}/\pi$.

Since the power dissipated in a load resistance R_L (see *Figures 9.9, 9.10* and *9.13* above) defines the r.m.s. value of the current and since the total power is the sum of the powers dissipated by the direct and alternating quantities (power is not a vector quantity, remember) we have

$$I^2 R_L = I_{dc}^2 R_L + I_{ac}^2 R_L$$
$$\therefore \quad I_{ac}^2 = I^2 - I_{dc}^2$$

from which

$$F_r = \frac{\sqrt{(I^2 - I_{dc}^2)}}{I_{dc}} = \sqrt{\left(\left[\frac{I}{I_{dc}}\right] - 1\right)}$$

A very low value of ripple factor is required to obtain a smooth conversion from alternating current to direct current.

Example (11) How could ripple factor be determined by using two meters, one, a moving-iron a.c. ammeter, the other a moving-coil d.c. ammeter?

Solution A moving-iron ammeter will read the r.m.s. value of the current, a moving-coil meter will indicate the average value of the current. By using these two meters in series, a measure of the ripple current can be made. For the half-wave output

$$\frac{I}{I_{dc}} = \frac{\hat{I}/2}{\hat{I}/\pi} = \frac{\pi}{2} = 1.57$$

---

and

$$F_r = \sqrt{([1.57]^2 - 1)} = 1.21 \ (121\%)$$

For the full-wave output

$$\frac{I}{I_{dc}} = \left(\frac{\hat{I}/\sqrt{2}}{2\hat{I}/\pi}\right) = \frac{\pi}{2\sqrt{2}} = 1.11$$

and

$$F_r = \sqrt{([1.11^2] - 1)} = 0.48 \ (48\%)$$

So although the use of full-wave rectification reduces the ripple component from 121 per cent to 48 per cent of the d.c. component, a moment's consideration will show that the output is far from being satisfactory for most electronic systems outside of battery charging and electroplating work.

Capacitor smoothing

Ripple factor can be greatly reduced by placing a capacitor across the load resistance. This capacitor is of large capacitance, generally within the range 100 to 5000 μF, depending in most cases upon the required working voltage and the physical dimensions of the component. A capacitor becomes bulky when high capacitance is associated with a high working voltage.

So in a qualitative view, the capacitor can be thought of as a low impedance path taken by the a.c. component of the rectified waveform, or it can be thought of as a 'reservoir' that stores charge during the short period when the diode is conducting and releases charge to the load when the diode is cut off.

Looking at *Figure 9.16*, consider what happens during the first positive half cycle after switching on. The capacitor will charge up to the peak value of the rectified output voltage, that is, to $\hat{V}$. During the interval represented by the missing negative half cycle of output no further charge is added to C and the voltage across its terminals will fall slightly as it discharges through the load R_L. However, provided the discharge is relatively slow, which in effect is the same thing as saying that the time constant CR_L is large compared to the time of one cycle of the waveform, the fall in the load p.d. before the arrival of the next positive half cycle will be very small. The capacitor is then 'topped-up' by the crest of this following voltage half cycle, and this process continues for the whole of the time that the circuit is operating. You will notice two very important points: the diode only switches on for a very small part of each alternate

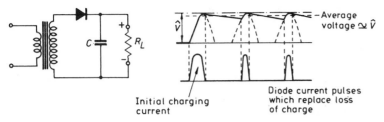

Figure 9.16

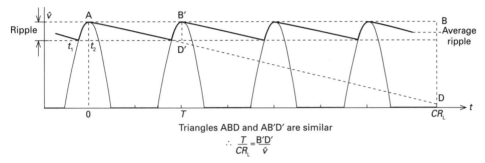

Triangles ABD and AB'D' are similar

$$\therefore \frac{T}{CR_L} = \frac{B'D'}{\hat{v}}$$

Figure 9.17

positive half cycle; it is prevented from being on for the complete period of each positive half cycle by the opposing voltage developed across R_L. Also, the average value of the d.c. output now approximates closely to the peak value of the alternating waveform, so providing an almost steady d.c. output level.

The best smoothing is obtained when CR_L is large; when this time constant is small the discharge *between* cycles becomes large and the ripple factor increases, giving a sawtoothed edge to the d.c. It is clear that, even in a poor case, the output is much better smoothed by the addition of the capacitor.

Let us expand for a moment on the conducting period of the diode. As we have seen, once the steady state has been reached, the output voltage is, on average, closely equal to the peak amplitude $\hat{V}$. Suppose for illustration that it is equal to 0.9 $\hat{V}$. This means that the voltage across R_L is acting in opposition to the positive peaks provided by the secondary of the transformer winding during which the diode tries normally to conduct. The diode cannot now conduct, however, until the voltage on its anode becomes greater than the voltage on its cathode. If the voltage on the cathode, then, is for our illustrative example 0.9 V, the diode remains switched off until the positive excursion at the upper transformer terminal exceeds this level. The lower graph of *Figure 9.16* shows the corresponding pulsations of diode current flowing into the load.

There are two dangers here for the unwary builder of a rectifier unit with capacitive smoothing. The

period during which the diode is conducting is a very small fraction of the cycle, and during this time it has to supply the capacitor with all of the charge lost over the bulk of the cycle. The current surge can therefore be very large, and hence the power dissipation in the diode's forward resistance is also large. Further, since the capacitor retains most of its charge between cycles, the voltage across R_L adds to the inverse voltage of the supply during the negative half cycles when the diode is reversed biased. The peak inverse voltage in a half wave rectifier circuit is therefore almost equal to *twice* the peak voltage, that is, almost double the value it reaches before the capacitor is added. The diode chosen for the job must be rated to withstand this.

Analysis of the operation

Figure 9.17 shows a more detailed representation of the smoothing operation. We can make the following assumptions about this representation which are justified on the grounds that the problem is seldom critical in practice:

(a) the time constant CR_L is sufficiently large so that the charging interval $t_2 - t_1$ is a very small fraction of the periodic time T;

(b) the diode switches on at the peak value of the supply voltage;

(c) the ripple segments can be approximated by straight lines.

Figure 9.17 shows all of these assumptions, and the variation in the ripple voltage. An estimation of the peak-to-peak value of this ripple can now be deduced. The discharge of the reservoir capacitor would follow an exponential path in reality, but using the assumption that this discharge follows a straight line for a time T which is small in relation to the time constant CR_L, the discharge rate remains at its initial value which is now considered constant; hence we can say

$$\frac{dV}{dt} = \frac{\hat{V}}{CR_L} \text{ volts per second}$$

Hence, since the length of time between recharging the capacitor is the periodic time T and the voltage will have changed by $(dV/dt)T$, the ripple voltage will be

$$\frac{dV}{dt} T = \frac{\hat{V}}{CR_L} T = \frac{\hat{V}}{CR_L f}$$

where f is the frequency, usually 50 Hz. Clearly, the greater the frequency, the capacitance and the load resistor, the smaller will be the ripple voltage.

Example (12) Given that the efficiency of a rectifier system is expressed as the ratio d.c. power out/total power input, find (a) the efficiency in terms of the ripple factor F_r, (b) the efficiency of a half-wave circuit.

Solution (a) We have

$$= \frac{\text{d.c. power}}{\text{output power}} = \frac{I_{dc}^2 R_L}{I^2 R_L} = \frac{I_{dc}^2}{I^2}$$

But
$$F_r = \sqrt{\left(\left[\frac{I^2}{I_{dc}^2}\right] - 1\right)}$$

and so
$$1 + F_r^2 = \frac{I^2}{I_{dc}^2}$$

Hence
$$\text{Efficiency} = \frac{1}{1 + F_r^2}$$

(b) For the half-wave circuit we have

$$= \frac{\hat{V}/\pi \cdot \hat{I}/\pi}{\hat{V}/2 \cdot \hat{I}/2} = \frac{4}{\pi^2} = 0.405 \ (40.5\%)$$

We will return to the problems of smoothing and smoothing systems in a later chapter since we are tending to anticipate those parts of the course which deal with power supplies and stabilized supplies. For the time being, then, we will carry on with some

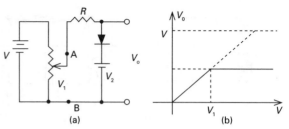

Figure 9.18

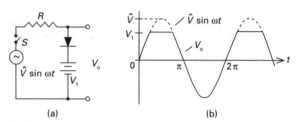

Figure 9.19

further diode applications and return to rectification and smoothing in the appropriate place.

Diode clipping

The voltage across a silicon diode will not exceed about 0.7 V for any level of current. In the circuit of *Figure 9.18(a)*, if a variable d.c. voltage is applied at terminals AB and advanced from 0 to V volts, the output will follow the characteristic of *Figure 9.19(b)*, being clipped or prevented from rising to the full output equal to V.

When the input is at 0 volts, the p.d. across the diode is $-V_2$ volts; it is therefore biased in the reverse direction and no current flows in the circuit (the output being unloaded). As the voltage at A is increased the diode remains off until V_1 exceeds V_2 by 0.7 V; the diode then becomes forward biased and remains so as V_1 is increased up to its maximum value of V. The current passed by the diode increases but the p.d. across it remains at about 0.7 V. The voltage V_1 is therefore clipped off at $(V_2 + 0.7)$ V. By adjustment of V_2 this clipping can be made to begin at any point intermediate between V and V_2.

If a sinusoidal source replaces the battery of *Figure 9.18*, a peak clipping circuit is obtained. In *Figure 9.19(a)*, if the switch S is closed at time $t = 0$, the diode remains non-conducting until $\hat{V} . \sin \omega t = (V + 0.7)$ V. The output V_0 (again assuming no external loading) will be equal to the generator voltage until this instant; thereafter V_0 remains at $(V + 0.7)$ V because there is no series resistor in the diode branch. When $\hat{V} . \sin \omega t$ falls below V_1 the diode is again

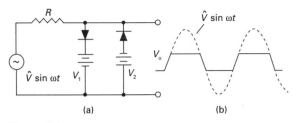

Figure 9.20

reverse biased and V_o again follows the generator voltage, as shown in diagram *(b)*.

By reversing the diode and the biasing battery, the applied voltage may be clipped on the peaks of the negative half cycles. In some applications both positive and negative peaks need to be clipped in this way, often asymmetrically, and a simple example of such a system is shown in *Figure 9.20*. By limiting the clipping levels to a very small fraction of the input peak value, a reasonable approximation to a rectangular wave may be obtained by this process.

(13)   Sketch the waveform (current against time ordinate) of the current through resistor R in the circuit of *Figure 9.20(a)*. What is the peak current if $R = 500\ \Omega$?

CLAMPING CIRCUITS

Another application for diodes are clamping circuits. In *Figure 9.21(a)* the diode is normally maintained in a conducting state by resistor R being returned to the positive supply rail. Considering an input rectangular waveform as shown at *(b)*, when V_1 increases, the diode current is increased by the amount of the capacitor charging current; the diode voltage, however, is kept at about 0.7 V. When V_1 falls, the voltage across C cannot change instantaneously, hence V_o falls with V_1. The diode is now reverse biased but a discharge current flows through R and the output fol-

lows an exponential curve until the point is reached where the diode again conducts. In this example the time constant is taken to be small compared to the half-period of the input wave; for a long time constant the exponential rise is small over this period and the output wave approximates very closely to the input. In all cases, the aiming voltage is the supply level V.

Notice that the input pulse is consequently *clamped* to a level a little above the earth line. This process is sometimes known as *d.c. restoration* and is employed where it may be necessary to transmit through a system the original reference levels of a waveform which have been shifted by couplings that have removed the d.c. component. We will return to this subject in the following chapter.

Example (14)   The input to the circuit of *Figure 9.22(a)* is the rectangular waveform shown at *(b)*. What will be the output waveform, assuming (i) that the diode is ideal, (ii) that the diode has a characteristic shown earlier in *Figure 9.15*.

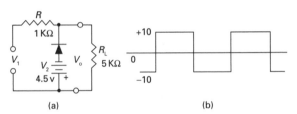

(a)   **(b)**

Figure 9.22

Solution   (i) During the positive half cycle of input, the diode will be reverse biased and non-conducting, hence the output will be given by

$$V_o = \frac{V_i R_L}{R_L + R} = \frac{5000 \times 10}{6000} = 8.33\ \text{V}$$

When the input is negative, the diode conducts and the output is clamped to the level of V_2, that is, -4.5 V. The output waveform and zero reference level are therefore as shown in *Figure 9.23*.

(ii) The positive half-cycle output will be un-affected by the presence of the diode forward

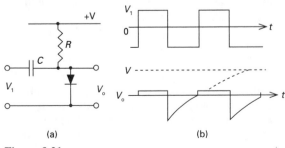

(a)   **(b)**

Figure 9.21

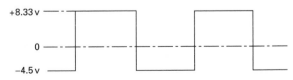

Figure 9.23

resistance since it will be reverse biased; the negative half-cycle, however, calls for us to make use of the diode characteristic, reproduced (over the relevant part of the range) in *Figure 9.24*, for convenience. We need the diode operating point and can obtain this by the use of the load line for the equivalent forward resistance.

Since the circuit to the left of the diode terminals has only linear elements, we can apply Thévenin to the diagram; we then get the equivalent of *Figure 9.25(a)* transformed into the Thévenin of *Figure 9.25(b)*. This is a circuit you should be familiar with by now. Applying the 833 Ω load line to the characteristic we see that the output is now clamped to the $(V_2 + 1)$ = 5.5 V level; the output waveform is now as shown in *Figure 9.26*.

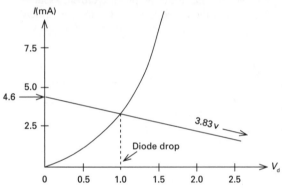

Figure 9.24

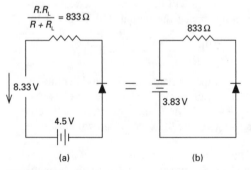

Figure 9.25

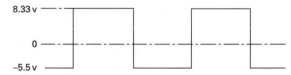

Figure 9.26

Example (15)   The waveform shown in *Figure 9.27(a)* is to be restored in such a way that the positive peaks are clamped approximately at earth (zero) potential. Explain how the circuit of *Figure 9.27(b)* will achieve this, assuming that any load resistance is very large.

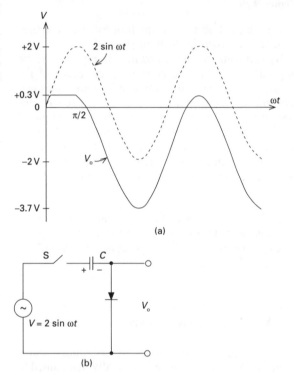

Figure 9.27

Solution   Taking the switch S to be closed at time = 0, the diode will not conduct until its terminal voltage reaches either about 0.7 V for silicon or 0.3 V for germanium. Taking a germanium diode, the output voltage will therefore be the same as the generator voltage until an input of about 0.3 V is reached. The diode will then conduct and V_o will remain at 0.3 V until the peak input point occurs where $\omega t = \pi/2$. At this time the generator voltage is 2 V and the capacitor voltage is 1.7 V with the polarity indicated. This voltage will remain constant (since the load resistance is assumed large); as V_i develops above the peak condition, V_o will be the generator voltage shifted downwards by 1.7 V. Hence the waveform is clamped with the positive peaks at about earth potential, as depicted.

(16)   What would be the effect of an appreciable load current flowing?

TRANSFER CHARACTERISTICS

A characteristic curve which relates the input current or voltage to the output current or voltage in an electronic device or system is known as a *transfer* characteristic. In *Figure 9.28*, for instance, a plot of output current I_o versus input current I_i would be a current transfer characteristic. Unless this characteristic is perfectly linear over the expected range of input current variation, the output current variation will not be a faithful reproduction of the input; we say that *distortion* has occurred.

Such distortion is often very undesirable, in high-fidelity audio amplifiers, for example; on the other hand, it is often introduced quite deliberately for a variety of reasons and purposes – obtaining a square wave output from a sine wave input, for example, as the clipper circuit of *Figure 9.20* illustrated. Non-linear elements, therefore, are harbingers of signal distortion unless precautions are taken.

We will not go into an analysis of distortion here, but illustrate the meaning of a transfer function from a couple of worked examples.

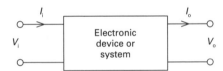

I_i I_o

Electronic device or system

V_i V_o

Figure 9.28

Example (17)   In the clipper circuit of *Figure 9.20*, let $V_1 = +6$ V, $V_2 = +4$ V (with the polarities shown), and $R = 5$ kΩ. If a load resistance of 40 kΩ is connected across the output terminals, find the equation representing the transfer characteristic for this circuit. Assume the diodes are ideal.

Solution   Both diodes are reverse biased when $-4 < V_o < +6$. For ideal diodes the branch currents will then be zero, hence the current around the outer loop will be

$$-V_i + 5000i + 40\,000i = 0$$

or

$$i = \frac{V_i}{45\,000}$$

The output voltage for this region of operation is

$$V_o = 40\,000i = \left[\frac{40\,000}{450\,000}\right]V_i = \left[\frac{8}{9}\right]V_i$$

Hence the overall transfer characteristic will have a gradient of 8/9 for the region $-4 < V_o < +6$. When $V_o = -4$ V, $V_i = -4.5$ V, and when $V_o = +6$ V, $V_i = 6.75$ V.

The characteristic can now be drawn as illustrated in *Figure 9.29*. The clipping of the input sine wave occurs when this wave swings beyond the linear portion of the transfer curve. Mathematically, there are discontinuities at the transition points.

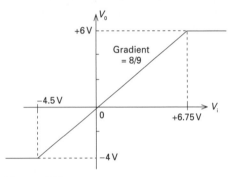

Figure 9.29

Example (18)   The transfer characteristic for a certain amplifying system is shown in *Figure 9.30*. What would be the effect at the output of (a) an input sinusoidal signal which swung between the limits of points A and B on the curve, (b) an input signal which swung between the limits of points P and Q on the curve?

Solution   The points A and B define a region of the curve which is for all practical purposes a straight line. Hence, a signal which operated over this part of the curve would appear in an undistorted form at the output. This is illustrated on the right of the diagram.

Over the region P to Q, however, the peaks of the input sine wave would be 'compressed' in the manner shown. There would not be true clipping in the sense that the characteristic of *Figure 9.30* would introduce it as no part of

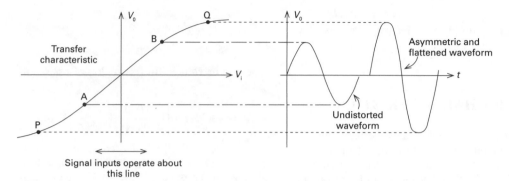

Figure 9.30

this present curve becomes zero gradient or horizontal over this range. The output wave is consequently distorted, nevertheless, by having flattened peaks and an asymmetrical displacement about the zero line. This kind of distortion is known as third-harmonic distortion and is equivalent to the addition of a wave of three times the frequency of the input sine wave, or fundamental frequency.

SUMMARY

- The important characteristic of a diode is the discrimination between forward and reverse voltages.
- If there is a single non-linear element in an otherwise linear circuit, a load line provides a simple graphical solution.
- A rectifier converts alternating current into unidirectional current. In a half-wave circuit with a load resistor R_L, $I_{dc} = \hat{V}/\pi R_L = \hat{I}/\pi$. A full-wave circuit with a load resistor R_L has $I_{dc} = 2\hat{I}/\pi$.
- Ripple factor F_r gives a figure for the effectiveness of rectification, where

$$F_r = \frac{V_{ac}}{V_{dc}} = \frac{I_{ac}}{I_{dc}} = \sqrt{\left(\left[\frac{I}{I_{dc}}\right]^2 - 1\right)}$$

- A clipping circuit provides an output voltage V_o proportional (or equal to) the input voltage V_1 up to a certain value V. Above V the wave is cut off or clipped.
- A clamping (or d.c. restoration) circuit holds

the peak values of certain signals to predetermined levels.

REVIEW QUESTIONS

1   Why is a diode model useful in certain calculations?
2   Why should a model be made up from *linear* elements?
3   Given an ideal diode, sketch the voltage developed across a load resistor in a half-wave circuit. A full-wave circuit.
4   Sketch the i,v characteristic of a real diode. Of an ideal diode.
5   What is meant by the efficiency of a rectifier system?
6   Explain how a full-wave rectifier operates. Use two different methods.
7   Write down, with respect to the peak input voltage V, (a) the average voltage, (b) the r.m.s. voltage, for the half-wave case. Do the same for the full-wave case.
8   Sketch a transfer characteristic for half- and full-wave rectifiers, assuming an input $v = 10 \sin \omega t$.
9   Explain the operation of a capacitor smoothing system connected to a full-wave rectifier.
10   During capacitor discharge, why doesn't some of the current flow through the diode?
11   Explain the operation of a diode clipper. Clamper.
12   In what way does the maximum diode current rating limit the value of the reservoir capacitor?

EXERCISES AND PROBLEMS

(19)   Divide the following list of component parts into linear and non-linear types, having a considered guess if you feel uncertain about any of them: carbon resistors, wirewound resistors, voltage-dependent resistors, capacitors, germanium diodes, inductors, field-effect transistors (FETs), light emitting diodes (LEDs), thermistors.

(20)   The three diodes of *Figure 9.31* can be assumed to have ideal linear gradients as shown. What are the voltages across each of the circuit resistors?

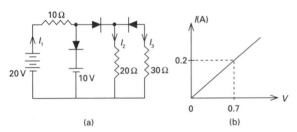

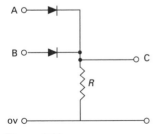

Figure 9.31

(21)   Diode D_2 in the previous problem is now reversed. What now will be the voltages across each resistor?

(22)   Inputs A and B in the circuit of *Figure 9.32* can have voltage levels of 0 V or +10 V. What will be the output level at C for (a) both A and B at 0 V, (b) either A or B at +10 V?

Figure 9.32

(23)   In a half-wave rectifier circuit, what effect on the current in the load would a resistance equal in value to the load have when (a) placed in parallel with the diode, (b) placed in series with the diode?

(24)   In a half-wave rectifier circuit the maximum permissible current is 2 A. Obtain a figure for the efficiency of the circuit, given that the load resistance is 50 Ω and the resistance of the remainder of the circuit is 10 Ω.

(25)   Plot the current/voltage characteristic of a power diode from the following table

V	0	10	20	30	40	50	60
I (mA)	0	2.0	6.5	13.5	21.5	30.0	38.5

For a supply of 75 V and a load of 3000 Ω connected in the diode anode circuit find (a) the load current, (b) the voltage across the load, (c) the power dissipated in the load, (d) the voltage across the diode.

(26)   In the clipping circuit of *Figure 9.20* (page 143), suppose that $V_1 = 8$ V, $V_2 = 6$ V and $R = 10$ kΩ. If a 100 kΩ load resistor is used, sketch the transfer characteristic for this circuit.

(27)   A voltage V_1 varies periodically from −2 V to +6 V. In a certain circuit it is necessary to clamp this voltage to a maximum positive excursion of +4 V. A silicon diode, a number of mixed resistors and a 6 V battery are to hand. Try and design a circuit that will fulfil this function.

(28)   Find the voltage V across the 10 Ω resistor when the non-linear resistance is expressed as $R = 20I$ ohms; see *Figure 9.33*.

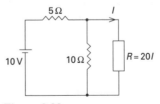

Figure 9.33

10

Small-signal amplifiers 1: bipolar transistors

The general amplifier
Fundamental mechanisms
Circuit configurations
Characteristic curves
The common-emitter amplifier
Leakage and bias stabilization

By definition, a 'small signal' is one to which an electronic device responds in a linear manner. In small signal amplifiers, therefore, we may limit our studies to small signals typically within the range of a few microvolts to a volt or so, superimposed on d.c. values, and assume linear relationships among signals in devices that are decidedly non-linear in their large signal behaviour.

THE GENERAL AMPLIFIER

An amplifier of any kind is essentially a box having two input terminals and two output terminals, as shown in *Figure 10.1*. This is known as a *two-port* network. Very often, one of the input terminals is commoned to one of the output terminals, and the arrangement is known as a *three-terminal* network. The box will contain one or more active devices, transistors or valves, together with some associated passive components such as resistors and capacitors. To operate, the box will also require some kind of d.c. power supply derived from batteries or rectifier units such as will be discussed in a later chapter.

In general (but not always) we expect an amplifier to fulfil two conditions:

1 The output signal will be greater in magnitude than the input signal.

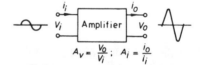

Figure 10.1

2 The output signal will be of exactly the same waveform (or shape) as the input signal.

The first of these conditions is a measure of the voltage or current amplification or gain provided by the system. We define

$$\text{Voltage amplification } A_v = \frac{\text{output signal voltage } v_o}{\text{input signal voltage } v_i}$$

and

$$\text{Current amplification } A_i = \frac{\text{output signal current } i_o}{\text{input signal current } i_i}$$

The signal voltages and currents will normally be measured in r.m.s. values, but for sine wave signals it is just as convenient in many cases to find the ratio of the output to input peak values; the resulting figures obtained for A_v and A_i are, of course, unaffected by this.

It is often necessary to know the power gain of an amplifier. This can be calculated from the product of voltage and current gain, so that

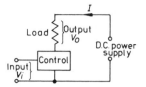

Figure 10.2

A_p = voltage gain × current gain = $A_v \cdot A_i$

The second of the conditions means that the signal waveform should not suffer an distortion during the process of amplification. It is not a simple matter to design an amplifier having negligible distortion, at least not without sophisticated and expensive circuit systems. For simple amplifiers of the kind of interest to us here some distortion would be inevitable, but it can be kept to a reasonably low level by careful attention to certain fundamental details.

In this chapter we shall investigate the basic amplifying properties of the bipolar transistor. No transistor is capable on its own of providing amplification, but when it is employed in conjunction with a load resistor (or a load impedance) amplification becomes possible. The conventional amplifier arrangement thus consists of a source of d.c. power (batteries etc.), a load (resistance or impedance), and a control device (transistor) connected as shown in *Figure 10.2*. From this diagram, we can deduce that the voltage amplification A_v is

$$\frac{\text{output signal voltage across the load}}{\text{signal voltage present at the input}} = \frac{v_o}{v_i}$$

We begin with a brief refresher reference to the fundamental operating principles of the bipolar transistor.

TRANSISTOR MECHANISMS

In *Figure 10.3* we have the representation of an *n-p-n* transistor with the appropriate forward and reverse biasing voltages applied to the respective electrodes. These voltage are, of course, the direct current supplies derived from batteries or d.c. power units, and along with the resulting currents are indicated in capital letters. The application of a negative potential to the emitter causes the majority electrons in the *n*-type material to be repelled from the emitter region. The emitter therefore acts as a source of electrons and these flow into the base region which is biased positively with respect to the emitter. We have, in other words, a forward-biased diode. In the base region the electrons drift towards the collector by a process of diffusion and are accepted by the collector which is

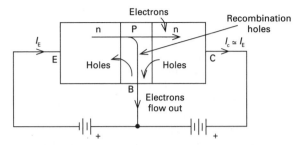

The indicated current direction is the true electronic flow. Conventional flow is opposite to this.

Figure 10.3

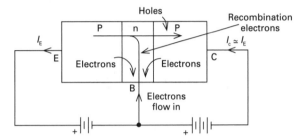

The indicated current direction is the true electronic flow. Conventional flow is opposite to this.

Figure 10.4

biased positively with respect to the base. Because the base is *p*-type material, electrons exist there only as minority carriers, so the electrons arriving at the collector are derived almost entirely from the emitter source. On their way across the base wafer, a small proportion, normally less than 1 per cent of the electrons, combine with holes in the base region and this loss of charge is made good by a flow of base current. It is to reduce this 'loss' of electrons that the base region is made very thin.

The effect seen from an external viewpoint is that of a fairly large flow of electrons from emitter across the base to the collector with a small flow of electrons *from* the base. For a small signal transistor, we might find such typical values as 1 mA for the emitter current I_E, 0.99 mA for the collector current I_c and the difference 0.01 mA for the base current I_B.

This is the mechanism of an *n-p-n* transistor.

For a *p-n-p* transistor, see *Figure 10.4*, the emitter acts as a source of positive charges or holes, which flow into the base when the base is biased negatively with respect to the emitter. The collector this time is biased negatively with respect to the base and so absorbs the positive charges diffusing across the base region. Here a small proportion of the holes leaving the emitter recombine with the majority electrons in the base and this loss of charge is made good by the flow of electrons *into* the base as base current.

Compare the *p-n-p* figure with that of the *n-p-n* paying particular attention to the direction of movement of the electron or hole carriers *inside* the transistor and the corresponding movement of *electron* carriers *only* in the external circuit. It is easier to comprehend the *n-p-n* example at first, since the carriers in all parts of the circuit are electrons. If you keep in mind that both battery connections are reversed in the *p-n-p* circuit and the external current (an electronic current) is also reversed, you should have no problems. We will, in most examples, refer to *n-p-n* devices.

Since the carriers originate at the emitter and distribute themselves between base and collector, the sum of the base and collector currents must always be equal to the emitter current, hence, as Kirchhoff will tell us, $I_E = I_B + I_C$.

CIRCUIT CONFIGURATIONS

There are three possible ways in which a transistor can be wired into an amplifier system, and these are indicated in *Figure 10.5*. As provision has to be made for a pair of input and output terminals in an amplifying system and the transistor has only three terminations, one or other of these terminations must be common to both input and output circuits. In the classification of the three possible connections, therefore, the circuit configuration is described in terms of this common terminal. Hence diagram *(a)* shows the common-base connection, *(b)* the common-emitter connection, and *(c)* the common-collector connection. The common electrode is, in general, treated as being at earth potential and the term 'earthed' or 'grounded' instead of 'common' is sometimes used to define the particular configuration; for instance, *(a)* might be referred to as a 'grounded-base' connection.

> (1) You have two batteries. Show how you would connect them to the three configurations shown in *Figure 10.5* in turn, so that the proper d.c. operating voltages were applied. Show also the directions of the currents (electronic, not conventional, here) in the external wires.

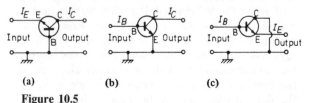

(a)   (b)   (c)

Figure 10.5

Now that you have attempted this last example, let us consider the relationship existing in each of the configurations between the input and output currents, assuming that we have the battery supplies appropriately connected to the various terminals. We may define the ratio

Current flowing in the output circuit
─────────────────────────────────────
Current flowing in the input circuit

as the *static current gain* of the transistor, the word static being used to imply that we are operating our devices solely from direct current sources, no other external components being involved.

Current gain in the common-base configuration is designated α^* and is the ratio of collector current I_C to emitter current I_E. Hence

$$\alpha = \frac{\text{Collector current}}{\text{Emitter current}} = \frac{I_C}{I_E}$$

In common-emitter configuration, current gain is designated β (sometimes α_E) and is the ratio of collector current I_C to base current I_B. So

$$\beta = \frac{\text{Collector current}}{\text{Base current}} = \frac{I_C}{I_B}$$

There is a relationship between α and β because for all the configurations the equation $I_E = I_B + I_C$ must hold. From the above definitions we have

$$\frac{\alpha}{\beta} = \frac{I_B}{I_E} = \frac{I_E - I_C}{I_E}$$

since

$$I_B = I_E - I_C$$

Hence

$$\frac{\alpha}{\beta} = 1 - \frac{I_C}{I_E} = 1 - \alpha$$

so rearranging

$$\beta = \frac{\alpha}{1 - \alpha}$$

and by transposition

$$\alpha = \frac{\beta}{1 + \beta}$$

* The symbols h_{FB} and h_{FE} are commonly used for static current gain and we will use these at the appropriate time.

(2)   An arrow on the emitter symbol that points towards the base indicates a(an) . . . transistor.

(3)   The circuit configuration in which the input is between base and emitter and the output between collector and emitter is the . . . connection.

(4)   A transistor has an emitter current of 2.7 mA and a collector current of 2.63 mA. What is its base current?

(5)   Calculate, for the previous transistor, the current gain for (a) common base, (b) common-emitter connection.
    What current ratio defines the gain of a common-collector configuration? Prove that this current gain is equivalent to $1 + \beta$.

You should have discovered one or two important facts from working these problems in conjunction with what has gone before. For instance, the ratio $I_C/I_E = \alpha$ is always less than unity for the reason that I_C is always less than I_E. In the same way, the ratio $I_C/I_B = \beta$ is always greater than unity for the reason that I_C is always greater than I_B.

Further, the current gain of the common-collector configuration is clearly always greater than unity since (if you have worked problem (5) above correctly) it is expressed as $\beta + 1$.

Typical figures for the current gains of common-base and common-emitter connections are 0.98 to 0.998 and 30 to 500, respectively.

(6)   A transistor has a common-emitter current gain of 350. What will be its gain in common base?

(7)   The common-base gain of a transistor is 0.985. What is its gain in common emitter?

From the point of view of current gain, the common-base connection seems to have no value as an amplifier, but this is because the word 'amplification' tends to become identified by an increase in something, often without knowledge of what that something is. The common-base connection, therefore, has no current gain as such a word is implied, but it nevertheless has other advantages which make it a relatively well-utilized configuration in circuit design.

CHARACTERISTICS

Because the transistor has three terminals and is effectively two interconnected diodes of opposing polarities, there are a number of voltage measurements which may be made on it in comparison with the simple relation between current and voltage which is made on the single semiconductor diode.

In the note which follows, we will use the conventional procedure for the direction of action of the various voltages and currents as this way is usually less confusing than when the electronic convention is used. In all cases, of course, Kirchhoff's laws must be obeyed at all junction points and around all loops, and all relevant quantities signed accordingly.

It is customary to fix the common rail which in most cases is taken as the earth line or chassis line, as a zero voltage reference and to measure all voltages relative to this. Voltages will therefore be identified by two subscripts if they act between two definite terminals of the transistor or if they are d.c. voltage sources such as battery supplies; the order of these subscripts will indicate the conventional direction of action of the voltages; that is, *the first subscript will indicate the point taken to be the more positive* with respect to the second subscript or to the common line. So, for example, treating an *n-p-n* transistor as a pair of diodes connected in opposition as *Figure 10.6* shows, the emitter-base diode being *forward* biased by battery V_{BB} and the base-collector diode being *reverse* biased by battery V_{CC}, then $V_{BE} + V_{CB} = V_{CE}$. An identical relationship will hold for a *p-n-p* transistor, since the voltages here are simply reversed in direction and the Kirchhoff summation will remain unaffected.

There are four quantities, or parameters as they are called, which can be held at a constant value or varied over a range of values, during the course of a measurements experiment; these are (a) input voltage, (b) input current, (c) output voltage and (d) output current. Four characteristic curves may then be drawn involving pair-combinations of these parameters so as to describe the d.c. performance of the transistor

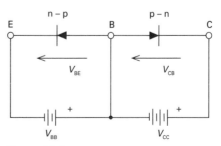

Figure 10.6

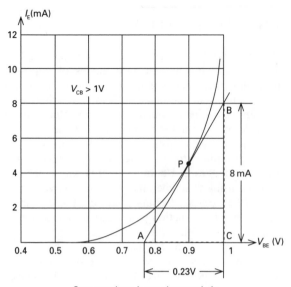

Common-base input characteristic

Figure 10.7

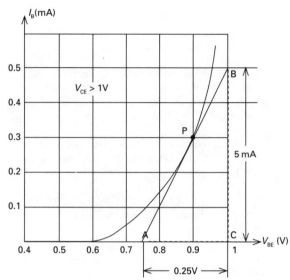

Common-emitter input characteristic

Figure 10.8

completely. The *static* characteristics can usually be found for particular transistors in the manufacturer's literature, and three of the more important of these curves will now be discussed.

The input characteristic

The input characteristic of a transistor is a plot of input current against input voltage, hence it concerns only the base-emitter junction which is the equivalent of a forward-biased diode. For the common-base configuration, where the 'live' input terminal is the emitter and the base is held at zero rail potential, the emitter current I_E is measured for a range of values of base-emitter voltage V_{BE}, the collector being maintained throughout the measurement at a constant voltage V_{CB}. The setting of V_{CB} does influence the position of the characteristic curve slightly, but this can for most purposes be ignored. A graph of I_E against V_{BE} can then be plotted in the manner illustrated in *Figure 10.7*. Notice that the value assigned to V_{CB} during the measurement is stated on the graph paper; in this case we have simply assumed that this voltage is greater than 1 volt.*

* In all graphs, voltage and current axes will be indicated positive, so that the first quadrant of the coordinate axes will contain the required curves. This is quite conventional, although strictly when negative values are involved, as in some *p-n-p* mode curves, other quadrants should be used to provide a true graphical representation. Some older books show such reversed axes. Our convention here does not affect the deductions drawn from the curves in any way.

For the common-emitter configuration, the 'live' input terminal is the base and the emitter is held at the zero rail potential. Hence base current I_B is measured for a range of values of base-emitter voltage V_{BE}, the collector again being maintained at a constant voltage V_{CE}. A typical curve is shown in *Figure 10.8*.

What information do the input characteristics give us? One of the chief interests is the input resistance of the transistor – what magnitude of resistance would be seen if we looked into the input terminals of the transistor? This parameter is very important in the study and design of transistor amplifiers. One thing we can be fairly certain about – the input resistance is likely to be relatively low because we are looking into a forward-biased diode. However, by the proper use of the input characteristic we can find out quite accurately what the actual working resistance will be for the two modes of connection.

Return to *Figure 10.7*, the common-base input characteristic. The graph gives us related values of I_E and V_{BE}; we could therefore select any point on the curve, read off the pair of related values represented by the point and by the division V_{BE}/I_E obtain the Ohm's law value of resistance at that point. However, the graph is markedly non-linear and we should clearly obtain a different value for the resistance at every point selected. What we are more concerned with here is the ratio

$$\frac{\text{the small change in } V_{BE}}{\text{the small change in } I_E} = \frac{\delta V_{BE}}{\delta I_E}$$

where δ simply means 'a small change in'. This is because when the transistor is employed as an amplifier, the input signal is *not* a d.c. level but a small *alternating* signal superimposed on a fixed d.c. level, this level being known as the *operating point*. This operating point depends for its position on the curve upon certain circuit requirements and once selected should not be allowed to change. The above ratio is then found as the *gradient* of the curve at the operating point P.

Unlike the *static* input resistance found by the simple division of V_{BE} by I_E the ratio $\delta V_{BE}/\delta I_E$ gives what is called the *dynamic* input resistance under actual signal conditions.

It is usual to denote this ratio in the symbolism

$$R_1 = \frac{\delta V_{BE}}{\delta I_E}\bigg|V_{CB}$$

where the subscript quantity outside of the vertical line is the constant quantity throughout that particular experiment.

In *Figure 10.7*, suppose the operating point P is fixed at an emitter current of about 4.5 mA. Through this point a tangent AB is drawn to the curve. The gradient of this tangent is the same as the gradient of the curve in the immediate neighbourhood of the point P. Now the ratio given above is the ratio of the horizontal change in V_{BE}, the length AC, to the corresponding change in I_E, the length BC. Hence the changes considered are 0.25 V in voltage for 8 mA in current. Hence

$$R_i = \frac{0.23}{8 \times 10^{-3}} = \frac{230}{8} = 29\,\Omega$$

In *Figure 10.8* we have the input characteristic of a common-emitter mode of connection. Since the vertical scale is now in fractions of a milli-amp, it is obvious that the input resistance will turn out to be much higher than it was in the common-base mode. Taking an operating point P corresponding to a base current of 0.3 mA and drawing the tangent AB as before, it is seen that

$$R_i = \frac{0.25}{0.3 \times 10^{-3}} = \frac{250}{0.3} = 833\,\Omega$$

The input resistance of the common-emitter mode is always much greater than that for common-base mode, the above figures being quite typical. Actual values of course depend on the transistor type, the working voltages and the selection of the operating point. The peak-to-peak magnitude of the input signal is also of importance since the object is to work along a linear part of the characteristic (that is, work along the tangent line) and this will be closely achieved if the input signal swing is small.

(8) A transistor has an emitter current of 1.5 mA and a base current of 10 µA. What will be its current gains in (a) common-base, (b) common-emitter mode?

(9) Referring to *Figure 10.7* and *10.8*, estimate the dynamic input resistance for each mode of connection, assuming that the operating point P is located at $I_E = 6$ mA for common base, and at $I_B = 0.2$ mA for common emitter. (Use a pencil to avoid spoiling your book diagrams.)

The output characteristic

The output characteristic possibly gives the most information of the graphical representations of transistor action because from this plot most other parameters may be derived. Here we plot the variation in collector current with a common V_{CB} (for common-base mode) and V_{CE} (for common-emitter mode). This is very dependent on the other variables, I_E in common base and I_B in common emitter; hence it is usual to plot a family of curves for a range of values for these quantities.

For the common-base mode, the emitter current is maintained at a constant value while a deliberate variation in the collector current is plotted against the resulting variation in the collector-base voltage. A suitable experimental set-up is shown in *Figure 10.9(a)*. I_E is fixed, say, at 2 mA by suitable adjustment of potentiometer R_1 and V_{CB} is then varied in a series of voltage steps using potentiometer R_2. Corresponding values of I_C are then recorded for each step in V_{CB} and a plot is made (for $I_E = 2$ mA) as shown in the lowest curve of *Figure 10.9(b)*. Notice that there is a rapid rise in the current to the saturation value with a small voltage applied, than the saturation region itself where I_C becomes relatively independent of V_{CB}.

This procedure is now repeated for other fixed values of I_E so that a complete family of curves may be plotted, each of these being annotated with the emitter current concerned. In the figure, these levels are taken in 2 mA increments up to a maximum of 10 mA. Each curve is seen to be basically the same shape with the saturation regions practically horizontal and parallel. Notice, as we should expect, that for each setting of I_E the collector current is almost identical to I_E, the small difference in each case being due to the flow out of base current.

In the common-base mode, the collector-base voltage can be reduced to zero and a collector current can

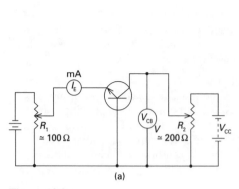

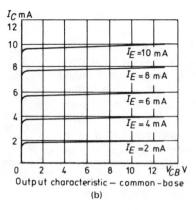

Output characteristic – common – base

(b)

(a)

Figure 10.9

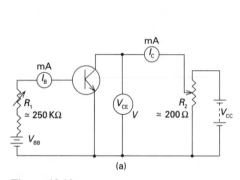

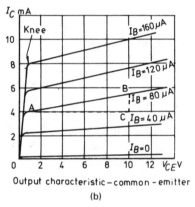

Output characteristic – common – emitter

(b)

(a)

Figure 10.10

still be present. Only by taking V_{CB} slightly *negative* can the collector current be reduced to zero.

> (10)   Can you explain the reason for the statement made in the last paragraph? (Don't worry if you can't; we shall be coming to this aspect in a short while.)

For the common-emitter mode, the output characteristic shows the variation in collector current against variations in the collector-emitter voltage, I_B being held constant. This time an experimental set-up similar to that shown in *Figure 10.10(a)* may be assembled.

Since the base-collector junction is a reverse-biased diode, a small leakage current will flow when $I_B = 0$. The lowest curve of *Figure 10.10(b)* shows this effect. By adjustment of potentiometer R_1 a small base current is set to flow, say, of $40\,\mu A$, and held constant while V_{CE} is adjusted throughout the desired range of voltage; this provides us with the next characteristic curve of the family. Going on in this

manner, increasing I_B in appropriate steps for each plot, the completed output characteristic is obtained. It is seen that, like the common-base example, the curves are basically of the same shape and reasonably flat in the saturation region, but not so flat as they were in the common-base mode. However, the collector current becomes substantially independent of collector voltage. This is because (as it is for the common-base mode) collector current is derived from the charge carriers originating in the emitter; increases in collector voltage, therefore, have little effect on collector current because it is not the collector voltage which is generating the carriers. As far as the output terminals are concerned, the transistor behaves as a constant current source, and this property has many useful applications in electronic circuit design.

Return now to *Figure 10.10(b)*. The curves give us related values of I_C and V_{CE} for various settings of I_B. The output resistance of the transistor is the effective resistance we should see by looking back into the output terminals, i.e. between collector and emitter; we should expect this to be relatively large because this time we are looking into a reverse-biased diode.

However, the value we measure is not simply the apparent resistance of the collector-base diode junction; working on the same principle as we did for the case of input resistance (and remembering the notes on such a subject from Chapter 2) we say

$$\text{Output resistance } R_o = \frac{\text{the small change in } V_{CE}}{\text{the small change in } I_C}$$

$$= \frac{\delta V_{CE}}{\delta I_C} \text{ for constant } I_B$$

and a triangle drawn on an appropriate characteristic enables R_o to be evaluated. Strictly, the hypotenuse of the triangle, AB, is a tangent to the curve, but as the lines are for all practical purposes straight and parallel, the characteristic can be considered as the actual hypotenuse. What we are saying is that the gradient of the lines is constant. From the diagram then

$$R_o = \frac{AC}{BC} = \frac{9}{1.5 \times 10^{-3}} = 6000 \ \Omega$$

Clearly, the output resistance increases as the characteristic lines become more horizontal, so referring to the curves for the common-base mode shown in *Figure 10.9(b)* it is apparent that the output resistance for this mode of connection will be very high. A typical value would be 500 kΩ. The common-emitter, on the other hand, would have a typical output resistance of 20 kΩ; the curves of *Figure 10.10(b)* were deliberately drawn to illustrate the calculation for R_o, and the value of 6 kΩ derived above is, in general terms, rather low.

The transfer characteristic

The transfer characteristic is a plot of the relationship between output current I_C and input current I_E or I_B depending on the mode of connection, at a specified collector voltage. The characteristic can be plotted directly from related values measured from the experimental circuits of *Figures 10.9(a)* or *10.9(b)*, or may be graphically derived from the respective output characteristics already discussed.

The output characteristics of *Figure 10.9(b)* can be used to determine the common-base transfer characteristic in the manner shown in *Figure 10.11*. For a specified collector-base voltage V_{CB}, in this example taken to be 8 V, values of I_E are projected across to the co-ordinate axes shown on the right. Notice that the vertical axis of I_C is common to both graphs. Where these horizontal projections meet the verticals for corresponding values of I_E, points lying on the transfer characteristic are established. Joining these intersections gives us the required characteristic.

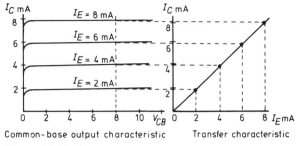

Common-base output characteristic   Transfer characteristic

Figure 10.11

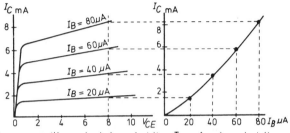

Common-emitter output characteristic   Transfer characteristic

Figure 10.12

The common-emitter transfer characteristic can be obtained similarly from an output characteristic of the form of *Figure 10.10(b)*, as shown in *Figure 10.12*. Here the specified collector-emitter voltage V_{CE} is again taken as 8 V. Values of I_B are projected across to the coordinate axes on the right of the figure, the vertical axis of I_C again being common to both graphs. Where these horizontal projections meet the verticals for corresponding values of I_B, points lying on the transfer characteristic are established, and the line joining them can be drawn in.

From the graphs, the gradients of the transfer characteristics give us

$$\frac{\delta I_C}{\delta I_E} = \alpha; \quad \frac{\delta I_C}{\delta I_B} = \beta$$

the current gains in common-base and common-emitter mode, respectively.

You will notice that the common-base transfer characteristic is much more linear than is the common emitter. Since I_E is always $I_C + I_B$ and I_C is plotted against I_E in *Figure 10.11*, the line will have a gradient which is only slightly less than 45°, α in other words is always slightly less than 1, since the tangent of 45° is 1.

(11) Use the transfer characteristic of *Figure 10.12* to obtain a rough estimation of the common-emitter current gain.

LEAKAGE CURRENT

Before going any further into general amplifier theory, it is necessary to look into the problem of leakage current. Look at *Figure 10.13*, which shows an *n-p-n* transistor in common-base configuration but with its emitter lead disconnected. Under this condition it might appear that the collector current I_C would be zero, since clearly I_E is zero, but this is not so. The collector circuit is still connected through the reverse-biased base-collector diode, and this diode must pass reverse current. This is where some of the facts you learned in Chapter 9 should stand you in good stead. The leakage current is due to the movement of minority carriers (holes in the *n-p-n* transistor) across the junction, and its direction is opposite to that of the main forward current which would flow if the diode was forward biased. But a movement of holes from collector to base inside the transistor is equivalent to a movement of electrons in the direction base to collector, hence in the external circuit the leakage current flows in the same sense as that due to collected electrons, the majority carriers. This leakage current is denoted by I_{CBO}, meaning that we are referring to the collector-base junction with 0 showing that the third electrode, the emitter, is left disconnected. This current still flows when the emitter is reconnected and the main forward current from the emitter is superimposed. So far we have taken the collector current to be $I_C = \alpha I_E$. But with the addition of the leakage current, the true total collector current becomes $I_C = \alpha I_E + I_{CBO}$ as *Figure 10.13* shows.

As I_{CBO} is very small at room temperatures (20–25°C), particularly with silicon material, it might seem that the small addition to the normal, relatively large forward current at the collector would be unimportant. This is true for the situation just described, but leakage current increases with increasing temperature and when we look at the problem in relation to the common-emitter configuration, such an increase leads to very undesirable effects in the transistor performance.

We illustrate the common-emitter situation in *Figure 10.14*. Here the base current is treated as the input. Since $I_E = I_C + I_B$ we can write the previous expression for I_C as

$$I_C = \alpha \cdot (I_C + I_B) + I_{CBO}$$

Then $I_C(1 - \alpha) = \alpha \cdot I_B + I_{CBO}$

$$\therefore \quad I_C = \frac{\alpha I_B}{1 - \alpha} + \frac{I_{CBO}}{1 - \alpha}$$

Now $\beta = \dfrac{\alpha}{1 - \alpha}$

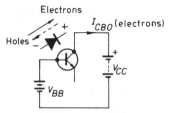

Figure 10.13

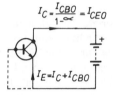

Figure 10.14

$$\therefore \quad I_C = \beta I_B + \frac{I_{CBO}}{1 - \alpha}$$

The first term here is the value of I_C we have so far taken as the output of the common-emitter amplifier with the input current equal to I_B. Refer back to page 150 if your memory has faded at this point. The second term must represent the leakage current when the base is disconnected; this is denoted I_{CEO}. Hence

$$I_{CEO} = \frac{I_{CBO} \times 1}{1 - \alpha} = (1 + \beta)I_{CBO}$$

Hence, for common-emitter connection

Total $I_C = \beta I_B + I_{CEO}$

This is identical in form to the expression for I_C in common-base mode, but the values of the parts are quite different. Suppose, for example, that $\alpha = 0.98$, so that $1/(1 - \alpha) = 50$. Then I_{CEO} is 50 times as large as I_{CBO}, which clearly aggravates the problem we mentioned earlier about the increase in leakage current with rise in temperature. What is happening, in fact, is that in common-emitter mode, the transistor is amplifying its own leakage current! This is definitely a most undesirable state of affairs and a matter of serious concern in transistor applications. Try the next two problems to see if your ideas are straight about this leakage problem.

(12)   If $I_C = 2.45$ mA, $I_{CBO} = 20\,\mu A$ and $I_E = 2.5$ mA, what is α?

(13)   If I_B is zero for the above transistor, what is I_C?

THE COMMON-EMITTER AMPLIFIER

For the two basic circuit configurations we discussed in the previous sections, we saw that the input resistance in both cases was the relatively low resistance of the forward-biased base-emitter junction; the output resistance was the relatively high resistance of the reverse-biased base-collector junction. It is this marked disparity between the input and output resistances, in conjunction with the external load resistor connected to the collector (output) circuit, which enables the transistor to act as an amplifier of alternating currents and voltages.

Figure 10.15 shows the basic circuit arrangement of a common-emitter amplifier. In the circuit, input takes place between base and emitter and the output is developed between collector and emitter. The emitter is the common electrode, and it is generally simpler to refer to the connection mode as base-input, collector-output amplification.

In the usual manner, the base is biased positively relative to the emitter (we are using an *n-p-n* transistor) by the battery V_{BB} and the collector is biased positively relative to the emitter (and the base) by battery V_{CC}. A load resistor R_L has been included in the collector lead. All this accords with what we have already discussed about basic transistor operation. The addition of R_L is the only major change in the general circuit arrangement. However, our previous work has

dealt with the transistor only from the point of view of the d.c. battery conditions and on this basis (with no load resistor in circuit) we defined the static current gain

$$\beta = \frac{\text{Change in } I_C}{\text{Change in } I_B \text{ with } V_{CE} \text{ constant}}$$

On this basis also, we drew the various characteristic curves relating the various electrode voltages and currents.

We are concerned now with what happens when an alternating signal is applied to the input terminals; this is the signal that we require to amplify, be it speech, music or a simple single-frequency tone. For simplicity we shall consider the alternating input to be a single frequency sine wave, expressed in the customary way as $i = \hat{I} \cdot \sin \omega t$. If the emitter-base voltage is allowed to alternate about a mean value, base current I_B also will vary about some mean value determined by battery V_{BB}. It is clear that the emitter current and hence the collector current will also vary about some mean value determined by V_{CC}. The load resistor R_L will have a p.d. developed across it by this alternating collector current, and this will represent the output alternating voltage from the amplifier. The processes involved are perhaps best brought out by way of a simple numeric example.

First, we recapitulate on the d.c. set-up, assuming that the a.c. signal input is zero. *Figure 10.16(a)* shows the arrangement. As usual we have the base-emitter junction biased in the forward direction by V_{BB}. This bias is taken, purely as an illustrative example, to be 1 V. This then fixes what we shall call the base-bias voltage. Suppose that this bias causes a base current $I_B = 0.1$ mA to flow; then this value of base current determines the mean d.c. level of base current about which the a.c. input signal will swing alternately positive and negative. This is the base current d.c. operating point.

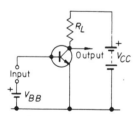

Figure 10.15

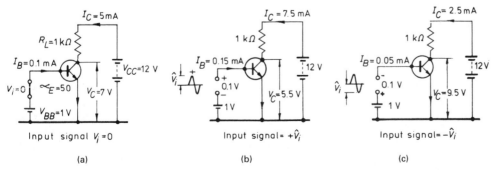

Figure 10.16

Now for this example the static current gain of the transistor has been taken to be 50. Since 0.1 mA is the steady base current, the collector current I_C will be given by $I_C = \beta \cdot I_B$, so

$$I_C = 50 \times 0.1 = 5 \text{ mA}$$

This current is shown in the diagram. When 5 mA flows through the load R_L, taken here to be 1 kΩ, there will be a steady voltage drop across R_L given by $I_C \cdot R_L = 5 \times 10^{-3} \times 1000 = 5$ V. The voltage at the collector V_{CE} will therefore be $V_{CC} - I_C R_L = 12 - 5 = 7$ V. This value of V_{CE} is the mean d.c. level about which the output a.c. voltage signal will swing alternately positive and negative. This is the collector current d.c. operating point. A transistor biased in this way is said to be operating in Class A conditions.

We have now fixed two quantities as operating points: the base current I_B at 0.1 mA, and the collector voltage V_{CE} at 7 V. This establishes the steady or static conditions. We now consider the working or dynamic conditions.

Turn to (b) of *Figure 10.16* and consider what happens when we apply the a.c. signal input in series with the base. The exact manner of doing this is unimportant at the moment. Suppose the input signal varies between the peak values +0.1 V and −0.1 V as shown, and imagine the condition at (b) to be such that the positive peak of the input is present at the input terminals. Then at this instant V_{BE} is increased by 0.1 V; suppose this causes I_B to increase by 0.05 mA, so that the instantaneous $I_B = 0.1 + 0.05$ mA = 0.15 mA. The collector current now increases to 0.15 $\times$ 50 = 7.5 mA and the voltage drop across R_L increases to $7.5 \times 10^{-3} \times 1000 = 7.5$ V. Hence the collector voltage V_{CE} falls to $12 - 7.5 = 5.5$ V. Notice now that for a change of 0.1 V at the base we have produced a change of 2.5 V across the collector load. If we consider the voltage across R_L as the output voltage, then

$$\text{Voltage gain} = \frac{\text{Voltage change across } R_L}{\text{Voltage change in } V_{EB}}$$

$$\therefore \quad A_v = \frac{2.5}{0.1} = 25$$

The input signal now changes from its positive peak value through zero to its negative peak value. At the instant it passes through zero, the circuit conditions have returned to those shown in *Figure 10.16(a)* and I_B, I_C and V_{CE} are momentarily at their mean operating values of 0.1 mA, 5 mA and 7 V, respectively. In (c) the input signal is at its most negative value, −0.1 V. At this instant V_{BE} is reduced by this amount and I_B decreases by 0.05 mA. The total I_B is now 0.1

− 0.05 mA. The collector current correspondingly falls to $0.05 \times 50 = 2.5$ mA and the voltage drop across R_L becomes $2.5 \times 10^{-3} \times 1000 = 2.5$ V. Hence the collector voltage V_{CE} rises to $12 - 2.5 = 9.5$ V.

The cyclic variation of V_{CE} about its mean value of 7 V goes on all the time the input signal varies I_B about its mean value of 0.1 mA. The transistor then provides us with a voltage amplification of 25, and the output variation is an exact replica of the input variation, i.e. a sine wave. Equal changes in I_B have caused equal changes in V_{CE}. There is therefore no distortion occurring in the process of amplification.

You will have noticed that we have interpreted the amplification provided by the transistor in the example as a voltage gain, symbolized A_V. There is also a current gain: I_B is changing by 0.05 mA on each input half cycle and the output current is changing by 2.5 mA correspondingly, so we have a current gain at 50, or β. In an actual amplifier circuit, for reasons which we will not pursue here, the dynamic working gain is always less than the static value of β, but an appreciable current gain is, nevertheless, provided by the common-emitter amplifier.

(14)   If the load resistor R_L is increased in value, there will be a greater voltage drop across it for a given collector current and hence a greater voltage gain. Is there any reason why R_L should not be made very large, say 100 kΩ, to obtain a large voltage gain?

We should note one other important point at this stage: when V_{BE} increased, V_{CE} decreased, and vice versa. The transistor therefore reverses the phase of the input voltage.

(15)   Does the amplifier reverse the phase of the input current?

USING THE CHARACTERISTIC CURVES

To have a reasonably good idea of the way an actual transistor will work as an amplifier, we have to know a number of the quantities mentioned in the previous illustrative example fairly accurately: the exact point, for example, to which I_B should be set, what amplitude of input signal we expect, what collector supply voltage is available and which transistor we are going

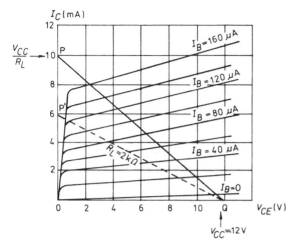

Figure 10.17

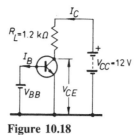

Figure 10.18

to use. All these things are inter-related and the choice of any one of them has to be made with an eye on the others.

Our starting point is the characteristic curves we discussed in the previous section. We are considering the common-emitter amplifier here, so the static characteristics relating to the common-emitter configuration will be our concern.

We look first at the output characteristic curves, a set of which were given earlier in *Figure 10.10(b)* and which are reproduced here as *Figure 10.17* for easy reference. These curves relate collector current I_C to collector voltage V_{CE} for various fixed values of I_B. Now we cannot use these individual curves to solve our problems about amplification, because as soon as a sinusoidal signal is applied at the base terminal, I_B varies over a range of values as our earlier example showed, and the individual curves relate to a constant I_B. It is necessary first of all to draw across the characteristic curves another line, known as the load line, which will show us the relationship existing between I_C and V_{CE} when I_B is changing. To draw the load line it is necessary for us to know

1 The value of the collector load resistor R_L.
2 The collector supply voltage V_{CC}.

Of course, these values themselves also depend to a certain degree on other factors, but we have to make a start somewhere. Consider the circuit of *Figure 10.18*, where $R_L = 1.2\,k\Omega$ and $V_{CC} = 12\,V$. These are typical figures for a small amplifier stage. The collector voltage is given by

$$V_{CE} = V_{CC} - I_C R_L$$

Now the supply voltage V_{CC} determines the value of the collector voltage when $I_C = 0$. Clearly, under

this condition there is no voltage drop across R_L and so

$$V_{CE} = V_{CC}$$

This establishes the position of point Q on the characteristic curve V_{CE} axis in *Figure 10.17*, and obviously V_{CE} can never exceed this value. Since $V_{CC} = 12\,V$, point Q represents this voltage limit.

Now suppose $V_{CE} = 0$. This can only happen if the voltage drop across R_L is exactly equal to V_{CC}. For the resistance given, $1.2\,k\Omega$, and a voltage drop equal to $12\,V$, the collector current

$$I_C = \frac{V_{CC}}{R_L} = \frac{12}{1200}\,A = 10\,mA$$

This condition establishes the position of point P on the I_C axis of the characteristic curves. By joining the points P and Q with a straight line we obtain the load line for the condition $R_L = 1.2\,k\Omega$.

For every given load resistance there will be a corresponding (and different) load line. If V_{CC} is kept at the same value, all the possible lines will start at the same point Q but will cut the I_C axis at different points P. Suppose, for example, R_L is increased to $2\,k\Omega$. When the voltage dropped across this load is $12\,V$, the current flowing through it will be $12/2000$ $A = 6\,mA$, hence the load line for $R_L = 2\,k\Omega$ will lie between Q and point P' as shown by the broken line. It is evident that the gradient of the load line (its steepness) decreases as R_L increases, and vice versa.

(16) Using the curves of *Figure 10.17*, draw load lines (lightly in pencil) for the following conditions:

(a) $V_{CC} = 10\,V$; $R_L = 1\,k\Omega$, $R_L = 2\,k\Omega$
(b) $V_{CC} = 8\,V$; $R_L = 1\,k\Omega$, $R_L = 1.5\,k\Omega$

Now check your answers and rub out the lines!

(17) We have plotted load lines using two extreme points P and Q. How do we know that the lines connecting these are, in fact, straight lines?

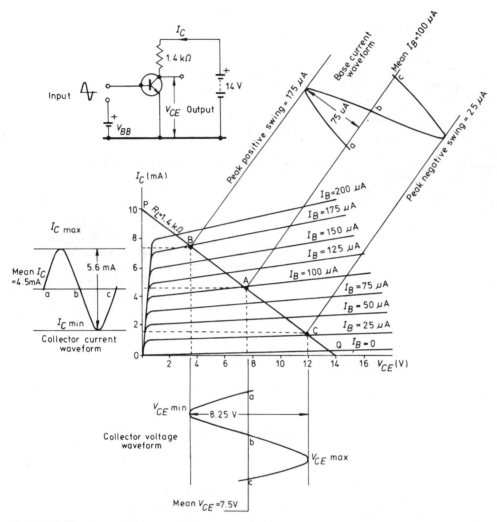

Figure 10.19

So far, so good. We have seen how a load line can be drawn to suit a particular value of load resistor. We have now to see how a particular load line will help us in establishing the proper mean values of I_B and V_{CE} about which the input and output signal alternation will respectively swing. We have to find a value of I_B so that, when the signal is superimposed upon it, the current and voltage variations at the collector lie within the limits determined, and imposed, by the extremities of the load line. Quite plainly the collector voltage can never exceed V_{CC} (point Q on the line) and the collector current can never be greater than that value which would make V_{CE} zero (point P on the line).

In *Figure 10.19* we have taken the output characteristics for a transistor amplifier which is to be used with a 1.4 kΩ load and a V_{CC} of 14 V. The load line

for $R_L = 1.4$ kΩ has been drawn in between points P and Q. The point A has been chosen as the base current operating point, i.e. $I_B = 100$ µA, because the excursions of I_B about this point which will result in proportional changes in I_C are very closely those between which the load line cuts through equal spacings between the static characteristics on either side of A. In other words, the distance from A to B and from A to C represents the peak permissible swings of base current under the stipulation that AB is very closely equal to AC. Any appreciable swing beyond these limits will exaggerate the discrepancy between AB and AC, or, worse still, will carry the transistor beyond collector current cut-off at end Q or work into the curved portions of the characteristics at end P. If this is allowed to happen, the output signal will no longer be identical in shape with the base current variations and distortion will be present. It is not usually possible to get A exactly midway between the limits B and C, particularly if a very large input swing is to be handled, so a small amount of distortion is tolerated as the price for a large output signal.

The instantaneous values of the base current waveform are now projected on to the load line to give the corresponding values of I_C; the instantaneous values of I_C (i_C) are likewise projected on to the load line to give the corresponding values of instantaneous output voltage (v_o). Notice that the output voltage waveform is in antiphase with the input current waveform; when base current changes in a positive direction from 100 µA to 175 µA, the collector voltage changes in a negative direction from 7.5 V to about 3.7 V. As base current is in phase with base voltage, the output voltage is antiphase to the input voltage, as we deduced from our earlier example.

From the diagram the maximum excursions of base current for closely equal spacings in either direction about A, where $I_B = 100$ µA, is 75 µA. The corresponding excursions of I_C are about 2.8 mA about the mean value of 4.5 mA. The swing in V_{CE} is then about 4.15 V about the mean value of 7.5 V. These mean values of voltage and current are often referred to as the quiescent values, since they represent the no-signal operating conditions. As a rough and ready rule, a mean operating value for V_{CE} is taken to be $V_{CC}/2$ for the purpose of quick calculations.

The signal variations shown in *Figure 10.19* represent the maximum we can allow for this particular circuit if serious distortion is to be avoided. If you measure AB and AC, you will find them approximately equal. Of course, there is nothing to prevent us from making the input signal smaller so that the swing in I_B might be only from 100 µA at A down to 75 µA on one peak and up to 125 µA on the other. The equality of distance along the load line about A would then be almost ideal. What we should get would simply be a smaller output, but with reduced distortion. The operating point must therefore be chosen with care, and once chosen, stabilized against any effects which might tend to move it. We shall come to this problem in the following section.

(18) What is the current gain, A_i, in the amplifier of *Figure 10.19*?

(19) Suppose the input current of this circuit flows in an input resistance of 1000 Ω. What is the approximate voltage gain, A_v?

BIAS STABILITY

It is necessary in good circuit design that once the operating point is selected, it should not significantly vary when the transistor characteristics change because of manufacturing differences, for example, or because of variations in the ambient temperature. As we have seen, there are a number of different coordinates involved in the specification of the operating point, i.e. I_B, I_C and V_{CE}, and a question arises as to which of these is most important in stabilizing the operating point.

As an example of the problem, consider *Figure 10.20*. This shows the effect of a temperature variation, and although the changes are exaggerated here for purposes of explanation, this does illustrate the situation clearly. The operating point for this particular transistor has been located on the $I_B = 125$ µA characteristic at point X. A varying base current of about 100 µA peak to peak is causing a collector current variation of about 7 mA peak to peak. Under the conditions shown, the collector current is not, however, a faithful reproduction of the base signal. When the temperature changes, the whole of the curve family moves either up or down relative to the axes, upwards for an increase in temperature and downwards for a decrease in temperature. The spacing between them also changes, though this is not shown on the diagrams. This action, superimposed on to the existing axes and load line background, will effectively shift the operating point one way or other along the load line, and although the operating point remains at $I_B = 125$ µA, the input base-current signal now results in clipped collector-current waveforms. The stage is driven either into the saturation region or into the I_C cut-off region depending on whether the temperature has risen or fallen, respectively. Replacing the transistor with another having a higher gain parameter or a lower gain parameter will produce the same effect.

How can the problem be overcome or minimized? A study of *Figure 10.20* shows that a stabilization of I_B is not the answer, since I_B was held constant in both examples. Relative to the other operating point coordinates, however, the operating point did shift. If some point along the curve axes was to have been held constant, such as a particular value of I_C, then more acceptable operation would have followed. For example, if the operating point had been maintained at about 6 mA instead of $I_B = 125$ µA, such clipping would not have occurred. The gain will not be stabilized by this suggestion, but gain stabilization is not the prime matter at this stage.

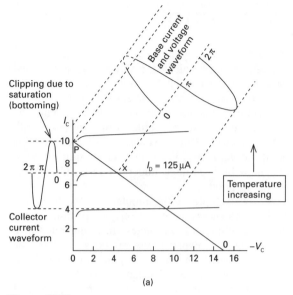

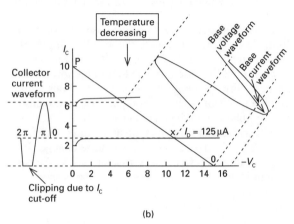

Figure 10.20

We look at a few practical suggestions for setting the bias current and see how that in turn can be made to stabilize the collector current at the operating point.

SETTING THE OPERATING POINT

We return now to the topics we were discussing in relation to operation along the load line PQ of *Figure 10.19*. Having selected the operating bias point A on a particular load line drawn on a particular set of characteristics, it is necessary to set the bias current of the transistor concerned to the required value and ensure that it stays there. In *Figure 10.19* and in the earlier diagrams we assumed that a separate bias battery V_{BB} was used for this purpose, but in practical designs it is not convenient to have this arrangement and in general the base bias is obtained from the same source as the collector supply, that is, a single battery (V_{CC}) provides all the necessary voltages throughout the amplifier. As the required bias voltage is much smaller than that required at the collector, the most simple modification is shown in *Figure 10.21*. Here the forward base bias is obtained by the insertion of a resistor R_B between the base and the positive terminal of the supply. The value of this resistor is easily calculated: in our example from *Figure 10.19*, the required I_B is 100 μA, hence

$$R_B = \frac{V_{CC}}{I_B} = \frac{14}{100 \times 10^{-6}} = 140 \text{ k}\Omega$$

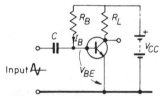

Figure 10.21

This value actually includes the internal base-to-emitter resistance, but as this is of the order of some few hundred ohms, it may be ignored in the calculation. R_B consequently limits the base current in the forward direction to the required value, in this case 100 μA. It is important to notice, however, that the bias is not developed across R_B but across the base-emitter junction as the result of the no-signal (d.c.) current through that junction. This action makes the base positive with respect to the emitter, so biasing the diode in the forward direction.

(20)   A transistor requires an operating base bias of 50 μA. The V_{CC} supply is 9 V. What value of resistor R_B is required to provide this bias?

(21)   The above transistor has β = 70, and is used with a collector load R_L of 2000 Ω. What will be the steady voltage at the collector?

Now the use of a single biasing resistor like this is not particularly good practice from the point of view

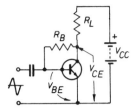

Figure 10.22

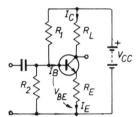

Figure 10.23

value of emitter current which will flow for a given base voltage at the junction of R_1 and R_2. Any increase in I_C will produce an increase in I_E and this in turn will increase the voltage drop across R_E. This reduces the forward bias voltage V_{BE} which then leads to a reduction in I_C, so partly compensating for the original increase. This argument applies equally well to changes in collector current resulting either from changes in β (which happen when a transistor is changed) or in the supply voltage. Thus this circuit gives better d.c. operating stability than one in which the emitter is connected directly to the earth line. It is usual to make R_E of such a value that a drop of about 0.5–1 V occurs across it, and to proportion the divider so that R_2 is about 5 to 10 times the value of R_E. The total current through the divider should normally be at least ten times the mean value of base current.

of maintaining stability in collector current. Suppose the temperature increases, then I_{CBO} also increases and I_{CEO} becomes β times this variation. Hence I_C increases, and V_{CE} and I_B are also influenced; for the d.c. collector current

$$I_C = \beta I_B + I_{CEO}$$

and the d.c. base current $= I_B - I_{CEO}$.

So, with any increase in leakage current, the total collector current increases and the total base current decreases. The operating point is consequently unstable, and the additional heating at the collector might lead to the effect of thermal runaway or at least to an increase in distortion.

An alternative method of base biasing is shown in *Figure 10.22*. Here the resistor R_B is returned, not to the V_{CC} line, but to the collector itself. If now the collector current increases for any reason, the collector voltage V_{CE} will fall. As the base bias resistor is taken from the collector the base current will also fall, since $I_B = V_{CE}/R_B$. Hence the collector current $I_C = \beta I_B$ will also fall and tend to restore itself to its original (prerise) value.

Figure 10.23 shows the most commonly used bias arrangement. It consists of a potential divider circuit made up from resistors R_1 and R_2 connected in series across the supply, and an emitter resistor R_E. If the potential divider is made up so that the voltage level at the centre point (the base connection) is that required to establish the proper base current, but at the same time the total value of $R_1 + R_2$ is such that the current flowing through the divider is very large compared with I_B, then the base current will remain substantially constant regardless of variations in collector current. The emitter resistor in turn determines the

Example (22)   A transistor with $\beta = 150$ is used in the circuit of *Figure 10.24*. The bias conditions are such that a collector current of 5 mA flows in the collector load. Ignoring leakage current and taking $V_{BE} = 0.65$ V, calculate a suitable value for R_B.

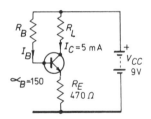

Figure 10.24

Solution

Emitter current $I_E = \dfrac{I_C}{\alpha} = \dfrac{I_C(\beta + 1)}{\beta}$

$$= \frac{151}{150} \times 5 \times 10^{-3}$$

$$= 5.033 \text{ mA}$$

Emitter voltage $= I_E R_E$

$$= 5.033 \times 10^{-3} \times 470$$

$$= 2.366 \text{ V}$$

$\therefore$ Base voltage $= 2.366 + 0.65$

$$= 3.02 \text{ V}$$

$\therefore$ Required voltage drop across $R_B = 9 - 3.02$
$= 5.98$ V

Now

Base current $I_B = \dfrac{I_C}{\beta} = \dfrac{5 \times 10^{-3}}{150} = 33.3$ μA

$\therefore$ $R_B = \dfrac{5.98}{33.3 \times 10^{-6}} = 180$ kΩ

Example (23) *Figure 10.25* shows a common-emitter amplifier with potential divider bias and an emitter resistor. The quiescent base current is 50 μA and the base-emitter voltage is 0.6 V. If the voltage drop across R_E is to be 1 V, assess suitable values for R_1, R_2, R_E and R_L.

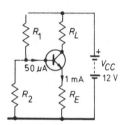

Figure 10.25

Solution For a 1 V drop across R_E at the indicated value of $I_E = 1$ mA, $R_E = 1$ kΩ. The current through the divider has to be large relative to I_B, so taking the current in R_2 to be $10I_B$ or 0.5 mA, the p.d. across $R_2 = V_{BE} + I_E R_E = 0.6 + 1.0 = 1.6$ V.

$\therefore$ $R_2 = \dfrac{1.6}{0.5 \times 10^{-3}} = 3200$ Ω

Current in $R_1 = 11I_B = 0.55$ mA

$\therefore$ p.d. across $R_1 = V_{CC} - 1.6 = 10.4$ V

$\therefore$ $R_1 = \dfrac{10.4}{0.55 \times 10^{-3}} = 19$ kΩ

As there is a 1 V drop across R_E, 11 V are available for the drop across R_L and V_{CE} in series. It is reasonable to take the mean V_{CE} at about the midpoint of this supply, hence $V_{CE} = 5.5$ V. So, taking $I_E = I_C = 1$ mA.

$R_L = \dfrac{5.5}{10^{-3}} = 5500$ Ω

In any practical circuit, the nearest preferred value of each of the resistors would be selected, together with a tolerance and a power rating to satisfy the circuit conditions. In the previous example, then, R_1 would be selected as 18 kΩ, R_2 as 3.3 kΩ and R_L as 5.6 kΩ.

(24) Now assess the minimum required power rating for each of R_1, R_2, R_E and R_L. Would a 0.25 W rating do for all these resistors?

MULTI-STAGE AMPLIFIERS

When a single stage of small signal amplification is not sufficient for its purpose, addition stages must be added. When the input signal is very small, it is generally better to get the required gain from two or three following (or *cascaded*) stages rather than to try to obtain the full gain from one stage.

The common-emitter amplifier is nearly always resistance-capacitance coupled between stages and, apart from the amplification itself, is an advantage it enjoys over the common-base circuit. This comes about from the higher input resistance of the common emitter, and also the high collector output resistance provides an ideal current source for the following base input.

Figure 10.26(a) shows a typical two-stage amplifier. Base bias for both stages is provided by potential dividers $R_1 R_2$ and $R_5 R_6$, and the emitter resistances R_3 and R_7 are bypassed with large value capacitors to eliminate signal negative feedback (of which more later). If n stages, each with a current gain β, are cascaded in this way, the overall gain is $\beta^n \cdot R_L/R_S$ where R_s is the input source resistance and R_L is the final output load.

An important point rises here; the intermediate collector load, R_4 in this example, is *not* simply the value of R_4. When we study equivalent circuits a little later on, we shall see that R_4 is effectively in parallel with the bias resistance R_5 and R_6 and the input resistance of transistor T_2. This reduces the value of R_4 *as far as the signal is concerned* to a value much lower than its stated value, hence the load line for that particular stage is not the one drawn under purely d.c. conditions. *Figure 10.26(b)* shows the effect of this; the load line on which the stage is actually working is the *a.c.* or *signal load line*, and because of the reduction in the effective load, this line has a different slope but still passes through the operating point.

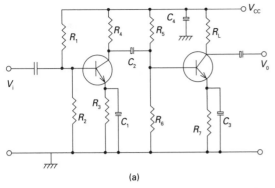
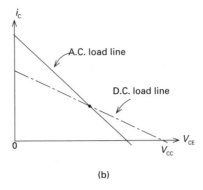

(a)

(b)

Figure 10.26

GAIN ESTIMATIONS

Although voltage and current gains can be estimated from the characteristic curves, it is sometimes sufficient to obtain values from simple formulae. The results are, of course, only approximations, but they are adequate for most general purposes. They are based on a knowledge of R_L, R_i and α_E. Input resistance R_i can be deduced from the input characteristic for a given V_{CE} as explained on page 152.

Assume a collector signal current of i_C so that the signal voltage across R_L is $i_C R_L$. If the current gain is β the base signal current $i_b = i_c/\beta$. Now for an input resistance R_i, the input voltage which must be applied across R_i to produce a base current of i_c/β is $i_c R_L/\beta$. Hence the voltage gain is given by the ratio

$$A_v = \frac{i_c R_L}{i_b R_i} = \beta\left(\frac{R_L}{R_i}\right)$$

The power gain of the amplifier can be evaluated similarly. The input signal power is given by $P_i = i_b^2 . R_i$ and the power delivered to the collector load by the transistor is $P_o = i_c^2 R_L$. Hence the power gain

$$A_p = \frac{P_o}{P_i} = \frac{i_c^2 . R_L}{i_b^2 . R_i} = \beta^2\left(\frac{R_L}{R_i}\right)$$

$$= \beta\left(\frac{v_o}{v_i}\right)$$

Example (25)   The input characteristic of the transistor used in the amplifier of *Figure 10.27(a)*

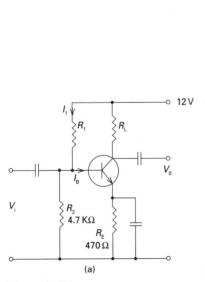

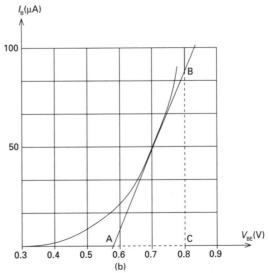

(a)

(b)

Figure 10.27

is shown at *(b)*. Under steady d.c. conditions the voltage across R_E is 2.5 V and the base current I_B is 50 μA. Estimate (a) the value of resistor R_1; (b) the voltage gain of the stage as it is shown; (c) the voltage gain of the stage when it is followed by an identical stage. Point out any approximations made.

Solution   This circuit has both potential-divider and emitter-resistor stabilization. From the input characteristic we see that for $I_B = 50$ μA, $V_{BE} = 0.7$ V. Hence the voltage at the junction of R_1 and R_2 is 2.5 + 0.7 = 3.2 V.

(a)   Current   through $R_1 = I_1 = \dfrac{3.2}{4.7 \times 10^3}$

= 0.68 mA

$$R_1 = \frac{12 - 3.2}{I_1 + I_B} = \frac{8.8}{0.73 \times 10^{-3}} = 12.05 \text{ k}\Omega$$

(b)   The *static* current gain $A_v = \dfrac{I_c}{I_B} \simeq \dfrac{I_E}{I_B}$;

$I_E = \dfrac{2.5}{470} = 5.32$ mA

$$\therefore \quad A_v = \frac{5.32 \times 10^{-3}}{50 \times 10^{-6}} = \frac{5320}{50} = 105$$

(c)   We need now to find the *dynamic* input resistance of the stage, so drawing a tangent to the input characteristic at the point where $I_B = 50$ μA we find

$$R_i = \frac{\text{change in } V_{BE}}{\text{change in } I_B} = \frac{AC}{BC} = \frac{0.21}{88 \times 10^{-6}}$$

= 2.39 kΩ

When this amplifier is connected to a further identical stage, the *effective a.c.* input resistance of this second stage will be R_L in parallel with R_2 and R_1. Working in kilohms

$$\frac{1}{RL'} = \frac{1}{4.7} + \frac{1}{12.05} + \frac{1}{2.39}$$

from which $R_L' = 1.4$ kΩ.

This resistance now appears to be in parallel with R_L of the first stage, hence the net resistance of the collector load of the first stage will no longer be 1.5 kΩ but will fall to R_L' where

$$\frac{1}{R_L'} = \frac{1}{1.5} + \frac{1}{1.4}$$

from which $R_L' = 0.72$ kΩ or 720 Ω.

To obtain the effective voltage gain of the stage, we can consider an arbitrary input voltage v_i and work through to the output v_o, the voltages now being signal voltages and hence measured in r.m.s.

Let $v_i = 0.03$ V r.m.s.

Then r.m.s. base current $= \dfrac{V_i}{R_i} = \dfrac{0.03}{2.39 \times 10^3}$ A

$$= 12.6 \text{ μA}$$

Taking the static *and* dynamic gains to be 105 (a reasonable approximation) the r.m.s. collector current will be

$i_c = 105 \times 12.6$ μA = 1323 μA = 1.323 mA

r.m.s. output voltage $v_o = i_c \cdot R_L'$

$$= 1.323 \times 10^{-3} \times 720$$

$$= 0.95 \text{ V}$$

$\therefore$ voltage gain $A_v = \dfrac{v_o}{v_i} = \dfrac{0.95}{0.03} = 32$

A shorter method would have been to use the approximate formula for gain given above, that is

$$A_v \simeq \frac{\beta R_L}{v_i}$$

which here $\simeq 105 \times \dfrac{720}{2390} = 32$

It might appear from the working of the last example that a number of approximations have been made and that the results obtained are also approximations. This is perfectly true, but in a practical circuit where transistor spreads and resistor tolerances are ever present, an exact analysis is impossible or, at best, extremely complicated. The approximations made are quite reasonable, and once having obtained the 'foundation' figures for the gain and so on of a particular circuit, some experimental adjustments with component values can be made on the basis of your calculations to bring the design to satisfactory fruition.

SUMMARY

- A small signal is one to which a given device responds linearly.

- Small signal transistors are three-terminal devices which can be connected in a circuit in one of three possible ways.
- In common-base connection, current gain is symbolized α (or α_B).
- In common-emitter connection, current gain is symbolized β (or α_E).
- Important parameters for transistors are input resistance, output resistance and current and voltage gain figures.
- Graphical methods can be used to construct load lines, operating points and determine current and voltage gains.
- Base bias must be stabilized and the operating point held fixed.
- Leakage currents are significant under certain operating conditions.

REVIEW QUESTIONS

1   What determines whether a signal is 'small' or 'large'?
2   Differentiate between the α and β values of transistor gain. Define α in terms of β; β in terms of α.
3   Why is emphasis placed on current gain in a bipolar transistor rather than voltage gain?
4   What distinguishes the 'static' gain of an amplifier from the 'dynamic' gain?
5   Explain why simply stabilizing base current is unsatisfactory.
6   Why is stability of the operating point necessary?
7   What is the purpose of biasing in a bipolar device?
8   Why must the current gain of a common-base connection always be less than unity? Why must the common-emitter always be greater than unity?
9   Distinguish between (a) the input characteristic, (b) the output characteristic, (c) the transfer characteristic. What properties of a transistor can be deduced from these characteristics?
10   How can a transfer characteristic be derived from an output characteristic?

EXERCISES AND PROBLEMS

(26)   What are the β values for a transistor having the following α values: (a) 0.938, (b) 0.975, (c) 0.962? Calculate the α values for a transistor having the following β values: (d) 35, (e) 75, (f) 220.

(27)   A transistor has $I_E = 2.50$ mA and $I_c = 2.47$ mA. What is its current gain in (a) common-base, (b) common-emitter mode?

(28)   If for a certain transistor $I_c = 4$ mA when $V_{CE} = 2$ V, and 5 mA when 8 V, I_E being held constant, what is the output resistance under this condition.

(29)   If $I_c = 5$ mA when $V_{CB} = 8$ V, and 5.02 mA when 12 V, I_E being held constant, what is the output resistance of the transistor?

(30)   In a common-emitter mode amplifier with V_{CE} held constant, a 65 µA change occurred in I_B when V_{BE} changed by 80 mV. What is the input resistance of the device under this condition?

(31)   State whether the following are true or false:

(a)   in common-base mode (i) I_E is dependent on I_c; (ii) V_{BE} is dependent on V_{CB}; (iii) I_c is independent of V_{CB}.
(b)   In common-emitter mode, no collector current flows if the base is disconnected.
(c)   Carriers injected by the emitter into the base region are of the same polarity as the collector.
(d)   Collector voltage influences the number of carriers injected into the base by the emitter.

(32)   The circuit of *Figure 10.28* uses a transistor in which $\beta = 115$ and $V_{BE} = 0.7$ V. Obtain estimations for the values of the bias resistor R_B

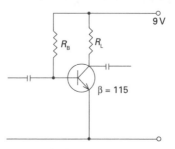

Figure 10.28

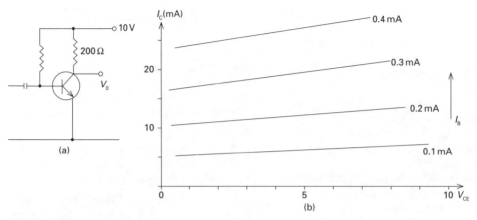

Figure 10.29

and the load resistor R_L, given that the operating point is located at the d.c. voltage and current values of 4.4 V and 9.2 mA, respectively.

(33) If the bias resistor of the previous example is returned to the collector terminal instead of the V_{cc} line to improve stabilization, what will be the value needed for R_B, assuming no changes in the base and collector currents.

(34) The following table gives the input characteristic parameters of a small signal transistor in common-emitter mode, for $V_{CE} = 4.5$ V

V_{BE}	75	100	125	150	175	200	(mV)
I_B	3	8	20	40	72	110	(μA)

Plot the input characteristic and from it evaluate the input resistance of the transistor when $I_B = 40$ μA, both static and dynamic.

(35) Using the characteristic given in *Figure 10.29(b)* in the circuit of *Figure 10.29(a)*, determine the current gain and the output voltage for an input current i = 0.1 sin ωt mA, given that the operating point is at $I_B = 0.2$ mA and $V_{CE} = 5$ V.

(36) An *n-p-n* transistor has the following output characteristics which may be assumed linear between the values of collector voltage given

Base current (μA)	Collector current (mA) for collector voltages of	
	1 V	5 V
30	1.4	1.6
50	3.0	3.5
70	4.6	5.2

This transistor is used as a common-emitter amplifier with load resistor $R_L = 1.2$ kΩ and $V_{cc} = 7$ V. The signal input resistance may be taken as 1 kΩ. Estimate the voltage gain A_v, the current gain A_i and the power gain A_p when an input current of 20 μA peak varies sinusoidally about a mean bias of 50 μA.

(37) Explain how leakage current inside a transistor affects the current flowing in the external circuit. Why is leakage current of more significance in common-emitter mode than in common-base?

(38) Prove that the equation of a load line can be expressed as $V_{CE} = V_{CC} - I_C R_L$. What is the significance of the negative sign? Why is this equation of a *straight* line?

11

Small-signal amplifiers 2: equivalent circuits

> *Common-base equivalent T-circuit*
> *Common-emitter equivalent T-circuit*
> *Hybrid parameters*
> *The general h-parameter model*
> *Thévenin and Norton equivalents*

We have already seen how complicated linear circuits can be reduced to simple equivalents by the application of Thévenin's and Norton's theorems. We have also dealt with the linear equivalents of otherwise non-linear devices such as diodes and zeners. We will now extend this kind of investigation to methods of simplifying electronic circuits by replacing non-linear devices such as transistors with linear elements, so that a relatively simple analysis can be made in terms of systems composed of such ideal elements.

It must be borne in mind, however, whenever circuit equivalents are used in this way, that any model is only an approximation to the actual behaviour of a device because an exact equivalent, if available, would be clumsy and not easily adapted to an interpretation of real practical conditions. Therefore, while some of these equivalent circuits are based on the physical processes going on within a semiconductor device, others do not necessarily represent in any way such actual physical interpretation. They do, nevertheless, represent linear models which are amenable to quantitative analysis where precision in the calculation of amplifier performance is unwarranted. We use, as far as possible, simplified models and their accompanying assumptions.

The actual character of a model depends on the frequency at which the circuit is operating and also on the magnitude of the signals expected; here we will be primarily concerned with small signal devices and what we might call 'mid-band' frequencies, where

neither extremely low nor extremely high frequencies are encountered.

BASIC CONSIDERATIONS

When circuits are designed and are known to be operating at some particular frequency or range of frequencies, the basic components of capacitance and inductance can be looked at as having either very low impedances or very high impedances at the frequencies concerned.

Capacitors are complete blocks, or open-circuits, to all d.c. supplies, but are chosen to have very low effective opposition at signal frequencies. Inductors have a very low opposition to direct currents, limited only by their self-resistance, but present an extremely large opposition at signal frequencies. A circuit such as an amplifier, therefore, presents a quite different picture to the a.c. signal than it does to the d.c. supplies.

In equivalent circuit analysis, the circuit that the a.c. signal 'sees' is the important one; the d.c. supplies rarely appear as such in circuit models. Resistors, of course, can be treated as resistances at all the frequencies we shall be interested in, though at very high frequencies these too have to suffer a modification.

Before we go on to have a look into models of the semiconductors themselves, we can supply the above observations to basic amplifier circuits to see how

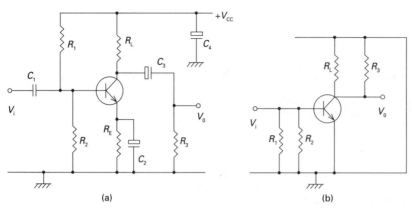

Figure 11.1

those observations affect the way the external circuit appears to the signal frequencies.

Figure 11.1(a) shows a simple common-emitter amplifier stage, similar to the first stage of the amplifier discussed on page 165. Here the output signal v_o is taken from across resistor R_3 instead of being passed to a following stage. Looking first at the various capacitors, these will behave as complete blocks to the d.c. supplies (and are included for that purpose) but they can be treated as short-circuits to the a.c. signal. The operating point of the stage and the d.c. load line are determined from the equivalent d.c. circuit which here shows the total resistance associated with the transistor to be $R_{dc} = R_L + R_E$. Since $V_{CE} = V_{CC} - R_L I_c$, we have

$$R_{dc} \simeq \frac{V_{CE}}{I_C}$$

The a.c. signal, however, sees a different situation. Capacitor C_2 is a short-circuit at the signal frequencies, therefore resistor R_E is effectively 'not there'. The V_{CC} line is a steady source of direct current but to the signal C_4 acts as a short-circuit, hence the supply line is effectively connected to the earth line. Then also the upper end of load R_L is effectively earthed and, because C_3 also acts as an a.c. short-circuit, R_L is placed in parallel with R_3 thereby effectively reducing the value of the load resistance. The a.c. load line must therefore differ from the d.c. load line in that it is now expressed as

$$R_{ac} = R_L \parallel R_3 = \frac{R_L R_3}{R_L + R_3} \quad \therefore \quad R_{dc} > R_{ac}$$

Hence the slope of the a.c. load line is different to that of the d.c. load line as *Figure 10.26* illustrated earlier on, and this in turn leads to the dynamic gain of an amplifier stage being less than that of the static

gain. *Figure 11.1(b)* shows the equivalent a.c. circuit to that given in *Figure 11.1(a)*.

(1)   Sketch the a.c. signal circuit for the two-stage amplifier of *Figure 10.26*.

Example (2)   In the amplifier stage of *Figure 11.2* the transistor has a current gain $\beta = 120$ and an a.c. input resistance of 1.5 kΩ. It is coupled to a following stage having an input resistance of 500 Ω by capacitor C_2. What will be the current flowing into the second stage, assuming that all capacitors have negligible reactance at the operating frequency?

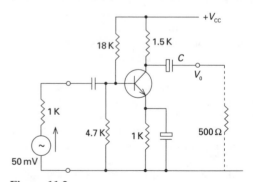

Figure 11.2

Solution   The a.c. equivalent input circuit is shown in *Figure 11.3(a)* where the Thévenin source feeds into three resistors in parallel. The effective input resistance R_i as far as the source is concerned is then found to be (working in kΩ)

$$\frac{1}{R_i} = \frac{1}{18} + \frac{1}{4.7} + \frac{1}{1.5}$$

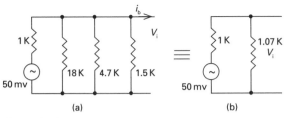

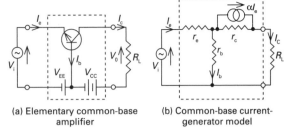

Figure 11.3

(a) Elementary common-base
amplifier

(b) Common-base current-
generator model

Figure 11.5

which, using conductances,

$$= 0.056 + 0.213 + 0.667 \text{ mS} = 0.936 \text{ mS}$$

Hence $R_i = 1.07 \text{ k}\Omega$, see *Figure 11.3(b)*

The amplifier input

$$v_i = \frac{1.07}{1 + 10.7} \times 50 \text{ mV} = 25.84 \text{ mV}$$

Now the a.c. base current

$$i_b = \frac{v_i}{R_i} = \frac{25.84 \text{ mV}}{1.5 \text{ k}\Omega} = 17.23 \text{ μA}$$

The a.c. collector current

$$i_c = \beta i_b = 120 \times 17.23 \times 10^{-3} = 2.07 \text{ mA}$$

As *Figure 11.4* now shows, part of this collector current is input i_o to the second stage. Hence

$$i_o = \frac{2.07 \times 1.5}{1.5 + 0.5} = 2.07 \times 0.81 \text{ mA}$$

$$= 1.55 \text{ mA}$$

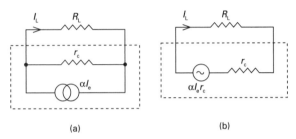

(a)

(b)

Figure 11.6

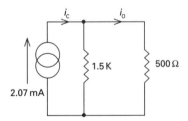

Figure 11.4

COMMON-BASE T-NETWORK MODEL

In what follows we will denote the equivalent-resistance elements of the transistor by lower case letter and lower case subscripts; the symbols denoting currents external to the transistor will have upper case letters. Remember that we are now dealing with signal components and that I_e, I_b and I_c are the r.m.s.

values of the a.c. components of currents at the emitter, base and collector terminals, respectively.

As the basic parameters of a transistor are these currents, the transistor resembles, and may be described in terms of, a three-terminal resistive network, that is, a two-port system with a common input and output line. The elementary circuit of a common-base configuration shown in *Figure 11.5(a)* may therefore be represented as a T-network equivalent, and this is shown in *Figure 11.5(b)*. Here r_e simulates the effective resistance of the forward-biased emitter-base diode, and r_c represents the resistance of the collector circuit to variations in collector current; this may be typically as high as 1 MΩ or more.

Since we have defined the current gain α in common base as I_c/I_e, the collector current $I_c = \alpha I_e$ and so the collector circuit may be depicted as a source of constant *current* αI_e in parallel with the collector a.c. resistance r_c; what we might recognize as a Norton equivalent. Some of the available αI_e generated current is 'lost' in the parallel r_c and this is another example of why the static and dynamic current gains are not the same, though the difference may be small. Only if $R_L = 0$, the static condition, does the wanted load current I_L equal αI_e.

By converting the Norton circuit into a Thévenin equivalent we can obtain an alternative form for the collector arm of the transistor. The constant current generator supplying the current αI_e shunted by r_c and having the load resistance R_L provides I_L through R_L which is, see *Figure 11.6(a)*

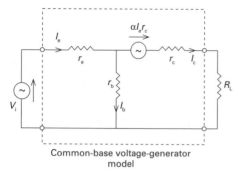

Common-base voltage-generator model

Figure 11.7

$$I_L = \frac{\alpha I_e \times r_c}{r_c + R_L} = \frac{\alpha I_e r_c}{r_c + R_L}$$

But this right-hand expression is the current I_L which would result if a constant voltage source of $\alpha I_e r_c$ were connected to a circuit made up of r_c and R_L in series, see *Figure 11.6(b)*. Therefore, so far as the load is concerned, circuits (*a*) and (*b*) are Norton–Thévenin equivalents and so therefore are the T-networks of *Figure 11.7*.

Applying Kirchhoff to the circuit of *Figure 11.7*, we obtain

$$I_e = I_b + I_c \qquad (11.1)$$

$$v_i = I_e r_e + I_b r_b \qquad (11.2)$$

$$\alpha I_e r_c = I_c(r_c + R_L) - I_b r_b \qquad (11.3)$$

From these equations we can obtain the current gain, the input resistance, the output resistance and the voltage gain ratio.

We will illustrate the use of them in solving problems with a couple of worked examples.

Example (3) A transistor having the following T-circuit parameters is used in a common-base configuration with a load resistance of 8 kΩ

$$\alpha = 0.97 \quad r_e = 20 \ \Omega \quad r_b = 400 \ \Omega \quad r_c = 1.2 \ M\Omega$$

Calculate (a) the current gain, (b) the input resistance, (c) the voltage gain for this circuit.

Solution Although the T-network equations may look clumsy, when the values of r_e, r_b r_c and α are known, direct substitution into the various equations together with some elementary rearrangement where necessary will provide the wanted solutions.

(a) The current gain $= \dfrac{I_c}{I_e}$, hence substituting values into equation (11.3) above:

$$0.97 I_e \times 1.2 \times 10^6$$

$$= I_c(1.2 \times 10^6 + 8000) - 400 I_b$$

$$= 1208 \times 10^3 I_c - 400 \ (I_e - I_c)$$

$$\therefore \quad 1164 \times 10^3 I_e + 400 I_c = 1208 \times 10^3 I_c + 400 I_c$$

$$11\,640 \times 10^2 I_e = 12\,080 \times 10^2 I_c$$

$$\therefore \quad \frac{I_c}{I_e} = \frac{11\,640 \times 10^2}{12\,080 \times 10^2} = 0.964$$

(b) The input resistance $= \dfrac{V_{eb}}{I_e} = \dfrac{v_i}{I_e}$ so from equation (11.2):

$$v_i = 20 I_e + 400 I_b$$

$$= 20 I_e + 400(I_e - I_c)$$

But $I_c = 0.964 I_e$ from answer (a)

$$\therefore \quad v_i = 420 I_e - (400 \times 0.964 I_e)$$

$$= 34.4 I_e$$

$$R_i = \frac{v_i}{I_e} = 34.4 \ \Omega$$

(c) Voltage gain $= \dfrac{V_c}{V_i} = \dfrac{I_c R_L}{I_e R_i}$

$$= \frac{0.964 \times 8000}{34.4} = 224$$

Notice from these results that the current gain worked out to be 0.964, a little less than the given α value of 0.97. The value of 0.97 is true only for static conditions when the load resistance is zero.

Example (4) A transistor having α = 0.98 is to operate in common-emitter mode with $I_e = 0.5$ mA. What is the value of the dynamic emitter resistance r_e? If now the collector load $R_L = 4.7$ kΩ and an output signal of 2 V r.m.s. is obtained, what will be the signal input current I_e and the voltage gain?

Solution For many low frequency applications, a sufficiently adequate model consists simply of an input resistance and a controlled current source αI_e, the shunting effect of r_c being ignored because of its very high value compared to the load R_L. We can then take the whole of the generated current to flow through the load. This example illustrates this.

Going forward to the next page we can find the dynamic emitter resistance by using the formula mentioned there:

$$r_e = \frac{25(\text{mA})}{I_c} = \frac{25}{0.5} = 50 \; \Omega$$

Replacing the transistor by the model of *Figure 11.5(b)* and ignoring r_c we have

$$v_o = \alpha I_e R_L$$

and the required signal current is then

$$I_e = \frac{v_o}{\alpha R_L} = \frac{2}{0.98 \times 4700} = 434 \; \mu\text{A}$$

The input voltage required is

$$v_i = r_e I_e = 50 \times 434 \times 10^{-6} = 0.0217 \; \text{V}$$

From this, the voltage gain $= \dfrac{v_o}{v_i} = \dfrac{2}{0.0217} = 92$

COMMON-EMITTER T-NETWORK MODEL

Whichever configuration a transistor is connected in, the device cannot 'distinguish' the way in which its external connections are actually made, since the physical relationships of the transistor inside its case are not changed. Hence, in the common-emitter configuration, the equivalent T-network has the same values as those for the common-base but are shifted from one part of the circuit to another.

The equivalent circuit for a transistor in common-emitter mode is drawn in *Figure 11.8(b)* which follows from the elementary amplifier circuit of *Figure 11.8(a)*. This equivalent is identical with the common-base network of *Figure 11.5(b)* except that r_e and r_b are interchanged; the base-emitter terminals now become the input port and the collector-emitter terminals the output port. This rearrangement, although appearing trivial at first encounter, leads to a complete change in the circuit performance.

Although the common-emitter equivalent of *Figure 11.7* could be derived, it is not a convenient model because the controlled source (the αI_e current source) is not now a function of the input current; we have to

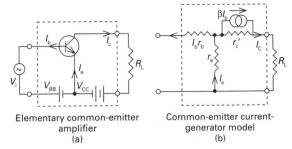

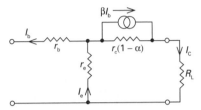

Figure 11.8

Elementary common-emitter
amplifier
(a)

Common-emitter current-
generator model
(b)

Figure 11.9

look at the collector generator as being dependent on the input quantity I_b rather than I_e. To change from the current source αI_e in shunt with r_c to one of βI_b in shunt with some new value of resistance, say r_c', we note that $I_e = I_c + I_b$ and that then making a source transformation of αI_e, the resulting voltage generator can be written as $\alpha(I_c + I_b)r_c$. But in series with this source is r_c and consequently there is a voltage drop of $I_c r_c$. If we combine the voltage rise $\alpha I_c r_c$ with the voltage drop $I_c r_c$ then the composite drop due to collector current is $I_c r_c (1 - \alpha)$ and the total rise is $\alpha I_b r_c$. Transforming these into a Norton current generator, then, gives us $\alpha I_b r_c / r_c(1 - \alpha) = I_b$ and the new resistance paralleling the source is $r_c' = r_c(1 - \alpha)$.

Figure 11.9 shows the completed equivalent circuit.

(5)   Deduce that an alternative constant-voltage form of the common-emitter equivalent circuit can be given by *Figure 11.10*. Why is this circuit not so useful as that given in *Figure 11.9*?

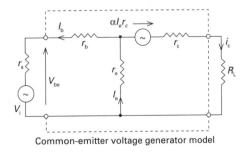

Common-emitter voltage generator model

Figure 11.10

(6)   Applying Kirchhoff to the equivalent circuit, show that:

$$I_e = I_c + I_b \qquad (11.4)$$

$$v_i = I_b r_b + I_e r_e \qquad (11.5)$$

$$I_e r_c = I_c(r_c + R_L) + I_e r \qquad (11.6)$$

Example (7)   If, in the equivalent circuit of *Figure 11.10*, the T-parameters are $r_e = 30\ \Omega$, $r_b = 400\ \Omega$, $r_c = 1\ M\Omega$, $\alpha = 0.98$, $r_s = 600\ \Omega$, find (a) the current and voltage gains of the amplifier, (b) the power gain, given that $R_L = 4.7\ k\Omega$.

Solution   Notice here that the input source has an internal resistance of $600\ \Omega$ which must be included in the input loop. In all problems of this sort, try to set down the various loop equations without referring back to those given above. In other words, having the equivalent circuit, treat it as a straightforward exercise using Kirchhoff's laws.

(a)   Going around the input loop we get

$$v_i = 1000I_b + 30I_e \qquad (i)$$

and for the output loop

$$0.98I_e \times 10^6$$
$$= I_c(10^6 + 4.7 \times 10^3) + 30I_e$$
$$= 1004.7 \times 10^3 I_c + 30I_e \qquad (ii)$$

or

$$0.98(I_c + I_b)10^6 - (1004.7 \times 10^3)I_c - 30I_e = 0$$

Collecting up and dividing through by 10^3 we get

$$980I_c + 980I_b - 1004.7I_c - 30I_e = 0$$

We can safely ignore the last term as this is negligible compared with the preceding two terms, hence

$$980I_c + 980I_b = 1004.7I_c$$

$$\therefore\ A_i = \frac{I_c}{I_b} = \frac{980}{24.7} \simeq 40$$

Going back to equation (i) and substituting $(I_c + I_b)$ for I_e, we get

$$v_i = 1030I_b + (30 \times 40I_b) = 2230I_b$$

Also

$$v_o = 4700I_c = 188\ 000I_b$$

$$\therefore\ A_v = \frac{v_o}{v_i} = \frac{188\ 000}{2230} = 84$$

(b)   Power gain $A_p = A_i \cdot A_v = 40 \times 84 = 3360$
This can be expressed in dB as 10 log 3360 $\simeq$ 35 dB.

(8)   Repeat the above example using the T-equivalent of *Figure 11.9*.

TRANSCONDUCTANCE

Because it is often useful to define the output current in terms of an input voltage, we can usefully introduce at this point a transconductance parameter g_m for a bipolar transistor. This parameter has particular features which simplify the work on the field-effect transistor also.

Since g_m is the ratio of input current to output voltage it has the dimensions of a conductance which is measured in siemens or sometimes as mA per volt, millisiemens (mS). Now the output current is βI_b and this is the input voltage multiplied by g_m.

$$\beta I_b = g_m v_i$$

where strictly $v_i = v_{be}$.

To a good approximation g_m is the reciprocal of the input resistance of the base-emitter diode which we have already seen to be expressed as $25/I_c$ (mA) ohms. Hence

$$g_m = \frac{\dfrac{1}{r_e \cdot I_e}\ (mA)}{25} = 40I_e\ mS$$

Since $I_c \simeq I_e$ we have $g_m \simeq 40I_c$ mA/V where I_c is in mA.

This is a transistor parameter which does not vary from one transistor to another but depends only on the collector current.

For many low frequency applications, a sufficiently adequate model consists only of an input resistance r_i and a controlled current source βI_b or $g_m v_{be}$.

HYBRID (H) PARAMETERS

There are three main disadvantages of T-parameters, although they do provide us with an introduction to what follows:

(a)   It is difficult to measure the T-parameters directly from relatively simple experimentation.
(b)   The Kirchhoff equations required to find currents and voltages gains and input and output

Figure 11.11

resistance tend to be cumbersome and lead to easily confused simultaneous equations.

(c)   The three circuit configurations require three different sets of equations.

What are known as hybrid or h-parameters have replaced the T-network equivalents of transistor action at low frequencies, and these parameters, which are based on standard two-port network theory, have the advantage that the expressions derived from them are applicable to all three transistor configurations. Further, the h-parameters are those most frequently quoted by manufacturers and the parameters themselves can be measured with little difficulty.

The 'black box' of *Figure 11.11* is typified by having two terminals which form the input and two others which form the output. What the box actually contains, whether transformer, amplifier or filter systems to name only three possibilities, they can be represented from the point of view of the terminals by four defining quantities. We have already covered such network 'boxes' in Chapter 2 where there was a common connection between one input and one output terminal, the internal assemblies there being passive resistive assemblies. The transistor being a 3-terminal device can here be represented in such a way, the common connection being illustrated by the broken internal line.

The four defining quantities are seen from the figure to be input and output currents i_1 and i_2, respectively, and input and output voltages v_1 and v_2, respectively. The directions in which these quantities are taken to act differ from those we have encountered so far in that there is a symmetrical arrangement of the four quantities with respect to the box terminals: currents i_1 and i_2 are both assumed to flow into the box, while voltages v_1 and v_2 are both taken to act 'upwards'. Such directions are nearly always followed in this branch of network theory. They will not make things difficult, however, because by taking proper note of the signs of the defining equations obtained relative to the assumed directions of action of the quantities, the algebra will take care of the rest. The difference between these 'new' conventions and those we used earlier for T-networks only arises because the former best suited the actual way a transistor performs. Whether a current actually flows into or out of certain

terminals or a voltage acts 'upwards' or 'downwards' in a practical case will come out with the washing as it were.

Those equivalent circuits already studied have in general been in the form of a Thévenin input circuit and a Norton output circuit. Although the correspondence is not exact, h-parameter networks bear the same sort of relation. The h-parameters are formed by the slopes of the static characteristics we looked at earlier and are used to express the output voltage and current in terms of the input voltage and current. The symbol for these characteristics is an 'h' with two letters in subscript. The first subscript refers to the particular characteristic function, e.g. an input resistance or a current gain; the second subscript designates the configuration in which the transistor is being used.

The parameters are

$$h_i = \frac{v_1}{i_1} \text{ which is the input resistance}$$

$$h_f = \frac{i_2}{i_1} \text{ which is the forward current gain or}$$
transfer ratio

$$h_o = \frac{i_2}{v_2} \text{ which is the output conductance (or}$$
admittance)

Strictly we are considering small voltage or current changes but example (9) later on will make this clear. It does not affect our present approach.

There is a fourth parameter h_r which is the reverse voltage feedback ratio v_1/v_2, but for our purpose this may be neglected as it is very small (about 10^{-5} in common emitter and common base) and has little effect upon the basic circuit behaviour.

When the second subscript is added, the configuration is established. Thus h_{ib}, h_{fb} and h_{ob} describe common-base performance, while h_{ie}, h_{fe} and h_{oe} are used for common-emitter connection.

As the actual ratios are formed, h_i will be a resistance measured in ohms, h_o will be reciprocal ohms measured in siemens, and h_f is a pure number. This is the reason these parameters, mixed as they are, are called hybrid.

The hybrid equivalent

An equivalent circuit, based on the h-parameters, is shown in *Figure 11.12(a)*. To describe completely this circuit, two equations are required: for the input loop we have a series circuit and the defining equation here consists of voltage drops and a voltage generator taken in order round the loop with due regard to

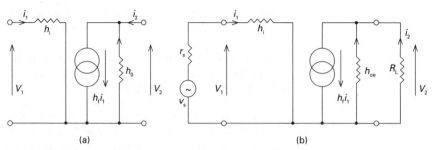

Figure 11.12

polarity. For the output loop we are concerned with currents, and here we assume that the current through h_o is in the same direction as the current produced by the generator $h_f i_1$.

Returning now to the input loop, the equation governing this loop is easily deduced:

$$v_1 = h_i i_1 \tag{11.7}$$

In the output loop we have a constant-current generator supplying current $h_f i_1$ and a parallel resistance $1/h_o$, hence the governing equation for this loop is

$$i_2 = h_f i_1 + h_o v_2 \tag{11.8}$$

Notice that these two equations differ in form from each other because the circuits they are derived from differ. In one we are dealing with loop voltages and in the other loop currents. In both circuits, however, the same two independent variables i_1 and v_2 are used.

We now add the necessary external components to the hybrid circuit, giving the equivalent shown in *Figure 11.12(b)*. This now includes the load resistance R_L, an input voltage v and its internal resistance r_s. From the basic equations above (which relate to the transistor only) we can now add further equations which relate to the external circuit:

$$v_1 = v_s - i_1 R_s \tag{11.9}$$

$$v_2 = -i_2 R_L \tag{11.10}$$

Notice the negative sign that has appeared in this last equation; this accords with the sign convention, the current flowing down through $h_f i_1$ but upwards through R_L. Also, the use of v_1, i_1, etc., instead of v_b or i_b, ensures that the equations will represent all three circuit configurations.

We can now derive a number of equations which express such things as current and voltage gains, input and output resistances, and power gain, all in terms of the signal voltages and currents and the h-parameters. For convenience, let us set down again the four basic relationships derived above:

$$v_1 = h_i i_1 \tag{i}$$

$$i_2 = h_f i_1 + h_o v_2 \tag{ii}$$

$$v_1 = v_s - i_1 R_s \tag{iii}$$

$$v_2 = -i_2 R_L \tag{iv}$$

To derive current gain, we eliminate i_2 between (ii) and (iv), this giving us

$$h_f i_1 + v_2 \left[\frac{1}{R_L} + h_o \right] = 0$$

$$v_2 = \frac{-h_f i_1}{h_o + \dfrac{1}{R_L}} = \frac{-i_1 h_f R_L}{h_o R_L + 1}$$

But

$$v_2 = -i_2 R_L$$

$$\therefore \quad -i_2 R_L = -i_1 \frac{h_f R_L}{h_o R_L + 1}$$

$$\therefore \quad \text{current gain} \quad \frac{i_2}{i_1} = A_i = -\frac{h_f}{h_o R_L + 1}$$

The current gain will always be slightly less than the direct value of either of the static values.

Now it is possible to carry on in this way and obtain expressions for voltage gain, input and output resistance, and so on, but this is not the best or desirable way to solve h-parameter problems. It leads to a tendency to rely solely on the fitting of given values into formulae and as a result nothing is learned about the manipulation of quantities which often depends on the form of wording the problem takes. It also involves slavishly remembering (or trying to remember) a number of formulae, and this procedure can never take the place of a proper understanding of the subject.

A number of worked examples follow: go through these carefully and see how progressive applications of the features and quantities of an equivalent circuit will lead to a solution without the necessity of having to remember formulae. And h-parameters have the advantage that one set of defining equations serves for all three configurations.

> Example (9)   Illustrate with simple diagrams how the h-parameters are related to the characteristic curves of a transistor, assuming a common-emitter configuration.
>
> *Solution*   Diagrams *(a)*, *(b)* and *(c)* of *Figure 11.13* show, respectively, on the static characteristic curves, the increments which define the h-parameters h_{ie}, h_{oe} and h_{fe}.

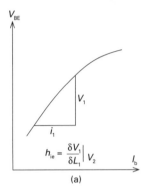

(a)

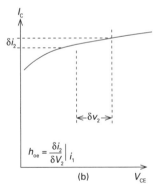

(b)

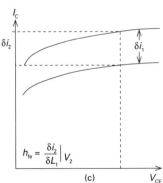

(c)

Figure 11.13

In diagram *(a)* h_{ie} is the input resistance of the transistor with the output short-circuited to the a.c. signal components, collector voltage v_2 being held constant; hence from the input characteristic shown

$$h_{ie} = \frac{\delta v_1}{\delta i_1} \text{ with } v_2 \text{ constant}$$

By a short-circuited output to the signal, we mean either that $R_L = 0$ or that the collector is returned to the common line by a large value capacitor. These methods are used under test conditions.

In diagram *(b)* h_{oe} is the output admittance with the input open-circuited to the signal components, the base current being constant; hence

$$h_{oe} = \frac{\delta i_2}{\delta v_2} \text{ with } i_1 \text{ constant}$$

The input can be open-circuited to the signal by the inclusion of a large resistor or a large reactive impedance in the form of a choke coil. Again, these methods are used under test conditions.

Finally, at *(c)*, h_{fe} is the forward current gain with the output short-circuited to the signal and v_2 held constant. Hence

$$h_{fe} = \frac{\delta i_2}{\delta i_1} \text{ with } v_2 \text{ constant}$$

Example (10)   The circuit of *Figure 11.14(a)* is a simple common-emitter amplifier with single

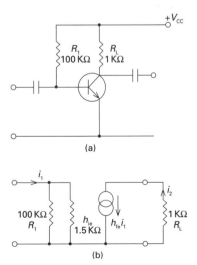

Figure 11.14

resistor base biasing. Draw an equivalent circuit in terms of the given h-parameters, neglecting the output h_{oe}. If $h_{ie} = 1.5$ kΩ, $h_{fe} = 120$ for the transistor, find (a) the a.c. input resistance, (b) the voltage gain.

Solution The equivalent circuit is shown at *(b)*. This is at its most simplified, since h_{oe} is neglected. This is justified if this parameter is assumed to be very much greater as a resistance $(1/h_{oe})$ than the load resistor R_L. However, the 100 kΩ bias resistor must be paralleled with the input resistance h_{ie}.

(a) Input resistance $\quad \dfrac{v_1}{i_1} = -\dfrac{h_{ie}R_1}{R_1 + h_{ie}}$

$$= \frac{100 \times 1.5}{100 + 1.5} = 1.48 \text{ k}\Omega$$

(b) Input voltage $\quad v_1 = i_1 r_i = 1.48 i_1$

Output voltage $\quad v_2 = -h_{fe}i_1 R_L$

$$= -120 \times 1000 i_1$$

$\therefore$ Voltage gain $A_v = \dfrac{v_2}{v_1}$

$$= -\frac{120 \times 1000 i_1}{1480 i_1}$$

$$= -81$$

(11) Does the input resistance of the above amplifier (or indeed any similar amplifier) change if V_{cc} changes?

Example (12) *Figure 11.15* and the given equations interrelate the signal currents and voltages at the terminals of a transistor in common-emitter configuration. If the signal input is 1 mV and the collector load a 1 kΩ resistor, calculate the output voltage.

$$v_1 = (10^3)i_1$$

$$i_2 = 80i_1 + (2 \times 10^{-5})v_2$$

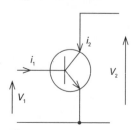

Figure 11.15

Solution From the defining equations we see that $h_{ie} = 10^3 \Omega$, $h_{fe} = 80$ and $h_{oe} = 2 \times 10^{-5}$ S.

Since 1 mV is applied to a resistance (h_{ie}) of $10^3 \Omega$, input current $i_1 = 10^{-6}$. Now

$$i_2 = -\frac{v_2}{R_L} = -\frac{v_2}{10^3} = 80i_1 + (2 \times 10^{-5})v_2$$

and substituting for i_1 gives us

$$-\frac{v}{1000} = 80 \times 10^{-6} + (2 \times 10^{-5})v$$

Multiplying through by 10^6 gives us

$$-10^3 v_2 = 80 + 20v_2$$

$\therefore \quad -1020v_2 = 80$ from which $v_2 = -\dfrac{80}{1020}$

$$= -78 \text{ mV}$$

Example (13) A transistor has the following h-parameters: $h_{ie} = 2.5$ kΩ, $h_{fe} = 50$, $h_{oe} = 20$ μS. It is used with a 5 kΩ collector load and the current in this load is 0.2 mA r.m.s. Estimate the current and voltage gains; hence find the power gain.

Solution The circuit is shown in *Figure 11.16*. Working in admittances we can write for the output loop:

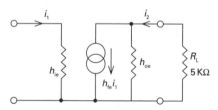

Figure 11.16

$$i_2 = -\frac{h_{fe}i_1}{R_L\left(h_{oe} + \dfrac{1}{R_L}\right)}$$

The current gain is

$$A_i = \frac{i_2}{i_1} = -\frac{\dfrac{h_{fe}}{R_L}}{h_{oe} + \dfrac{1}{R_L}} = -\frac{50/5000}{(20 + 200)10^{-6}}$$

The input signal required to develop 0.2 mA in the load is

$$i_1 = -\frac{i_2}{45.5} = -\frac{0.2 \times 10^{-3}}{45.5} = -4.4 \,\mu A$$

In the input loop

$$v_1 = i_1 h_{ie} = 4.4 \times 10^{-6} \times 2500 = 11 \text{ mV}$$

$$v_2 = -i_2 R_L = -0.2 \times 10^{-3} \times 5000 = 1 \text{ V}$$

$$\therefore \quad A_v = \frac{v_2}{v_1} = -\frac{-1}{11 \times 10^{-3}} = -91$$

Power gain is the product of current and voltage gains, hence

$$A_p = 45.5 \times 91 = 4140$$

Can you express in decibels?

Example (14)   The amplifier of *Figure 11.17(a)* uses a transistor with hybrid parameters $h_{ie} = 2 \text{ k}\Omega$, $h_{fe} = 95$, h_{oe} being negligible. What is the voltage gain between input (v_1) and output (v_o), assuming that the bias resistors and coupling capacitors have no effect on the signal component, when the transformer secondary terminals are (a) on open-circuit, (b) looking into a 100 Ω resistive load?

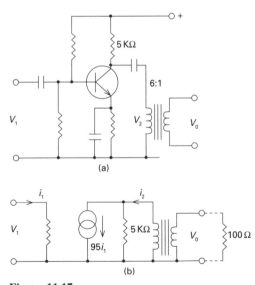

(a)

(b)

Figure 11.17

Solution   Figure 11.17(b) shows the equivalent circuit.

When the output terminals are on open-circuit, the primary winding of the transformer acts as a very high inductive impedance and has no shunting on the 5 kΩ collector load, in other words the transformer may be ignored from this point of view. It will, however, step down the voltage which appears across its primary terminals.

(a)   Setting up the appropriate equations, we get

$$v_1 = h_{ie} i_1 = 2000 i_1$$

$$i_2 = h_{fe} i_1 = 95 i_1$$

$$v_2 = 5000 i_2 = 5000 \times 95 i_1$$

$$\frac{v_2}{v_1} = \frac{5000 \times 95 i_2}{2000 i_1} = 238$$

$$\therefore \quad v_o = \frac{1}{6} \times 238 = 39.7 v_i$$

(b)   When the 100 Ω load is connected to the transformer secondary, the effective resistance seen at the primary terminals will be $6^2 \times 100 = 6400 \,\Omega$. Hence the total effective collector load will be

$$\frac{5000 \times 6400}{5000 + 6400} = 21.8 \text{ k}\Omega$$

Then $\dfrac{v_2}{v_1} = \dfrac{2.81 \times 10^3 \times 95 i_1}{2000 i_1} = 133$

$$\therefore \quad v_o = \frac{133}{6} = 16.6 v_1$$

It may appear at first that by introducing the transformer we have sacrificed a lot of the gain we could have got directly into the 5000 Ω load; but the true loading is actually the 100 Ω which, if placed directly in shunt with the 5000 Ω, would have reduced the gain to a very low value. As we saw earlier, a transformer is an impedance matching device, and in this example it was doing exactly that. Of course, by a proper choice of the turns ratio, any resistance can be matched to the requirements of the transistor output circuit.

However, there is an electronic way of matching high resistance outputs to low resistance inputs, and this will now be discussed.

The emitter-follower

In the circuit of *Figure 11.18*, the load resistor R_L is taken from its so far familiar position in the collector circuit of the common-emitter and common-base amplifiers and placed in series with the emitter. As the collector is then returned directly to V_{cc} (an a.c. earth), the configuration resulting is known as common-collector mode. This connection goes under the more familiar name of emitter-follower.

To bias an emitter-follower stage, a constant base current is required as *Figure 11.18* indicates, where this is obtained from a single resistor R_1, though the more customary potential divider method may be employed. It must be noticed, however, that in this circuit the base terminal is above the zero or earth line potential by $V_{BE} + I_E R_L$. Therefore, R_1 will have to be smaller in value than those we would have used in previous circuits and this is not desirable if a high level of input resistance is to be maintained.

Suppose a sinusoidal signal to be applied at the input terminals. As v_1 rises towards its positive peak value, the emitter current increases and the emitter voltage rises in step, relative to the earth line. The converse occurs when the input signal falls towards its negative peak value. Hence the emitter voltage 'follows' the input signal variations, and both base and emitter variations are in phase at all times. The effective base-emitter input signal to the stage, however, is the difference between v_1 and the drop across load R_L (v_2), and taken around the loop these voltages are in phase opposition. Notice particularly that v_2 is in phase with v_1 and hence the voltage gain v_2/v_1 is positive; but v_1 is phase opposed to v_2 relative to the emitter, and so the voltage gain is going to be very small.

A simple way of looking at what is happening is to consider the input and output circuits to be linked through the forward-biased base-emitter diode. We have seen earlier that the value of this resistance is given by $25/I_E$ Ω; this will always work out to be a relatively low value, generally less than 25 Ω. This

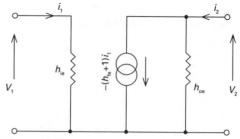

Figure 11.19

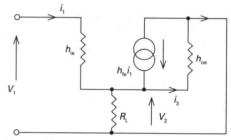

Figure 11.20

value is consequently small relative to the load R_L which may be several thousands of ohms. Since the diode resistance and R_L are effectively in series and so constitute a potential divider for the signal, the voltage gain must always be less than unity, and in practice lies between 0.95 and 0.99. So the emitter follows the base very closely.

h-parameters for the emitter-follower

For the common-collector configuration, the h-parameters are similar to those of the common-emitter. These can therefore be used directly in the equivalent circuit for common-emitter with the proviso that we replace h_{fc} by $-(h_{fe} + 1)$. *Figure 11.19* shows how we can do this, using the common-emitter parameters. It is better, however, to use the original circuit with the emitter load resistor 'lifting' this current above the earth line as *Figure 11.20* shows. We can then use the common-emitter parameters throughout and write the defining equations in terms of these.

Examples will perhaps illustrate the procedures needed.

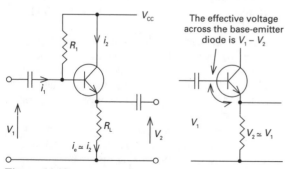

The effective voltage across the base-emitter diode is $V_1 - V_2$

Figure 11.18

Example (15)   A transistor with $h_{ie} = 1.5$ kΩ, $h_{fe} = 100$, $h_{oe} = 33$ μS, is used as an emitter-follower with an emitter load resistor $R_L = 22$ kΩ. The circuit is fed from a voltage source of internal resistance 50 Ω. Calculate the current and voltage gain.

Solution   From the circuit of *Figure 11.21* we can write the defining equations:

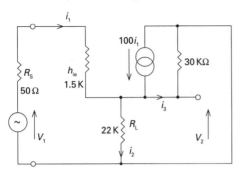

Figure 11.21

$$i_2 = 100i_1 - i_3 \tag{i}$$

$$v_1 = 1.55i_1 + 22i_2 \tag{ii}$$

$$\therefore \quad (22 + 30)i_2 - (30 \times 100)i_1 = 0$$

$$\text{or} \quad 52i_2 - 3000i_1 = 0 \tag{iii}$$

Make a note that we are working in $k\Omega$ and that $1/h_{oe} = 30 \ k\Omega$. We can now obtain the current gain straight away from equation (iii):

$$A_i = \frac{i_2}{i_1} = \frac{3000}{52} = 57.7$$

Using this result, we can write $i_1 = i_2/57.7$ and then, since $A_v = v_2/v_1$

$$A_v = \frac{i_2 R_L}{1.55i_1 + 22i_2} = \frac{R_L}{\left[\left(\frac{1.55}{57.7}\right) + 22\right]}$$

$$= \frac{22}{22.03} = 0.998$$

The emitter-follower has a high current gain and a voltage gain less than unity. The current gain is roughly identical with that we would expect from a common-emitter configuration for the same transistor; in fact, the current gain of the common-collector is $1 + h_{fe}$. Can you deduce this for yourself? Think about the very basic relationships.

Additionally, the emitter-follower has a very high input resistance, typically a megohm or so (though the shunting effect of bias resistors must be taken into account), and a very low output resistance, typically below $100 \ \Omega$. These properties make it useful as a matching device. We have seen earlier that maximum power is transferred from a source to a load if the

resistance of the load is equal to the resistance of the source. Consequently, the emitter-follower is usually preferred to the use of transformers as such resistance matching units, particularly at frequencies above some 25 kHz. Many signal sources, such as pick-ups, microphones and recorder heads, require load resistances of a high order, often a megohm or more, and feeding into a common-emitter stage does not match to this very well.

It can be shown that to a good approximation the input resistance $R_i \simeq h_{fe}R_L$ and the output resistance $R_o \simeq R_s/h_{fe}$, where R_s is the resistance of the signal source feeding the amplifier. This value of R_o has R_L effectively in shunt with it.

Bootstrapping

It is the shunting of the bias resistor(s) in the emitter-follower configuration that degrades the otherwise high input resistance that the circuit provides. By arranging the circuit in the manner shown in *Figure 11.22* the shunting effect of the bias resistors R_1 and R_2 upon the input is practically eliminated. This technique goes under the name of bootstrapping.

Capacitor C acts as a d.c. block between the emitter and base circuits but is selected to have a negligible reactance at the signal frequency. Hence there is signal feedback between the output at the emitter terminal and the input at the base. Because the voltage gain is practically unity, the signal amplitude across R_L is almost of equal amplitude to the incoming signal; in addition, it is in phase with the incoming signal. Hence the potentials at either end of resistor R_3 are approximately equal. This is equivalent to R_3 having an extremely high resistance to the signal frequency but not to the d.c. bias current flowing into the base. Hence the shunting effect of R and R_2 are isolated from the input terminals.

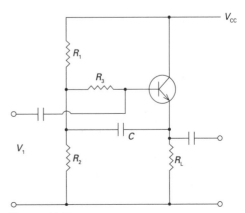

Figure 11.22

(16)   Would there be any additional advantage in making R_3 very large in the bootstrapped circuit, say several megohms?

SUMMARY

- Equivalent circuits replace non-linear devices with linear elements.
- Linear models are used to predict the performance of amplifiers in which signals are small in relation to bias values.
- Equivalent circuits represent the way the circuit system appears to signal voltages and currents, all d.c. supplies being ignored.
- For many low frequency applications, a sufficiently adequate model consists of an input resistance and a controlled current source.
- T-equivalent circuits require three different sets of equations to suit the three configurations.
- Hybrid parameters have the advantage that a standard form can be used to cover all three configurations.
- Comparisons between the configurations are as follows:

	Common-base	Common-emitter	Common-collector
Input resistance	10–300 Ω	500–2000 Ω	100 Ω–1 MΩ
Output resistance	100 kΩ–1 MΩ	5 kΩ–50 kΩ	50 kΩ–1 kΩ
Current gain	1	10–300	50–250
Voltage gain	10–300	10–300	1
Phase gain	0°	180°	0°

REVIEW QUESTIONS

1   Explain the difference between V/I and $\delta V/\delta I$.
2   In which major principle does an equivalent circuit differ from a real circuit?
3   What is the effective impedance of a d.c. power supply?

4   How can capacitors be ignored in equivalent circuits? Is there a situation where a capacitor could not be ignored?
5   What is meant by mid-band frequencies when applied to amplifiers?
6   Why are static and dynamic gain values different? Is it reasonable to treat them as being identical in certain cases?
7   Why are h-parameter equivalents more useful than T-parameters?
8   What transistor characteristic does not vary from one specimen to another but depends only on collector current?
9   Why are h-parameters so named?
10   Why is an emitter-follower possibly preferable to an iron-cored transformer for resistance matching purposes?

EXERCISES AND PROBLEMS

(17)   Deduce the Thévenin and Norton equivalent circuits for terminals P and Q in the network of *Figure 11.23*.

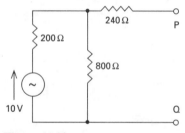

Figure 11.23

A 100 Ω load resistor is connected across P-Q. Use Thévenin to calculate the load current and Norton to calculate the load voltage. What is the load power? Would a one-eighth watt rating do for this resistor?

(18)   What restrictions are placed on the use of equivalent circuits for transistors? Draw an equivalent T-circuit for a transistor and show that if the base resistance r_b is negligible

$$i_c = \alpha i_e - \frac{v_e}{r_e}$$

(19) Show that from the h-parameter equivalent circuit of *Figure 11.12(b)* that the voltage gain A_v can be expressed by

$$\frac{-h_{fe}}{(1 + h_{oe}R_L) \times \dfrac{R_L}{h_{oe}}}$$

(20) By using Kirchhoff's relationship between i_c, i_b and i_e, show that the circuits of *Figure 11.24* are identical.

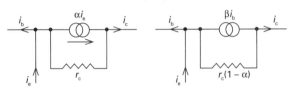

Figure 11.24

(21) In the circuit of *Figure 11.14(a)*, if R_L is reduced to 680 Ω, $h_{ie} = 1$ kΩ, $h_{fe} = 85$, find (a) the input resistance, (b) the voltage gain.

(22) In the emitter-follower circuit of *Figure 11.18*, $R_1 = 470$ kΩ, $R_L = 1$ kΩ. The parameters for the transistor are given in the defining equations:

common-emitter: $h_{ie} = 600$ Ω, $h_{fe} = 99$

common-collector: $h_{ic} = 600$ Ω, $h_{fe} = -100$

If h_{oe} and h_{oc} may be neglected, sketch the equivalent circuits for both configurations and find approximately (a) the input resistance, (b) the output resistance for each case.

(23) The small signal voltages and currents for the transistor in *Figure 11.25* are related by the equations:

$$v_1 = 2000i_1$$

$$i_2 = 40i_1 + 10^{-4}v_2$$

for a particular bias point.

The transistor is used under the same bias conditions as an collector-emitter amplifier with a collector load of 5000 Ω. A signal source having an e.m.f. of 1 mV and an internal resistance of 400 Ω is connected to the base input.

Draw an equivalent circuit, calculate the base and collector signal currents and the output signal voltage across the collector load.

(24) Given the emitter-follower circuit of *Figure 11.26* where the transistor parameters are $h_{ie} = 1.8$ kΩ, $h_{fe} = 80$, $h_{oe} = 40$ μS, calculate the mid-band current gain $A_i = v_2/v_1$, assuming that the effect of the bias resistor is negligible. Calculate also the mid-band power gain where the output power is that developed in the 10 kΩ external load, and the input power is that to the base of the transistor.

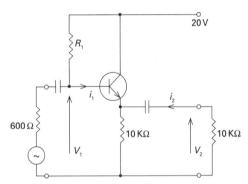

Figure 11.26

(25) Draw a complete equivalent circuit for the two-stage amplifier of *Figure 11.27*. Taking the h_{ie} value for both transistors as 800 Ω, estimate the turns ratio required of the inter-stage transformer. What approximations do you make?

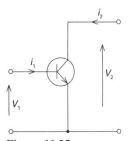

Figure 11.25

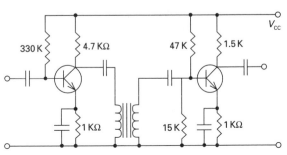

Figure 11.27

12

The unipolar transistor

A comparison of the unipolar with the bi-polar device
The construction and operation of the unipolar transistor
Depletion and enhancement modes
Output and transfer characteristics
The common-source amplifier

We turn now to the unipolar or field-effect transistor (FET). This is a device which exhibits certain characteristics that are markedly superior to those of bipolar junction transistors and which operates on the principle that the effective cross-sectional area and hence the resistance of a conducting rod of semiconductor material can be controlled by the magnitude of a *voltage* applied at the input terminals.

The FET operates upon a completely different principle from bipolar transistors. In these, the junction has been in series with the main current path from emitter to collector, and the operation of the transistor has depended on the injection of majority carriers from the emitter into the base region. There is no such injection in FETs which depend only upon one effective *p-n* junction and only one type of charge carrier. For this reason FETs are known as *unipolar* transistors.

As an amplifier the FET has a very high input impedance comparable with that of a thermionic valve, generates less noise than the ordinary transistor, has high power gain and a good high frequency performance. In addition it has a large signal handling capability, voltage swings at the input being measured in volts. At the best, base voltage swings on bipolar transistors are measured in fractions of a volt.

There are several forms of FET and these are discussed below.

THE JUNCTION GATE FET

In its simplest form, the junction gate FET (or JUGFET) is constructed as shown in *Figure 12.1*. Here

a length of semiconductor material, which may be either *p*- or *n*-type crystal, has ohmic (non-rectifying) contacts made at each end. The length of semiconductor is known as the *channel* and the end connections form the *source* and the *drain*. We shall assume throughout this discussion that the channel is *n*-type material as this form of construction is the most commonly used in practical designs.

With no voltages applied to the end connections, the resistance of the channel $R = \rho l/A$ where ρ is the resistivity of the material and l and A are the length and cross-sectional area of the channel, respectively. For example, if $\rho = 5$ ohm-cm, $l = 0.1$ cm and $A = 0.001$ cm^2, the channel resistance is 500 Ω. If the source end of the channel is effectively earthed and the drain end is taken to a positive potential, a current will flow along the channel (conventionally) from source to drain. This is drain current I_D. Clearly, if the effective cross-sectional area of the channel can be varied by some means, its resistance and hence the drain current will be brought under external control.

A means of varying the resistance of the channel is shown in *Figure 12.2(a)*. Two *p*-type regions, known as *gates*, are positioned one on each side of the channel. If these two gates are connected to each other and to the source terminal, they form reverse-biased diode *p-n* junctions with the channel crystal and

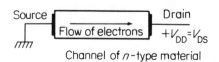

Figure 12.1

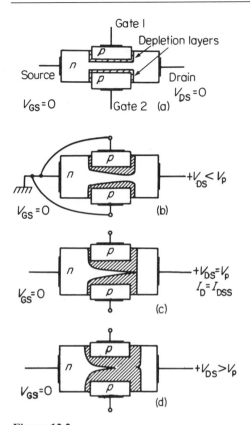

Figure 12.2

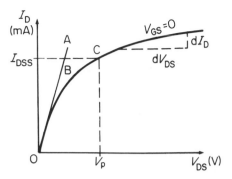

Figure 12.3

depletion layers will be established as shown in *Figure 12.2(b)*. As the channel has finite conductivity, there will be an approximately linear fall in potential along the channel from the positive drain terminal to the zero (earth) potential at the source terminal, hence the contours of the depletion layers will take the form shown in *Figure 12.2(b)*. As the channel has finite conductivity, the reverse bias voltage between channel and gate is greatest at that end. The flow of electrons from source to drain is now restricted to the wedge-shaped path shown which represents a channel of reduced cross-sectional area compared with the normal condition of diagram *(a)*.

The effective area of the channel is clearly dependent on the drain-source potential, V_{DS}, because if this potential is increased, the depletion regions will grow and eventually meet; the channel conduction area is then reduced to zero at a point towards the drain end of the channel as *Figure 12.2(c)* depicts. The channel is now said to be *pinched-off*, and the value of V_{DS} at which this occurs is known as the *pinch-off voltage*, V_p.

It is important to take note of the fact that the drain current does not cease when the drain voltage reaches pinch-off because a voltage equal to V_p still exists

between the pinch-off point and the source, and the electric field along the channel causes the carriers (electrons) in the channel to drift from source to drain. However, because of the high effective resistance of the channel, the drain current does become substantially independent of the drain voltage.

As V_{DS} is increased beyond V_p the depletion layer thickens between the gate and the drain as shown in *Figure 12.2(d)* and the additional drain voltage is effectively absorbed by the increased field in the wider pinched-off region. The electric field between the original pinch-off point and the source remains substantially unaffected, hence the channel current and so the drain current remains constant. Electrons which arrive at the pinch-off point find themselves faced by a positive potential and are swept through the depletion layer region in exactly the same way as electrons are swept from base to collector in a bipolar *n-p-n* transistor.

It is now possible to represent the relationship between the drain current I_D and the drain-source voltage V_{DS} in graphical form. In *Figure 12.3* the line OA represents the behaviour of the channel acting simply as a semiconductor resistor; the drain current I_D would follow this line for increases in V_{DS} if the resistance of the channel was constant, unaffected, that is, by the effect of the increased drain voltage upon the width of the depletion regions. At point B the action of the depletion layers begins to take effect and there is a departure from the linear characteristic. At point C the applied V_{DS} reaches V_p and pinch-off occurs. For values of V_{DS} greater than V_p the channel remains pinched off and I_D becomes virtually constant and independent of V_{DS}. This characteristic curve is for the particular case when the gate-source potential is zero and is so indicated on the diagram. The drain current which flows when $V_{GS} = 0$ and $V_{DS} = V_p$ is indicated as I_{DSS}, representing the saturated (constant) drain current with the input short-circuited, i.e. gates connected to source.

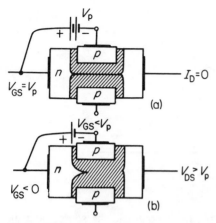

Figure 12.4

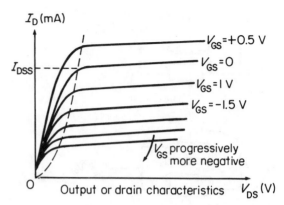

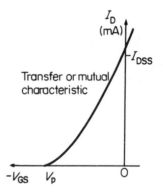

Figure 12.5

Suppose now that the gates are negatively biased relative to the source instead of being at zero potential. The depletion regions will clearly be thicker for a given value of V_{DS} than they were when there was no such negative bias on the gates. As a result, pinch-off and saturation drain current will occur at lower values of V_{DS}, and when V_{GS} is sufficiently negative, the electric field between the source and the original pinch-off point will be eliminated and drain current will cease to flow. This situation is illustrated in *Figure 12.4(a)*, and this diagram should be compared to the pinch-off condition of $V_{GS} = 0$ in *Figure 12.2(c)*.

In *n*-channel FETs the gate must be negative with respect to the source to achieve pinch-off. The reverse polarity is necessary for a *p*-channel device. The sign of V_p is therefore as follows:

n-channel $V_p < 0$
p-channel $V_p > 0$

In operation as an amplifier the FET is biased so that $V_{DS} > V_p$ and $V_{GS} < V_p$. This condition is illustrated in *Figure 12.4(b)*.

Characteristic curves

There are two characteristic curves of interest to use here, the *output* or *drain characteristic* and the *transfer* or *mutual characteristic*.

We have already touched on the form of the output characteristic in discussing *Figure 12.3* and we can now extend that diagram to include the effect of an increasing negative bias on the gate of the transistor. This has been done in *Figure 12.5* and a family of curves relating I_D to V_{DS} for different values of V_{GS} has been obtained. Notice from this diagram that it is possible to operate the junction FET with a small positive gate bias. The superficial similarity of these

Figure 12.6

curves to the output characteristics of bipolar transistors should be noted.

The broken line in the diagram represents the locus of a point passing through the respective pinch-off locations on each separate curve. To the left of this line the FET behaves as a variable resistor and this region is known as the *ohmic* region. To the right of the line the FET behaves as a constant current generator and exhibits a very high output resistance, just as the bipolar transistor does. As mentioned, the FET is normally operated in this saturation region where values of V_{DS} exceed V_p.

The transfer or mutual characteristic relates the dependence of drain current I_D upon the gate-source voltage V_{GS} and a typical characteristic is shown in *Figure 12.6*. There is actually a family of these curves, each relating to a particular value of V_{DS}, but as I_D is practically constant beyond the pinch-off point, the curves are almost coincident with one another. To avoid confusion, only a single curve is shown in the figure. Notice that the particular value of I_D for which V_{GS} is zero is I_{DSS}, in accord with the output characteristics of *Figure 12.3* and *Figure 12.4*.

The pinch-off voltage V_p appears on this characteristic as well as on the output characteristics because the channel can be pinched-off by applying V_p between gate and source in the reverse direction. The depletion layers then meet along the whole length of the channel and $I_D = 0$. Strictly, a small leakage current continues to flow along the channel which varies with temperature; this temperature dependence of V_p is due entirely to the variation of barrier potential at the p-n junctions, the same effect that causes variation of V_{BE} with temperature in junction transistors. We shall return to problems of temperature effects a little later on.

FET parameters

In order to analyse the amplifying properties of the junction FET, we borrow two parameters from thermionic valve theory. These parameters are mutual conductance and drain slope resistance.

1. Mutual conductance g_{fs} (or g_m) is defined

$$g_{fs} = \frac{\text{the small change in } I_D}{\text{the small change in } V_{GS}} \text{ for a constant } V_{DS}$$

$$= \left. \frac{\delta I_D}{\delta V_{GS}} \right| V_{DS} = k$$

This parameter, relating the dependence of drain current I_D to the gate-source voltage V_{GS}, represents the gradient of the transfer characteristic of the FET.

As g_{fs} represents the ratio of a current to a voltage, its unit of measurement will be the siemen. For a junction FET, g_{fs} will range typically from 10^{-2} to 10^{-3} S.

The equation of the characteristic is complicated, but to a good approximation I_D varies as the square of V_{GS} so that the curve is closely parabolic. It can be shown experimentally that the relationship between I_D and V_{GS} can be expressed in the form

$$I_D \simeq I_{DSS} \left[1 - \frac{V_{GS}}{V_p} \right]^2 \tag{12.1}$$

Notice from this that when V_{GS} is zero, $I_D = I_{DSS}$ as it should.

Example (1)   Using the equation just stated, derive an expression for the mutual conductance of a FET. What is the mutual conductance for the case where $V_{GS} = 0$?

Solution   Since $g_{fs} = dI_D/dV_{GS}$, the mutual conductance can be obtained by differentiating equation (12.1), noting that V_p is constant.

Then

$$g_{fs} = 2I_{DSS} \left[1 - \frac{V_{GS}}{V_p} \right] \left[-\frac{1}{V_p} \right]$$

$$= \frac{-2I_{DSS}}{V_p} \left[1 - \frac{V_{GS}}{V_p} \right] \text{ siemens}$$

The value of g_{fs} obtained at zero bias, or when $V_{GS} = 0$, is denoted by g_{fso} and this follows at once from the previous result:

$$g_{fso} = -\frac{2I_{DSS}}{V_p} \text{ siemens}$$

Notice that g_{fs} or g_{fso} will *always* be positive since either V_p or I will be negative, whether the transistor is n- or p-channel. The gradient of the mutual characteristic is clearly positive as a glance will show.

(2)   Can you see why the expression for g_{fso} above would be useful in a practical method of measuring the g_{fs} of a transistor; (Hint: substitute g_{fso} back into the equation for g_{fs}, and then think about how you would measure the various parameters.)

Example (3)   The output characteristics for the 2N5457 FET are shown in *Figure 12.7*. Using these curves, derive the mutual characteristic for this transistor for $V_{DS} = 10$ V. Hence estimate the mutual conductance of the 2N5457 at a gate bias of -0.5 V.

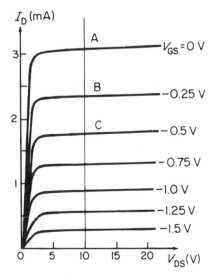

Figure 12.7

Solution   This example will illustrate how the mutual characteristic can be obtained from the output characteristics. From *Figure 12.7* we notice that V_{GS} ranges from zero to −1.5 V while I_D ranges from zero to just over 3 mA. Accordingly we draw the axes for the required mutual characteristic to cover the same ranges. These are shown in *Figure 12.8*.

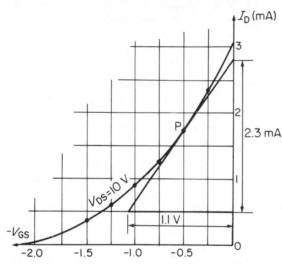

Figure 12.8

Referring now to the $V_{DS} = 10$ V point on the output horizontal axis, we make a note of the I_D values where a line erected vertically from this point cuts the V curves; these are the points A, B, C, etc. These I_D values are then plotted on the mutual characteristic axes against the corresponding values of V_{GS}. Connecting the points so obtained, the required characteristic curve appears as the diagram of *Figure 12.8* shows. To find g_{fs} at the point where $V_{GS} = -0.5$ V (point P), we draw a tangent to the curve, and then

$$g_{fs} = \frac{\delta I_D}{\delta V_{GS}} = \frac{2.3 \times 10^{-3}}{1.1} = 2.1 \times 10^{-3} \text{ S}$$

$$= 2.1 \text{ mS}$$

As in valve theory, g_{fs} is sometimes expressed in mA per volt; the above solution might equally well be written as 2.1 mA/V. This then tells us that the drain current changes by 2.1 mA when the gate voltage changes by 1 V on a bias of −0.5 V.

We come now to the second of the FET parameters:

2. Drain slope resistance r_d (or r_{ds}) is defined

$$r_d = \frac{\text{the small change in } V_{DS}}{\text{the small change in } I_D} \text{ for a constant } V_{GS}$$

$$= \left. \frac{\delta V_{DS}}{\delta I_D} \right|_{V_{GS} = k}$$

This parameter, relating the dependence of drain current I_D on the drain-source voltage V_{DS}, represents the reciprocal of the gradient of the output characteristic in the saturation region. *Figure 12.3* shows the meaning of drain slope resistance. You should observe that the gradients of the drain characteristics are, like the output curves of a bipolar transistor, very flat in the saturation region and hence the output resistance is very high.

THE JUGFET AS AN AMPLIFIER

Except for the rather different biasing required, the junction FET described above can be used in any of the transistor circuits already discussed. The drain, gate and source are loosely equivalent to the collector, base and emitter of the bipolar transistor, or to the anode, grid and cathode of a thermionic valve. There are, however, two very important differences between the FET and the ordinary transistor which must be emphasized.

First, the control of current flow through the FET is by way of V_{GS}, that is, a *voltage* control and not a current control. Because the input junctions are reverse biased, the gate leakage current is negligible and the input resistance of the FET is correspondingly very high, of the order of 10^3 to 10^6 megohms. Although the input resistance decreases rapidly when the junctions become forward biased, the input resistance remains relatively high (a megohm or more) in a silicon device provided the forward bias does not exceed some 0.5 V at ordinary temperatures. The FET can therefore be operated as a small signal amplifier with $V_{GS} = 0$.

Secondly, the FET is a majority carrier device; in the *n*-channel form only *electrons* are the carriers drifting from source to drain. In the *p*-channel form only *holes* are the carriers drifting from source to drain, V_{DS} now being of reversed polarity. In both respects, the *n*-channel FET resembles the thermionic valve much more closely than it does a bipolar transistor.

We shall be interested here only in the *common-source* configuration which is the circuit equivalent of the common-emitter connection.

It is useful at this stage to relate the two FET parameters g_{fs} and r_d to a third parameter. We have noted that as far as I_D is concerned, its control can be brought about either by varying V_{DS} (the output characteristic) or V_{GS} (the mutual characteristic). The gate voltage V_{GS} exerts a much greater influence on I_D than does the drain voltage V_{DS} for a given variation of potential. Suppose a small increase in V_{DS} causes an increase in I_D, and that I_D can then be restored to its original value by a small negative change in V_{GS}. The ratio of these two changes in the drain and gate voltages which produce *the same change* in I_D is called the *amplification factor* of the FET and is symbolized μ. So

μ = the small change in V_{DS}/the small change in V_{GS} producing the same change in I_D

$$= \frac{\delta V_{DS}}{\delta V_{GS}} \bigg|\ I_D = k$$

Since μ is the ratio of two voltages, it is simply expressed as a number. Amplification factor is related to g_{fs} and r_d because the product

$$g_{fs} \times r_d = \frac{\delta I_D}{\delta V_{GS}} \times \frac{\delta V_{DS}}{\delta I_D} = \frac{\delta V_{DS}}{\delta V_{GS}} = \mu$$

Hence

$$\mu = g_{fs} \cdot r_d \tag{12.2}$$

The circuit, and the circuit symbol, for an *n*-channel FET common-source amplifier is shown in *Figure 12.9*. The input signal V_i is applied between gate and earth, and the drain circuit contains a load resistor R_L across which the output voltage V_o is developed by the flow of drain current. A resistor R_S is included in the source lead and the drain current also flows through this. A voltage equal to $R_S \cdot I_D$ is therefore developed across R_S and the source terminal is raised by this potential above the earth line. As the gate is connected

to earth through resistor R_G the source is effectively positive with respect to the gate, or, what is the same thing, the gate is *biased negatively* with respect to the source. There is no d.c. voltage developed across R_G because the gate current is negligible; at the same time R_G presents a high impedance to the a.c. input signal. Notice that the drain supply voltage is designated V_{DD}.

Suppose the gate voltage to move positively by a small amount due to the input signal; then I_D will increase and there will be an increased voltage drop across R_L. Hence the drain voltage V_D will fall. Like the bipolar common-emitter amplifier, the common-source FET amplifier introduces a 180° voltage phase change. This negative change in drain voltage in turn causes a further change in drain current, *separate from* but *simultaneous* with the change in drain current due to the original change in gate voltage. The *total change* in I_D is therefore the sum of two simultaneous changes in V_i and V_D. Let a change d V_i in input voltage cause a change dI_D in drain current. Then by definition of g_{fs}

$$\delta I_D = g_{fs} \cdot \delta V_i$$

This change in I_D will in turn produce a change dV_D in drain voltage, so by definition of r_d

$$\delta I_D = \frac{1}{r_d} \cdot \delta V_D$$

But $\delta V_D = -I_D \cdot R_L$, the negative sign indicating that dV_D falls as I_D increases. The total change in I_D is therefore

$$\delta I_D = g_{fs} \cdot \delta V_i - \frac{1}{r_d} \cdot \delta I_D \cdot R_L$$

Rearranging:

$$\delta I_D = \frac{g_{fs} \cdot r_d}{r_d + R_L} \delta V_i = \frac{\mu}{r_d + R_L} \cdot \delta V_i$$

But the output voltage change across R_L is d$V_o = -dI_D \cdot R_L$ hence

$$\delta V_o = \frac{\mu R}{r_d + R_L} \cdot \delta V_i$$

$$\therefore \quad \frac{\delta V_o}{\delta V_i} = A_v = -\frac{\mu R_L}{r_d + R_L} \tag{12.3}$$

which is an expression for the voltage gain of the FET common-source amplifier.

If we assume that r_d is very much *greater* than R_L (which is true in many amplifiers), the expression for voltage gain approximates to

$$A_v \simeq -\frac{\mu R_L}{r_d} \simeq -g_m \cdot R_L$$

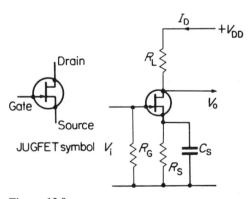

Figure 12.9

Since the signal input current to a FET is negligible, the current gain is very high but of little importance, and we shall not pursue it further.

Example (4)   In a common-source amplifier, the parameters for the FET are $g_{fs} = 2.5$ mS, $r_d = 100$ kΩ. If the load resistor is 22 kΩ, calculate the voltage gain of the amplifier.

Solution   From equation (6.3)

$$A_v = -\frac{\mu R_L}{r_d + R_L}$$

But

$$\mu = g_{fs} \times r_d = 2.5 \times 10^{-3} \times 100 \times 10^3$$

$$= 250$$

$$\therefore \; A_v = -\frac{250 \times 2200}{12\,2000}$$

$$= -45$$

It is now possible to derive an equivalent circuit for the FET amplifier, for equation (12.3) connects voltage output V_o, a constant voltage source of e.m.f. $\mu \cdot V_i$ volts, and a resistance made up of r_d and R_L in series. Rewriting the equation gives us

$$V_o = -\mu \cdot V_i \times \frac{R_L}{r_d + R_L}$$

and this can be represented by the circuit shown in *Figure 12.10(a)*. The FET is replaced by a *constant-voltage* generator whose e.m.f. is $-\mu \cdot V_i$ and this sends current through a resistance made up of r_d and R_L in series, the output voltage V_o being that fraction of $-\mu \cdot V_i$ developed across R_L.

As the FET behaves as a constant-current generator, however, it is often better to use a constant-current

form of equivalent circuit and this can be done by expressing equation (12.3) in the form

$$V_o = -g_{fs} \cdot V_i \frac{r_d \cdot R_L}{r_d + R_L}$$

where μ has been replaced by the product $r_d \cdot g_{fs}$. This now expresses V_o in terms of a constant current $g_{fs} \cdot V_i$ flowing in a circuit made up of r_d and R_L in parallel. This circuit replaces the FET by a generator which provides a *constant-current* output feeding a parallel arrangement of r_d and R_L. This circuit is shown in *Figure 12.10(b)*. We have assumed for both forms of equivalent circuit that the input resistance of the FET is infinitely high and that the device capacitances are negligible. That is, we are considering a low-frequency equivalent circuit representation.

(5)   A voltage gain of 20 is required from a FET having $g_{fs} = 2.0$ mS and $r_d = 50$ kΩ. What should be the value of the drain load resistor to provide this gain?

Example (6)   In the amplifier of *Figure 12.9*, $V_p = -2$ V and $I_{DSS} = 2.0$ mA. The circuit is to be biased so that $I_D = 1.2$ mA, the drain supply V_{DD} being 20 V. Estimate (a) V_{GS}; (b) mutual conductance g_{fs}; (c) source bias resistor R_S; (d) the required value of R_L to give a voltage gain of 15.

Solution
(a)   From our basic equation

$$I_D \simeq I_{DSS} \left[1 - \frac{V_{GS}}{V_p} \right]^2$$

we have by arrangement

$$V_{GS} \simeq V_p \left\{ 1 - \left[\frac{I_D}{I_{DSS}} \right]^{\frac{1}{2}} \right\}$$

Check this for yourself before going on. Now inserting the given values we obtain

$$V_{GS} \simeq 2 \left\{ 1 - \left[\frac{1.2}{2} \right]^{\frac{1}{2}} \right\} \simeq -0.45 \text{ V}$$

(b)   Here we can use the equation

$$g_{fs} = -\frac{2 I_{DSS}}{V_p} \left[1 - \frac{V_{GS}}{V_p} \right]$$

and inserting values

$$g_{fs} = -\frac{2 \times 2}{-2} \left[1 - \frac{0.45}{2} \right] = 1.55 \text{ mS}$$

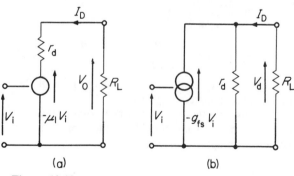

(a)   (b)

Figure 12.10

LOAD LINE ANALYSIS

(c)   By Ohm's law

$$R_S = \frac{-V_{GS}}{I_D} = \frac{0.45 \times 10^3}{1.2}$$

$$= 375 \ \Omega$$

(d)   We have to use the approximate expression for voltage gain here as we do not know the values of r_d and μ. Assuming that r_d is very large compared with R_L, we have

$$A_v = -g_{fs} \cdot R_L$$

or

$$-15 = -1.55 \times 10^{-3} \times R_L$$

from which

$$R_L = 9.7 \ k\Omega$$

A 10 kΩ resistor would be used here.

(7)   *Figure 12.11* shows a mutual characteristic curve for a FET. Prove that if the gradient of the curve at the point where V_{GS} is 0 is continued as the diagram shows, it will cut the horizontal axis at the point $V_p/2$.

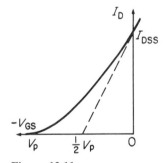

Figure 12.11

(8)   The following values are taken from the linear portions of the static characteristics of a FET:

V_{DS}	15	15	10	volts
I_D	13.5	10.5	12.7	mA
V_{GS}	−0.5	−1.0	−0.5	volts

Calculate the parameters r_d, g_{fs} and μ for this FET.

When a load resistor is placed in the drain output circuit to provide a signal voltage output, a load line can be drawn across the output characteristics as it was for the bipolar transistor amplifier. *Figure 12.12* shows a typical graph of this sort, the gradient of the load line depending on whether an a.c. or a d.c. load condition is being considered. The d.c. load line has a gradient given by $-1/(R_L + R_S)$ and cuts the horizontal axis at V_{DD}, since for $I_D = 0$ the drain voltage must rise to the applied voltage. The other end of the line assumes that the volts drop across the FET is zero at saturation, hence the drain current will be $V_{DD}/(R_L + R_S)$, R_L and R_S being effectively in series across V_{DD}.

The a.c. load line has a gradient given by $-1/R_L$ since R_S is effectively short-circuited by capacitor C_S (see *Figure 12.9*) at signal frequencies, and so cuts the horizontal axis at $(V_{DD} - V_S)$. For Class A operation both load lines must pass through the operating point P which lies on the V_{GS} characteristic corresponding to the d.c. gate-to-source voltage developed across R_S. The procedure for determining the load lines is, in other words, similar to that already described for bipolar transistors and the method of using them to obtain information on the amplifier performance is also identical. The procedure can therefore be summarized here as follows:

(a)   The input signal excursions applied to the gate terminal of the amplifier centred on the operating point P must be such that the peaks do not intrude beyond the knee of the $V_{GS} = 0$ characteristic (saturation) or below the horizontal axis limit (cut-off).

(b)   The d.c. output drain voltage and drain current at quiescence determine the position of P and this must be chosen with regard to the conditions outlined in (a) above. P will normally be

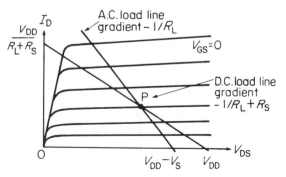

Figure 12.12

positioned so that the signal swings over characteristic curves which are linear and equally spaced.

THE INSULATED GATE FET

It is the high input resistance of the junction FET which makes it a particularly attractive device in many applications. If an extremely high input resistance is necessary, another type of FET may be employed. This is the metal-oxide semiconductor FET or MOSFET, sometimes also referred to as a MOST or insulated-gate FET, or IGFET.

This device differs from the junction FET in that the gate is actually insulated from the channel by a very thin (about 100 nm) layer of oxide insulation, usually silicon dioxide (glass). The input resistance is then typically in the range 10^6 to 10^8 megohms. MOSFETs are described under two general types: the *depletion* type and the *enhancement* type.

Figures 12.13(a) and (b) illustrate the constructional features of the MOSFET. In both forms the gate and channel form the two plates of a capacitor separated by the thin silicon dioxide layer. Because the gate is insulated in this way, V_{GS} can be either positive or negative with respect to the channel without conduction taking place through the gate-channel circuit. Any potential applied to the gate establishes a charge on the gate and this induces an equal but opposite charge in the channel.

In *Figure 12.13(a)*, when the gate potential is positive, a negative charge is induced in the *p*-type substrate at its interface with the silicon dioxide dielectric. This charge repels the majority carriers (holes) from the surface of the substrate and the minority carriers (electrons) that remain form an *n*-channel 'bridge' which connects together the existing *n*-type source and drain electrodes. Consequently, when the drain is connected to a positive supply voltage V_{DD}, electrons flow from source to drain by way of the induced *n*-channel. Increasing the gate-source voltage V_{GS} *positively* widens or *enhances* the induced channel and the flow of drain current increases. For this reason this form of MOSFET is known as an *n*-channel enhancement FET.

When drain current flows along the channel, a voltage drop is established and this tends to cancel the field set up across the dielectric by the positive gate bias. (Think about the workings of the JUGFET at this point.) When the cancellation is sufficient almost to eliminate the induced *n*-channel layer, the channel pinches off and the drain current saturates at a value which is substantially independent of any increases in drain voltage. The output and mutual characteristic curves for an *n*-channel enhancement mode transistor are shown in *Figure 12.14* together

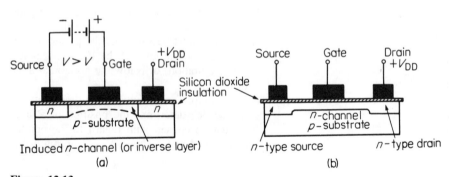

Figure 12.13

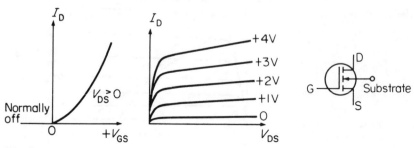

Figure 12.14

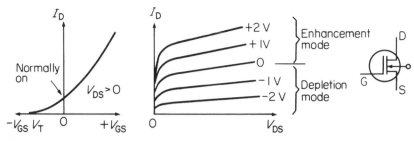

Figure 12.15

with the circuit symbol. On the transfer characteristic the value of the gate-source voltage at which drain current falls to zero is called the threshold voltage V_T. It is that voltage which *just* induces the n-channel at the surface of the p-type substrate. The curve does not cross the I_D axis. The enhancement FET is a normally OFF device.

---

(9) Sketch on appropriately marked axes the characteristic curves of a p-channel enhancement type MOSFET.

Figure 12.13(b) shows the construction of an n-channel depletion mode MOSFET. Here the n-channel is introduced during manufacture by n-type doping of the surface layer of the substrate between the n-type source and drain regions. The construction is then essentially similar to the JUGFET except that the gate is insulated from the channel by the oxide layer. As a result, current flows from source to drain with zero gate voltage when a positive voltage is applied at the drain. If the gate is made *negative* with respect to the channel, the n-channel width is reduced (or depleted) as electrons are expelled from its interface with the oxide dielectric, and drain current decreases. The transistor is then operating in the depletion mode, just as the JUGFET does. Here, however, the gate voltage may be made positive, in which case the channel width is enhanced and drain current increases. Hence this type of MOSFET will operate in either the depletion or enhancement modes. The depletion FET is a normally ON device.

Figure 12.15 shows the output and mutual characteristics of a transistor of this kind. Notice that the threshold voltage now has a negative value (as for the JUGFET) and that the transfer curve crosses the I_D axis, where $V_{GS} = 0$ and $I_D = I_{DSS}$. The symbol for the depletion MOSFET is shown to the right of the diagram. The channel is now represented by a full line.

SUMMARY

- The field-effect transistor is a three-terminal device, these terminals being known as the source, gate and drain.
- FETs are characterized by their high input resistance.
- In the FET, the channel conductivity is controlled by the gate voltage.
- FETs operate with only one type of charge carrier.
- In the constant current region, the mutual characteristic has a parabolic form.
- In the MOSFET device, the gate is insulated from the channel by a thin oxide layer.
- The density of the mobile carriers in the channel may be depleted or enhanced.
- MOSFETs are described as being either depletion or enhancement types.

REVIEW QUESTIONS

1 What are the functions of the source, gate and drain in a JFET?
2 Sketch a cross-section of a JFET at pinch off.
3 What is meant by pinch off in an FET?
4 In the name Field-Effect Transistor, what kind of field is actually concerned and what is its effect?
5 Why does drain current still flow even when the FET is pinched off?
6 Why are FETs called unipolar transistors?
7 Why is the resistance of an FET very high compared with a bipolar transistor?
8 What is (a) the mutual conductance, (b) the amplification factor of an FET in its

application as an amplifier? How are these characteristics related?

9 Make sketches of the form of the mutual characteristic for (a) the JFET, (b) the depletion MOSFET, (c) the enhancement MOSFET.

10 Explain the operation of an enhancement type FET.

EXERCISES AND PROBLEMS

(10) Sketch the construction and describe the principle of operation of a junction field-effect transistor. Explain what is meant by pinch-off and show its effect on the drain characteristics.

Draw a circuit diagram of an automatically biased junction FET used in a Class A amplifier in the common-source mode. (C. & G.)

(11) Describe the principle and operation of either of the following insulated-gate field effect transistors: (a) depletion type, (b) enhancement type. Sketch the symbols for these devices and draw typical gate/drain characteristic curves for both types.

(12) Define the parameters: (a) mutual conductance g_{fs}; (b) drain slope resistance r_d; (c) amplification factor μ. Given the relevant characteristic curves of a FET, how would you determine these parameters?

(13) A FET has a transconductance (mutual conductance) of 3 mS. If the gate-source voltage changes by 1 V, what will be the change in the drain current? What assumption have you made in obtaining an answer?

(14) When the drain voltage of a FET is reduced from 20 V to 10 V the drain current falls from 2.5 mA to 2.35 mA. Assuming the gate-source voltage remains unchanged, what is the drain slope resistance of the FET?

(15) The following values are taken from the linear portions of the static characteristics of a certain FET:

V_{DS}	12	12	8	volts
I_D	6.2	2.9	5.9	mA
V_{GS}	−1	−2	−1	volts

Estimate the parameters r_d, g_{fs} and μ.

(16) Complete the following statements:

(a) In general terms, FETs have properties most similar to . . .

(b) A JUGFET can operate only in the . . . mode.

(c) The V_{GS} polarity for an enhancement mode FET is dependent on. . . .

(d) I_{DSS} is the drain current which flows when V_{GS} is. . . .

(e) In a JUGFET majority carriers flow from . . . to. . . .

(17) A FET has parameters g_{fs} = 2.0 mS, r_d = 50 kΩ. What is its amplification factor? If this FET is used as a common-source amplifier with a drain load of 10 kΩ, calculate the voltage gain.

(18) A FET used in common-source mode has g_{fs} = 2.0 mS, r_d = 100 kΩ. What value of load resistor will provide a voltage gain of 28 dB?

(19) In the circuit of Figure 12.9, V_p = −2 V and I_{DSS} = 2 mA. It is desired to bias the circuit to a drain current of 1.1 mA. If the drain supply is 15 V, estimate (a) V_{GS}, (b) g_{fs}, (c) source resistor R_S, (d) the required value of R_L to provide a voltage gain of 10.

(20) The output characteristics of a FET are given in the table. Plot these characteristics and determine the drain-source resistance from the characteristic for V_{GS} = −2 V.

V_{DS} (V)	I_D (mA)			
	V_{GS} = 0	−1.0	−2.0	−3.0
2	5.6	4.15	3.0	1.90
6	7.0	5.25	3.62	1.91
9	8.1	6.0	4.1	1.92

Use the curves to estimate the g_{fs} of the FET for V_{DS} = 5 V.

(21) In the common-source amplifier of Figure 12.16, the quiescent conditions are such that V_{GS} = −1.0 V and the drain current I_D = 4 mA. Assuming the input resistance of the FET itself is infinite, obtain values of R_1 and R_2 such that the effective input resistance of the amplifier is 0.75 MΩ. What will be the approximate gain of the stage if g_{fs} = 5 mS? (Hint: R_1 and R_2 are effectively in parallel with the input terminals.)

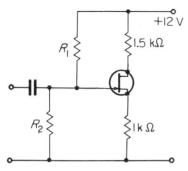

+12 V

R_1

1.5 kΩ

R_2

1kΩ

Figure 12.16

(22)   The equation for the drain current I_D of a FET is

$$I_D = 7.5\left[1 - \frac{V_{GS}}{-2.5}\right]^2 \text{ mA}$$

for a V_{DS} of 9 V. Find I_D for each of the following values of V_{GS}: 0, −0.5, −1, −1.5, −2 and −2.5 V. Plot the mutual characteristic for this FET and estimate g_{fs} when $V_{GS} = -1.5$ V.

(23)   A FET is used as a common-source amplifier with a parallel-tuned resonant circuit as the drain load. The tuned circuit consists of an inductor of 400 μH and resistance 20 Ω in parallel with a 300 pF capacitor. Calculate the resonant frequency.

If the transistor parameters are $g_{fs} = 1.5$ mS, $r_d = 20$ kΩ, calculate the voltage gain of the amplifier.

13

D.C. power supplies

We have covered the rectifying properties of diodes and their basic applications in an earlier chapter and seen how a d.c. power supply can be obtained from an alternating source, in general the domestic mains supply. The ripple from the rectified outputs of the various diode arrangements is reduced to a very low value by the smoothing circuits discussed and for many pieces of electrical and electronic equipment such a supply is adequate. *Figure 13.1* will remind us of the nature of this very basic d.c. unit.

However, many electronic systems impose more stringent conditions for their operating supplies than those given by the circuit of *Figure 13.1*. For example, computer systems, d.c. amplifiers and stable oscillators require some additional features and requirements to the provision of their operating power supplies.

The reasons for this can be summarized as follows:

(a) The output voltage changes if the alternating supply input varies;
(b) The output voltage changes if the loading varies;
(c) The 100 Hz ripple component in the output (which remains finite in spite of the smoothing system) tends to increase as the current drawn by the loading increases.

The basic circuit, then, of *Figure 13.1* is said to be *unstabilized* against these usually inevitable variations. A *stabilized* or *regulated* power supply unit, on the other hand, is designed to reduce any remaining ripple

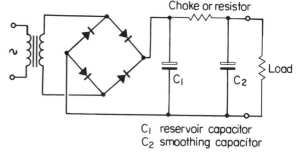

Figure 13.1

components to a minimum and to provide a stable voltage output regardless of mains voltage and load current variations. These conditions bring to mind the theoretically desirable constant-voltage generator discussed in Chapter 2, where a source of e.m.f. was available having a zero internal resistance.

There are two main forms of voltage regulation which provide us with very good approximations to this ideal: shunt regulation and series regulation, and we will investigate these and the component systems they require in this chapter. For the most fundamental form of shunt regulation we return to the humble diode.

THE ZENER DIODE

If the reverse voltage applied to an ordinary diode is increased sufficiently, a sudden increase in reverse

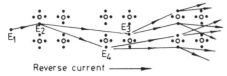

Figure 13.2

current will be observed. The diode is no longer a one-way device but becomes at this point capable of conduction in the reverse direction. This comes about because of two mechanisms associated with the atomic structure and the degree of impurity doping of the semiconductor material from which the diode is fabricated; these two conditions are known as the *avalanche* effect and the *zener* effect. The zener effect occurs at voltages *lower* than those causing the avalanche effect and is actually controllable by adjustment of the impurity doping of the material during manufacture. We look at these two effects in turn.

The avalanche effect is caused by the minority carriers being accelerated by the applied field. *Figure 13.2* shows part of the crystal lattice of the semiconductor material with four atomic nuclei each with their four valence electrons. When a sufficiently high reverse voltage is applied, an electron E_1, say, is speeded up through the lattice with enough energy to dislodge a valence electron E_2. This free electron now comes under the influence of the applied potential and is similarly accelerated. In turn, these two electrons collide with further valences E_3 and E_4 and break the covalent bonds. There are now four electrons available to continue the effect of dislodging further electrons in the same way; this kind of chain reaction builds up in an extremely short period of time to an avalanche of electrons throughout the lattice and so into the external circuit. This is the breakdown point associated with ordinary diodes and, in general terms, occurs at quite high voltage levels.

The so-called zener effect (named after C. Zener, who first explained the phenomenon in 1934!) is caused by the intensity of the electric field established across the diode junction, much as an electric field is set up across the dielectric of a charged capacitor and places a stress across the dielectric. The covalent bonds of the semiconductor materials are then broken and the reverse current becomes very large. This is the breakdown point in zener diodes. As mentioned above, doping reduces the thickness of the depletion layer in these diodes and the breakdown point occurs at a lower voltage than does the avalanche effect.

In either form of breakdown the reverse current is large and practically independent of the applied voltage; if the power then dissipated within the diode is

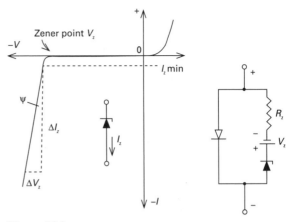

Figure 13.3

beyond its power rating, the junction will be destroyed. However, a diode will survive this treatment so long as its intrinsic power handling capacity is not exceeded, and this can be ensured by placing a safety resistor in series with the junction.

Figure 13.3 shows a typical characteristic of the reverse region of a diode; you will see from this that the curve follows the reverse voltage axis until the zener point is reached. This region is the normal reverse saturation (or leakage) current of any diode; the point at which the breakdown occurs in zener diodes can be as low as 2.5 V or as high as 100 V or more. They are inexpensive components and available in a scale of preferred voltages similar to that used for resistor values.

Let us now examine the characteristic of *Figure 13.3* in more detail. Once the breakdown is established the voltage across the diode remains substantially constant at the level V_z irrespective of a large variation of the current through it, that is, the ratio $\Delta V_z / \Delta I_z = \tan \psi$ is very small, and the system behaves as a linear resistor R_z in series with a source of e.m.f. This is indicated in the circuit model on the right of the figure where the voltage source shows that reverse current does not flow until the negative p.d. across the diode exceeds V_z. A parallel diode, shown in broken lines, covers the normal forward bias conditions which normally have no significance in the practical applications of zener diodes. The ratio $\Delta V_z / \Delta I_z$ is known as the *slope resistance* of the diode and is represented in the model as resistor R_z in series with the circuit.

The slope resistance therefore represents the internal resistance of a source whose e.m.f. is V_b. Notice that because the slope resistance is finite, that is, ΔV_z is not zero, the voltage will not be absolutely constant for variations in current I_z.

THE ZENER REGULATOR

The circuit of the most simple shunt regulator using a zener diode is shown in *Figure 13.4*. Here the diode is connected to the output of a power supply of the type shown in *Figure 13.1* at terminals P and Q. Resistor R_s is connected in series with the zener and provides the necessary protection against excessive current flow at the breakdown point. The load is represented as usual by resistor R_L across the output terminals R and S. Notice how the diode is connected; its *cathode* goes to the positive terminal of the d.c. supply, so that it is reverse biased and under proper application will be operating in the breakdown condition.

To understand the action of the circuit, the zener can be looked on as a current reservoir so long as it remains broken down. It then responds to variations in both the supply voltage at terminals P and Q, and the load current I_L flowing through R_L as follows:

(a) Suppose the d.c. voltage at P-Q increases for some reason, then the current through the zener will increase while the voltage increase will be developed across resistor R_s – *not* across the zener. The zener voltage, remember, remains at the breakdown value V_z, irrespective of the current through it. Similarly, if the voltage at P-Q decreases, the zener will surrender the extra current and the voltage across R_s will fall. So the variation is 'absorbed' as it were by the series resistor R_s and the output voltage at R-S remains constant.

(b) If the load current I_L increases for any reason, the zener current will decrease by the same amount. Similarly, if the load current decreases, the zener current will increase by the same amount. This time, the zener takes up any excess current and sheds any current difference demanded by the load, so acting as a current reservoir while maintaining a constant voltage at the load terminals.

There is a minimum zener current for which the voltage regulation described is effective and this is

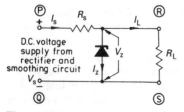

Figure 13.4

indicated in *Figure 13.3*. If the zener current falls below this level, the breakdown fails and the stabilizing effect no longer holds. Zener diodes can regulate satisfactorily down to currents of the order of 0.5 mA or so; the upper limit of current is, of course, dependent on the power rating of the device. Nominal ratings range from 400 mW up to some 25 W.

The following worked examples will illustrate the general action of zener regulators. You need nothing more than Ohm's law as your mathematical equipment.

Example (1) A change in current from 2 mA to 5 mA in a zener diode resulted in a change in the p.d. across the diode of 20 mV. What was the slope resistance of this diode?

Solution The change in voltage

$$\Delta V_z = 20 \times 10^{-3} \text{ V}$$

The change in the current

$$\Delta I_z = (5 - 2) \times 10^{-3} \text{ A}$$

Then the slope resistance

$$R_z = \frac{\Delta V_z}{\Delta I_z} = \frac{20}{3} = 6.67 \ \Omega$$

Example (2) The circuit of *Figure 13.5* provides a 12 V regulated output from an unregulated input of 20 V. The zener is rated at 1.3 W power dissipation; what value of series resistor would you use?

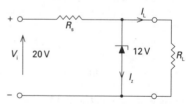

Figure 13.5

Solution The various currents and their distribution are illustrated in the figure. The power dissipated in the zener must not exceed 1.3 W, and since the zener current I_z will be greatest when $I_L = 0$, i.e. with the load disconnected, the maximum permissible level of I_z will be

$$\frac{\text{Maximum power rating}}{\text{Zener voltage}} = \frac{1.3}{12} = 0.108 \text{ A}$$

The drop across R_s will be $(20 - 12) = 8$ V,

hence $R_s = \dfrac{8}{0.108} = 74\ \Omega$

The nearest preferred value to this is 75 Ω, though in practice it would probably be wisest to go up to 82 Ω to provide a small margin of safety without materially affecting the performance of the circuit.

Example (3)   If the zener of the previous example regulates down to a current of 3 mA, estimate the greatest and the least supply voltages between which the regulation will be effective, given that the load resistance $R_L = 1500\ \Omega$.

Solution   For the *least* value of the supply voltage, the zener must still be passing its minimum current of 3 mA. The load current $I_L = 12/1500 = 0.008$ A or 8 mA, hence the supply current $(I_L + I_z)$ will be $(8 + 3) = 11$ mA.

$$\therefore\quad V_{min} = (11 \times 10^{-3} \times 75) + 12 = 12.83\ \text{V}$$

For the *greatest* value of the supply voltage, the zener current must not exceed 0.108 A or 108 mA as we have already established. Therefore the supply current will be $(I_L + I_z) = (108 + 8) = 116$ mA and

$$V_{max} = (116 \times 10^{-3} \times 75) + 12 = 20.7\ \text{V}$$

Example (4)   A zener diode having the reverse characteristic of *Figure 13.6(a)* is used in the circuit of diagram *(b)*, where the minimum value of the input voltage supply is 15 V. If the diode power rating is 500 mW, what is the greatest input voltage permitted for regulation to take place and what degree of regulation will be actually achieved?

Solution   The zener appears to be designed to operate at a nominal 9.1 V, a value in the preferred range. Consequently, the current in the load resistor R_s is $9.1/1000 = 9.1$ mA. So that the load voltage will remain substantially constant, the diode reverse current should not encroach on the bend in the characteristic, that is, below about 3 mA. So with the supply voltage V_i at its least value of 15 V, R_s should be about $(15 - 9.1)\ \text{V}/(9.1 + 3)\ \text{mA} = 487\ \Omega$. What preferred value would you choose here?

As the zener maximum dissipation is 500 mW, the greatest permissible zener current will be 500 mW/9.1 V = 55 mA. Under this condition the current in R_s will be $(55 + 9.1) = 64.1$ mA and the input V_i will then be $(64.1 \times 10^{-3} \times 487) + 9.1 = 40.3$ V.

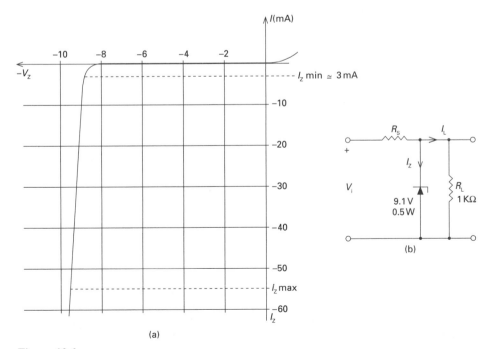

(a)

(b)

Figure 13.6

From the characteristic we see that the voltage across the load (and across the zener) is about 9.4 V, thus a change in the input voltage from 15 V to 40.3 V produces a change in the load voltage from 9.1 to 9.4 V, only 0.3 V.

(5)   Determine, with values, the elements in the model of the zener used in the previous example.

(6)   Sketch a model for a zener diode having the following characteristics:

Nominal voltage at 5 mA = 12 V
Maximum slope resistance at 5 mA = 10 Ω
A linear gradient above 1 mA

This zener stabilizes a load current of 20 mA at 12 V from a nominal 24 V d.c. source. If this supply varies by 2 V, what will be the load voltage change?

ZENER PROTECTION

Another application of zener diodes is shown in *Figure 13.7* which illustrates a method of protecting a sensitive meter or similar device against accidental overload. Suppose the meter to have a full-scale deflection (f.s.d.) at a current I_m and that at this current the voltage developed across the instrument is V_m. Resistors R_1 and R_2 are then selected in conjunction with the zener characteristic so that if the applied voltage V exceeds V_m, the excess current resulting will be bypassed through the diode and not through the meter. This can be arranged by ensuring that when V equal V_m the zener just breaks down but draws only a negligible current. At this same time, of course, the meter must just be at its f.s.d.

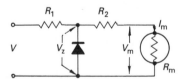

Figure 13.7

Always work problems of the kind we are encountering in this section (as in any others) by the direct application of Ohm's law and common sense.

Example (7)   A meter has a f.s.d. of 0.5 mA and a resistance of 1000 Ω. It is used as a volt-

meter by the addition of a series resistance and is scaled 0–10 V. Design a suitable protection circuit using a 6.8 V zener diode so that the meter will not be overloaded by connection to voltage sources greater than 10 V.

Solution   There are two parts to this problem: we first need to find the value of the series resistance which converts the 0.5 mA f.s.d. meter into a 0–10 V voltmeter. Looking at *Figure 13.8(a)* we see that when 10 V is applied to the series arrangement, the meter current should be 0.5 mA. Therefore

$$R + 1000 = \frac{10}{0.5 \times 10^{-3}} = 20\ 000\ \Omega$$

$$\therefore \quad R = 19\ 000\ \Omega$$

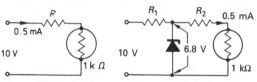

Figure 13.8

The protection arrangement is shown in *Figure 13.8(b)*, and here $R_1 + R_2$ must be made equal to 19 000 Ω. When the applied voltage is 10 V, the zener must just break down, but the current through it will be negligible. At this time the meter must just indicate f.s.d. The current through R_1, R_2 and the meter in series is therefore 0.5 mA. The situation, in other words, is identical with the circuit in diagram (*a*). Then for 6.8 V across the zener

$$R_1 = \frac{10 - 6.8}{0.5 \times 10^{-3}} = 6400\ \Omega$$

Hence

$$R_2 = 19\ 000 - 6400 = 12\ 600\ \Omega$$

In a practical situation some experimenting with the values would be necessary. Zener diodes have a tolerance on their nominal voltages, usually 5 per cent, and the same with resistors. Allowing for this, the calculations performed above are unlikely to be precise in actual circumstances. Always use your calculations, especially when you are applying them to bench-top experimentation, simply as a good guide – that's why hands-on work is always necessary.

SERIES REGULATION

The basic zener regulator circuits we have so far discussed have all been forms of shunt regulation, the regulating device being placed in parallel, or shunt, with the load.

In the series form of regulator, the control element is connected in series with the supply lead feeding to the load. If the input voltage V_i or the load current I_L change, the voltage across the control element adjusts itself in such a way that the output voltage remains constant.

A basic form of series regulation is shown in *Figure 13.9*. This circuit is essentially a transistor current amplifier used in conjunction with a zener diode which acts now as a fixed voltage reference V_{ref}. Strictly, we have a shunt regulator (the zener) acting through a series controller (the transistor). This transistor is usually a power type of device, capable of carrying and controlling a relatively large current.

The base of this transistor is maintained at a constant voltage by the zener diode which is supplied with current through resistor R_B exactly as we have already described. This transistor then acts as a current amplifier, so enabling the circuit to provide much higher currents than can be handled by a zener diode alone. If, for example, the current gain of the transistor is 50, then a base current of 20 mA will provide an output current of 1 A, and such a base current is now the effective zener load.

Since the transistor is connected as an emitter-follower, the output load R_L being in the emitter circuit lead, the output voltage follows the reference closely, the difference being the base-emitter voltage drop, approximately 0.7 V. This difference can be easily compensated for by using a zener having a voltage rating about 0.7 V above the required V_O: for instance, for a 10 V output, an 11 V zener could be used.

The regulation takes place like this: let us assume both the input voltage and the reference voltage to be constant, then the current I through R_B is constant. Suppose now that the load current I_L falls; this will reduce the base current I_B since $I_B = I_L/(h_{FE} + 1)$, but the reference current I_z will increase because the base voltage will rise. Thus the sum of the currents ($I_B + I_z$) remains essentially constant. The converse action takes place when the load current tends to increase; hence the output tends to remain constant. If there is any variation in the input current I_i, current I through R_B changes and this current change will be absorbed by the zener diode in the manner we have seen for the basic regulator. Hence, although this circuit gives a

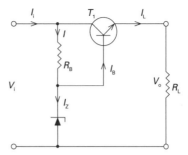

Figure 13.9

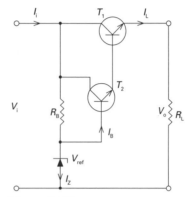

Figure 13.10

much improved load regulation, its line regulation compared with the single zener is no better.

This circuit has a particular disadvantage in that if the load is disconnected or the input voltage goes sufficiently high, the zener calls for a high dissipation device, the series base resistor R_B being relatively small in this usage. So, although such a circuit will be found adequate for supplying constant load currents at a fixed voltage, some degree of sophistication is called for in more stringent cases.

One such improvement is the introduction of a high gain amplifier between the zener diode and the series transistor. By an appropriate connection, a second transistor can be introduced so that a Darlington pair is formed; this arrangement is seen in *Figure 13.10*. The maximum base current I_B is then $I_L/(h_{FE1} + 1)(h_{FE2} + 1)$ and this is usually very small, typically of the order of a milliampere or so. In this way a low-power reference diode can be used, with minimum currents of the order of a few milliamperes at most. Further, resistor R_B can be made very much larger than that used in the circuit of *Figure 13.9*.

Closed loop regulators

Although the emitter-follower regulators give satisfactory performances for many general-purpose

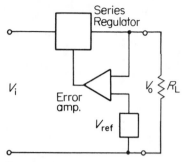

Figure 13.11

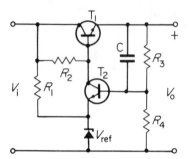

Figure 13.12

applications, the desirable property of a constant-voltage source, that is, a very low output impedance, is not attainable from these circuits. The basic principle of a so-called closed-loop regulator, sometimes known as a comparator or error detection system, is illustrated in the block diagram of *Figure 13.11*. A fraction of the output voltage V_o is compared with a reference voltage V_{ref}, obtained as before from a zener diode, and the difference between these voltages is amplified and used to control a series regulator. This in turn controls V_o and acts in such a direction that the initial difference is minimized, irrespective of which way the output has shifted in the first place. The feedback amplifier is known as an error amplifier since it senses the small difference, or error, existing between the change in V_o and the reference.

A practical form of this circuit is shown in *Figure 13.12*. In low current systems the series regulator may be a single transistor T_1 as seen here, though a Darlington connected pair is often encountered. Transistor T_2 with collector load R_2 is the error amplifier; the emitter of T_2 is held at a constant voltage determined by the zener diode and fed, in the usual manner, through resistor R_1. Resistors R_3 and R_4 are selected so that their ratio provides the desired ratio between V_{ref} and V_o, while their total sum provides a bleed current that is large compared to the base current of T_2. You should be able to say why this is necessary. Under these conditions:

$$V_{ref} = \frac{V_o R_3}{R_3 + R_4} \qquad (13.1)$$

and the bleed current through R_3 and R_4 in series is

$$I = \frac{V_o}{R_3 + R_4} \qquad (13.2)$$

Solving these simultaneously (do this for yourself) we find the values of R_3 and R_4 to be

$$R_4 = \frac{V_{ref}}{I} \quad \text{and} \quad R_3 = \frac{V_o}{I} - R_4$$

Resistor R_1 is selected so that the diode current is sufficient to establish a stable breakdown condition so that V_{ref} is constant. This then fixes the emitter of the error amplifier at a constant voltage. The voltage drop across R_2 due to the conduction of T_2 is fed directly to the base of the series regulator T_1 where it provides forward bias for this device.

To follow the operation of this circuit, suppose that the output voltage V_o changes for some reason, either a variation in load current or a variation in the input V_i. This change is sensed at the base of the error amplifier T_2 and, because its emitter potential is fixed, the base variation will either increase or decrease the collector current of T_2 via R_2 accordingly, as the output change is an increase or a decrease in V_o, respectively. In other words, the variation is compared with the zener reference and T_2 senses in which direction its base potential has moved relative to that reference. The current variation in the collector of T_2 causes the drop across R_2 to change, and so the forward bias of the series regulator is also changed. This change in bias changes V_c and the output current is therefore adjusted in such a way that the original change in V_o is minimized. This system cannot, of course, compensate absolutely, since an error must be present for the circuit to operate.

You will probably spot that this is an example of negative feedback, and the circuit can be analysed in terms of feedback theory. What is essential, of course, is that a polarity reversal must take place in the error amplifier so that the feedback is negative. Go around the circuit with a specific direction for the change in V_o and check that this requirement is so.

In circuits of this sort, the gain of T_2 should be as large as possible, and T_1 must be capable of dealing with the full-load output current and generally with a high power capability. The voltage drop across T_1 must also be large enough at all times to ensure that it conducts at all times. You may wonder about the presence of capacitor C connected across R_3. This is not always provided but when it is its purpose is to

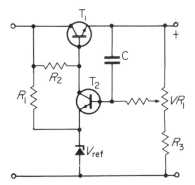

Figure 13.13

assist the circuit to cope with rapid voltage changes, particularly ripple variations. R_3 is effectively short-circuited by C at such frequencies and the full output variations are imposed on the base of T_2. For slow output variations, attenuation occurs across the potential divider R_3, R_4; since the base and emitter of T_2 are practically at the same potential, the attenuation is in fact that expressed by equation (13.1) above, for by rearrangement

$$\frac{V_o}{V_{ref}} = \frac{R_3 + R_4}{R_4}$$

From this relationship it can be seen that

$$V_o = \left[\frac{R_3 + R_4}{R_4}\right] V_{ref} = \left[\frac{R_3}{R_4} + 1\right] V_{ref}$$

so that V_o can be adjusted by manual variation of the ratio R_3/R_4.

A circuit having an adjustable output voltage control is shown in *Figure 13.13*. Potentiometer VR_1 replaces the fixed divider chain and permits a limited range of control, but the output voltage can never be less than that of the reference. If it is required to control the output voltage down to zero, a negative supply rail has to be provided.

(8)   Why cannot the output voltage be reduced below the reference level in the above circuit?

(9)   What would happen to the output voltage, if anything, if it did drop below the reference level?

(10)   There is an additional resistor placed between potentiometer VR_1 and the base of the error amplifier T_2 in *Figure 13.13*. Why do you think it is placed there?

Example (11)   *Figure 13.14* shows a series regulated power supply using two transistors T_2 and T_3 in the amplifier feedback path. Deduce how this circuit operates to maintain a constant output voltage.

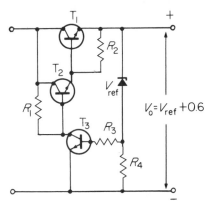

Figure 13.14

Solution   This circuit looks rather different in some respects from those discussed above. The zener diode, for example, is on the 'other side' of the series regulator and its feed resistor R_4 is returned to the negative supply line. However, taking the circuit a step at a time soon reveals that there is quite a lot we can recognize. Clearly, it is immaterial whether the zener has its feed resistor on the positive side of the supply or the negative. The thing that matters is that the diode is reverse biased. The zener will break down at its rated voltage V_z and thereafter will behave as a constant voltage source. Ignoring R_3 for the moment, the output voltage will be the sum of V_z and the base-emitter voltage of T_3 (about 0.7 V), so the zener rating determines the order of output voltage we are going to get.

We have already seen that the resistor in series with a zener takes up any voltage changes occurring across the combination; hence if the output voltage tends to change, the potential across R_4 will change accordingly and the base input to T_3 also changes. The output of T_3 drives the base of T_2; this transistor is in emitter-follower configuration and the load impedance output at the emitter controls the series regulator T_1 forming the emitter load of T_2.

Resistor R_2 is included to shunt away the collector-base leakage current of T_1, while R_3 is selected to limit the base current of T_2 in the event of a fault condition arising.

> (12)   Check the polarity variations around the circuit of *Figure 13.14* for yourself and verify that the control exercised by T_1 is in the proper direction to reduce an original output voltage change.

USING OPERATIONAL AMPLIFIERS

Integrated-circuit operational amplifiers can be employed in the design of a regulated power supply to great advantage over discrete transistors. Such amplifiers can provide open-loop gain figures of several thousand and many different regulator configurations can be devised from their employment. When used in a typical circuit of the form shown in *Figure 13.15*, excellent regulation characteristics are obtained.

It is necessary to ensure in the use of integrated amplifiers in this way that the output voltage V_o and the reference V_{ref} do not exceed the permissible output voltage swing of the operational amplifier, otherwise regulation will cease. For this reason, the non-inverting input to the amplifier should be positioned at a level of about $V_o/2$, obtained from the zener reference. Transistor T_1 is a common-emitter base driver for the series regulator T_1 and provides a base potential that is the desired output voltage for the operational amplifier. The voltage gain of T_2 is low because the collector load resistor R_1 is effectively in parallel with the input resistance of T_1 which is small. An adjustable output voltage is achieved by variation of the voltage on the inverting input to the operational amplifier by potentiometer VR_1. In most designs of this kind, feedback around the operational amplifier is necessary to reduce the gain and broaden its bandwidth to ensure stability.

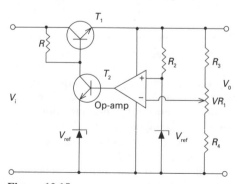

Figure 13.15

Example (13)   In the circuit of *Figure 13.9* suppose T_1 has a static gain (h_{FE}) = 40 and V_{BE} = 0.7 V at I_E = 20 mA. By making any necessary assumptions as you proceed, design a power supply of this type to provide an output voltage of 20 V at a maximum load current of 500 mA.

Solution   This solution is not intended to be comprehensive but it will provide you with the rudiments of tackling a design without pure guesswork or hit-and-miss methodology.

Looking at the circuit, we can make the following notes: (a) the reference zener current I_z is least when the load current is greatest; (b) we must ensure that the minimum diode current does not allow the zener to go out of the breakdown region; (c) we must choose a minimum value for the voltage across R_B, bearing in mind that the higher this voltage, the higher the required input voltage V_i must be to make sure that observation (b) is maintained.

From these considerations, let us assume that a zener current of 5 mA is an acceptable level assuring a stable breakdown. The zener voltage should then be $(V_o + V_{BE})$ = 20 + 0.7 = 20.7 V. Either a 20 V or a 22 V zener could be employed here, with I_z at 5 mA. Now choose 2 V as a minimum voltage drop across R_B; then the value of R_B will be found from the following:

$$I_B = \frac{I_L}{h_{FE} + 1} \text{ and } R_B = \frac{V_B}{I_Z + I_B}$$

$$\therefore \quad R_B = \frac{V_B}{I_Z + \dfrac{I_L}{h_{FE} + 1}} = \frac{2}{5 \text{ mA} + \dfrac{500}{41} \text{ mA}}$$

$$= \frac{2}{17.2} \text{ mA} = 116 \ \Omega$$

The minimum voltage input V_i to the circuit is then 20.7 + 2 = 22.7 V.

PROTECTION METHODS

Semiconductor devices have a failing in one sense; if an accidental overload occurs, for example if the output terminals of any of the previously discussed power supplies are short-circuited, any ordinary fuses used as protection devices act too slowly in most cases to prevent the semiconductor elements from being irreparably damaged. Integrated circuits are often the

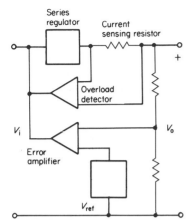

Figure 13.16

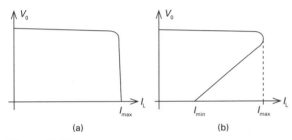

Figure 13.17

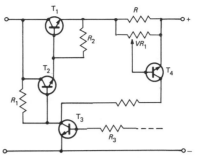

Figure 13.18

Again, a practical form of such a system is shown in *Figure 13.18*. This appears even more complicated, but what we have is the circuit of *Figure 13.14* with one additional transistor, T_4. This forms the overload detector. When a current drawn by the load exceeds a certain level, the voltage across the current-sensing resistor R rises sufficiently to exceed the nominal switch-on level of 0.7 V for T_4. This transistor then switches on and takes control away from the error detector and shunts down the series regulator to a point where the maximum current output is maintained, but not exceeded. The point at which T_4 switches on (in terms of the voltage across R) is decided by the setting of VR_1, so a range of current-limiting values is available. The current-sensing resistor is usually of 1 or 2 ohms in value, with VR_1 a few hundred ohms.

The power developed in T_1 is dissipated in the form of heat (as it is in any series regulator) and this transistor is almost always mounted on a suitable heat sink. The error amplifier is often also treated in the same way.

most vulnerable in this respect. The answer to this problem can be found by the use of some technique that automatically limits the maximum current available from the supply to a level that will not prove excessive to semiconductor devices in either the power unit itself or in the circuits which the power unit may be feeding.

We will first consider the block diagram of *Figure 13.16* where some additional circuitry has been added to the general form of power supplies already investigated. You will recognize the familiar series regulator, the error amplifier and the reference voltage box; the additions are a current-sensing resistor and an overload detector. The principle of this protection system (of which many circuit variations are possible) is that when an excessive load current flows through the current-sensing resistor, the voltage developed across it exceeds a predetermined level and switches on the overload detector. This then takes control away from the error amplifier and biases the series regulator back to a safe current condition; this is sometimes designed simply to limit the output to its normal maximum current, or to utilize a 'fold back' system which actually reduces the current to a low value, much below its maximum rating. *Figures 13.17(a)* and *(b)* show the output characteristics for both these measures, respectively.

(14)   The value of the current-sensing resistor in a circuit similar to that of *Figure 13.18* is 1.5 Ω and is shunted with a pre-set control (as VR_1) of 100 Ω. When this control is set to its midway position, what will be the greatest output current obtainable from this circuit?

(15)   Do you think the introduction of the current-sensing resistor affects the regulation of the circuit? Give reasons for your answer. Would it make things better to put it on the unregulated side of T_1?

Example (16)   In the regulator shown in *Figure 13.18* the voltage gains of T_1 and T_2 are, respectively, 12 and 120. If the input voltage V_i changes by 3 V, what will be the change in output voltage V_o?

Solution   We will work this one in general symbols, leaving the actual evaluation until the end. Let the output voltage change be ΔV_o. Then the change at the base of T_2 will be $\Delta V_o/6$.

The control voltage at the collector of T_2 will be $A_{v2} \cdot \Delta V_o/6$ where A_{v2} is the voltage gain of T_2. Hence the base-emitter voltage of T_1 will be

$$\frac{A_v \Delta V_o}{6} - \Delta V_o = 20\Delta V_o - \Delta V_o = 19\Delta V_o$$

This will produce a voltage across T_1 equal to $A_{v1} \cdot 19\Delta V_o$. But the voltage across T_1 is also equal to $3 - \Delta V_o$. Hence $3 - \Delta V_o = 228\Delta V_o$ and $\Delta V_o = 3/229 = 0.013$ V or 13 mV.

Fold back protection

In the previous protection circuits which depend on a maximum output current limitation should the output become shorted, the power dissipation in the series regulator transistor, as mentioned above, becomes high, since practically the whole of the input voltage V appears across T_1. For currents of several amperes, a protective heat sink becomes physically large (as well as expensive), and in any case, the high dissipation represents a waste of energy in unnecessary heating.

In the circuit of *Figure 13.19* this problem does not arise because the load current is folded back (as in *Figure 13.17(b)*) to a level which is a small fraction of the maximum value as soon as the maximum value is exceeded. Make a note that the circuit given is the current limiter only; the usual series regulator can be placed between this limiter and the load terminals. We assume here for purposes of analysis that the load is placed directly on the limiter output as the broken lines indicate.

When the load is connected, I_L flows through resistor R_1 which is of low value and produces a voltage

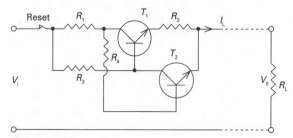

Figure 13.19

drop of a volt or so. Resistor R_3 is selected to be smaller than $R_1 \times$ (current gain of T_1) so that transistor T_1 is switched hard on with a very small emitter-collector voltage drop, V_{ce1}. Now transistor T_2 will not conduct until $V_{ce1} + I_L R_2$ rises to about 0.7 V. At this point, T_2 switches on and its collector current takes base drive away from T_1 and brings it out of the saturated state. This action increases V_{ce1} and thus increases the forward bias of T_2; the process is then cumulative so that in a very short time T_1 is cut off and the only current getting through to the load is by way of resistors R_3 and R_4. Hence I_L is folded back to a very much smaller level than the design maximum.

It might appear that T_2 would now take the brunt of excessive current, but this is not so, since it is in saturation during the relevant operating time and so its dissipation will be relatively small. It must, however, be capable of carrying the current V_1/R_3. Unlike the previous circuits, this one remains cut off even when the short circuit is removed; the system can be reset by switching off the input V_i momentarily.

INTEGRATED REGULATORS

Integrated regulator systems are available in a great variety of dissipations and output voltages. These integrated packages contain most of the basic components of a regulator, that is, the zener reference, error amplifier, a series transistor and protection circuit, and offer the advantages of excellent regulation, compact size and ease of use.

All of the circuits using discrete component parts we have looked at in this chapter can be fabricated on a single silicon chip. A large range of such integrated regulators can be looked up in manufacturers' catalogues and are readily obtained from electronic supply sources. Many of these units are three or four terminal devices and are usually mounted in the standard T0220 style packages; others are found in the circular T100 packages, and in the 14-pin DIL form. Both fixed voltage and variable voltage types are available. The smaller fixed voltage types can often be used throughout a piece of electronic equipment to provide voltage regulation at a number of points rather than a single regulator looking after a large and complicated circuit. This technique helps overcome stability problems by preventing interaction between sections of a common supply line, and enables a range of voltage options to be applied to different sections of the circuitry, rather than using a resistance divider system which is wasteful of energy.

FIXED VOLTAGE REGULATORS

The fixed voltage types of integrated regulators are available in voltages which cover the most common applications, that is, 5 V for TTL logic systems, 12 V and 24 V for CMOS and operational amplifier circuits. Intermediate values are obtainable, namely, 8 V and 15 V; in all of these regulators adjustments can be made by simple external additions to increase the nominal voltage rating by up to 20 per cent and more. Current ratings range from 100 mA through 1, 2 and 3 A. They are thus extremely versatile and useful units, containing up to a dozen transistors and resistors as well as one or two zener references.

Although apart from the transformer and reservoir capacitor, the whole regulating circuit can be found inside the integrated packages, a few precautions are necessary whenever these devices are put into operation. Two basic circuits using the time honoured 78 or 79 series, available in T0220 packages, are given in *Figures 13.20(a)* and *(b)* respectively. The simplicity of these circuits is apparent at once, the only external components being a protective diode in the event of reversed voltage being applied to the output terminals (such as could happen if an inductive circuit was suddenly switched on or off), a small smoothing capacitor and a return resistor for the emitter of the internal series regulator. Stability is also helped by capacitor C_2 which must be connected directly to the relevant pins of the integrated regulator, not wired simply in parallel with the reservoir capacitor C_1.

The 78 series 7805, 7812, 7815 and 7824, providing outputs of 5, 12, 15 and 24 V, respectively, all supply maximum currents of 1 A when the case is mounted on a small heat sink. The 79 series provides the same voltage and current range but gives a negative output polarity. When higher current is required, the 78T series will provide up to 3 A. In these regulators, it is necessary to maintain the input from the reservoir capacitor at least 3 V above the nominal output voltage so that the regulating action is operational. There is a series of these regulators which have the capability of maintaining their action with a much smaller input–output differential, nominally about 1 volt; these can be obtained under the LM29 series. The differential for either series can, of course, be greater than the design minimum; in general it may be anything up to some 25 V but this means that the drop across the internal series regulator is unnecessarily high and the power dissipation is consequently excessive for a given load current. The input voltage should be chosen to suit the required output voltage, with the recommended differential in mind.

There are a number of ways of adjusting the nominal fixed voltage output from these regulators. The circuit of *Figure 13.21(a)* being useful in cases where only a small increase is needed, perhaps to make up a nominal voltage which is at the lower tolerance end of the range. This can be as much as 0.2 V, so a 5 V device which gives us only 4.8 V can be brought closer or a bit above the actual required level of 5 V. The introduction of a diode placed in the common line of the regulator can modify the internal reference by

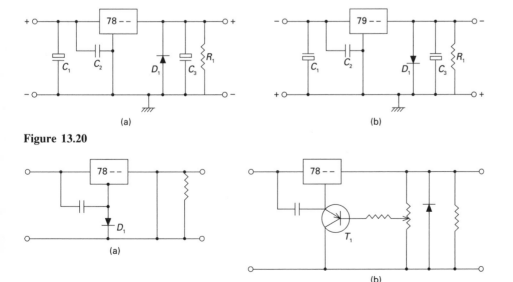

Figure 13.20

Figure 13.21

about 0.7 V for a silicon diode or about 0.2 V for a germanium diode, so that a 4.8 V regulator can provide a 5.0 V output with a germanium diode, while a good approximation to 9 V can be obtained by using the silicon diode with an 8 V regulator. The same effect can be brought about by using a resistor of about 50 to 100 Ω in place of a diode, but this does tend to affect the regulation slightly.

Substantial increases in the output voltage can be obtained by the method shown in *Figure 13.21(b)*. This not only improves the regulation over that obtained from a diode but enables a 5 V regulator to operate with an output, say, of 9 V or more. This is a useful thing if you have only a few 5 V regulators but wish to try an experiment that calls for a higher range of voltages. Transistor T_1 has to be a *p-n-p* type for positive regulators or *n-p-n* for negative regulators; the parallel protective diode must also be reversed to suit the output polarity.

VARIABLE VOLTAGE REGULATORS

Integrated regulators are available which are designed to provide an adjustable output over a wide range and at high load currents. These are generally found in T05 packages which have to be mounted on suitable heat sinks in the same way as power transistors. There are a great variety of such regulators to choose from, and we shall mention only two in any detail here. The most common types have been around for some years and have stood the test of time: these are the LM317K and the 338K, with the 317K providing an output adjustable from about 1.5 V up to 35 V at 1.5 A, and the 338K providing a similar output range at a current of 5 A. A more recent addition is the LM396K which gives us a load current up to 10 A.

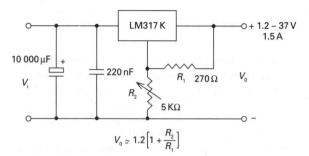

$$V_0 \approx 1.2\left[1 + \frac{R_2}{R_1}\right]$$

Figure 13.22

A basic practical circuit for the 317K is shown in *Figure 13.22* where the voltage range covered is from about 1.5 V to 35 V at a maximum current for 1.5 A. It is possible to get down to zero volts by providing a negative supply rail of a few volts for the return of adjustment potentiometer VR_2.

Figure 13.23 shows a circuit using the 338K regulator, this providing an output adjustable from about 1.5 V to 25 V at a maximum current of 5 A. This circuit is similar to that for the 317K except that a range switch S_1 has been introduced to break the output into two ranges: 1.5 V to 15 V and 15 V to 25 V. The purpose of this range switching might not be obvious at first glance, but it is necessary to avoid excessive power dissipation in the regulator which would come about if large load currents were being drawn at low output voltages. For example, suppose the input voltage V_i to be 30 V; if the control potentiometer VR_1 is set, say, for a 5 V output V_o, then there is a voltage drop across the regulator of 25 V which, at a current of 5 A, represents a dissipation of 125 W. This dissipation would call for a massive heat sink. By switching the transformer output so that the input voltage V_i is about 17 V and by simultaneously switching out the series resistor to VR_1, the output can be restricted to the range 1.5 V to 15 V and the dissipation considerably reduced.

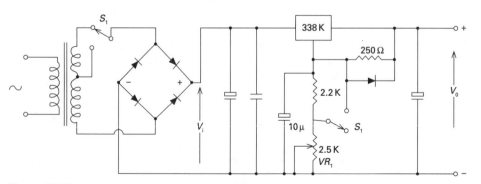

Figure 13.23

MULTI-PURPOSE REGULATORS

Integrated regulators are available which, by providing a number of useful external points of connection, can be used to construct a large number of different regulated power circuits. The 723 regulator chip is a well-tested example of such a regulator which can be obtained in either the circular ten-lead T100 package or in the 14-pin DIL package. The former has a larger power dissipation, 800 mW as against 660 mW for the DIL, and a higher ambient temperature rating but otherwise they are similar. The basic circuit of the regulator having the DIL package is shown in *Figure 13.24*. You will recognize the various parts of this circuit from what we have already covered about discrete transistor systems, but there are one or two additional innovations here.

The package contains the following elements:

(a) A zener diode reference which is fed from a constant-current source and included in a feedback amplifier which serves to lower its slope resistance and hence improve its stabilization;

(b) An error amplifier with both inverting and non-inverting inputs brought out for external connection at pins 4 and 5, respectively, as well as the power supply terminals at pins 7 and 8;

(c) A current limiting transistor T_2 in which both base and emitter connections are brought to pins 2 and 3, respectively, for connection to an external current-sensing resistor;

(d) A medium power emitter-follower series control regulator T_1; the possible power dissipation of the whole circuit hinges upon this part of the system.

Figure 13.25 shows a basic example of how externally added components can be added to the integrated package, illustrating here an adjustable supply covering the output range 6–25 V approximately. In this example, the zener reference voltage is connected directly to the non-inverting input of the error amplifier, and the supply to this amplifier is derived from the same rail as the normal input voltage. A current-sensing resistor R_s is selected so that, when the maximum permissible load current is reached (or a fault condition occurs), the p.d. across it is 0.7 V and switches on the limiting transistor T_2; this short-circuits the base-emitter of regulator T_1 and prevents any further increase in load current. Adjustable output voltage control is effected by varying the voltage fed to the inverting pin of the error amplifier and capacitor C is added to reduce the high frequency gain of this amplifier and so avoid a risk of high frequency instability.

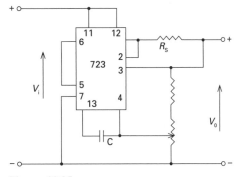

Figure 13.25

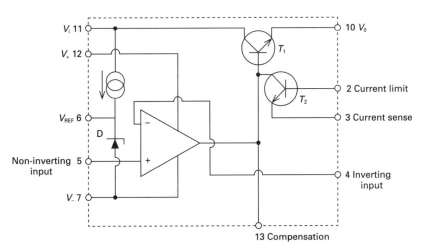

Figure 13.24

Example (17)   The following table gives the absolute maximum ratings of a 723 integrated regulator in 14-pin DIL format:

Continuous voltage from $V+$ to $V-$	40 V
Input–output voltage differential	40 V
Maximum output current	150 mA
Maximum power dissipation	660 mW

You are using a current limiting resistor of 5 Ω in the circuit of *Figure 13.25*; approximately at what voltage will the output be limited, and what will be the greatest permissible input–output differential under this condition?

Solution   The maximum output current is stated to be 150 mA but this must be considered in conjunction with the maximum power dissipation of 660 mW. When resistor R_s is chosen as 5 Ω, then the limiter will switch on at a current of about 0.7/5 = 0.014 A or 140 mA; this current, however, will only permit an input–output differential of 660 mW/140 = 4.7 V, since the maximum dissipation is 660 mW. For a 40 V maximum input, therefore, the maximum output will be 40 − 4.7 = 35.3 V.

(18)   Do the previous calculation, assuming that R_s = 7 Ω and the input voltage is 30 V.

Example (17) above may give the impression that the 723 has very restricted applications. This is not so; besides a very wide variety of fixed voltage outputs covering all the standard levels, and variable outputs adjustable over wide or restricted ranges, increased current outputs can be catered for by using the 723 as the control element of an auxiliary power transistor which takes over as the series regulator. The addition of the components shown in *Figure 13.26* to the basic arrangement of *Figure 13.25* enables load currents up to several amperes to be obtained, depending on the

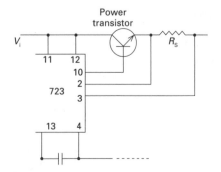

Figure 13.26

power transistor employed. The compensating capacitor C needs to be increased to around 470–1000 pF here because of the additional phase shift at high frequencies which the power transistor introduces.

SUMMARY

- A voltage regulator minimizes changes in d.c. load voltage.
- In zener diodes, small voltage changes cause large current changes that produce voltage drops to compensate for changes in input voltage or output voltage.
- Zener diodes can be used as shunt voltage regulators.
- Zener diodes can be used as instrument protection devices.
- Integrated operational amplifiers can be used as error detectors in voltage regulators.
- Regulators are available in completely integrated form.

REVIEW QUESTIONS

1   What factors affect the output voltage of an unregulated d.c. power supply?
2   Differentiate between a regulated and an unregulated supply.
3   'After breakdown, a zener behaves as a linear resistance.' What do you understand by this statement?
4   What is meant by the slope resistance of a zener diode?
5   Is zener diode regulation a series or a shunt regulator?
6   What determines the point at which a diode suffers a reverse voltage breakdown?
7   Is it true that zener breakdown occurs in wide junctions at very high voltages?
8   Is there any difference between zener breakdown and avalanche breakdown?
9   Make sketches, and comment on, a regulator output characteristic with (a) current limiting, (b) fold-back.
10   List any advantages/disadvantages you might find in (a) shunt regulators, (b) series regulators.

11   Describe any form of short-circuit protection used in a power supply unit.

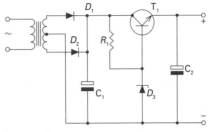

Figure 13.27

EXERCISES AND PROBLEMS

(19)   Draw a circuit diagram of a simple zener diode regulating supply and explain how it operates to maintain a constant voltage across a load resistance.

(20)   The voltage across a zener diode changes from 6.15 V to 6.2 V when the diode current rises from 20 mA to 35 mA. What is the slope resistance of this zener?

(21)   In the circuit of *Figure 13.18(b)*, show that the power dissipated in the zener is $I_z I_L R_L$.

(22)   A 20 V regulated d.c. supply is required from a 50 V d.c. input. A 20 V zener is employed having a power rating of 2 W. What series resistor is needed?

(23)   If the zener of the previous problem regulates down to a current of 5 mA, calculate the greatest and least supply voltage between which regulation will be obtained, given that the load resistance is 1 kΩ.

(24)   A 68 V zener diode is used to supply a variable load from a 100 V d.c. source. If the series resistor used is 730 Ω, what minimum wattage rating for the zener would you select?

(25)   A zener diode is to provide a 16 V regulated output from a 20 V unregulated source. The load resistance $R_L = 2$ kΩ and the zener operating current is 8 mA. What series resistor is required and what power will it dissipate? What preferred values for resistance and power rating would you select in practice?

(26)   Explain the operation of the power supply shown in *Figure 13.27*. If an output of 15 V is required from this circuit, suggest (a) a suitable input voltage across C_1, (b) a voltage and power rating for zener diode D_3, (c) a value for R_1.

(27)   Given that the stabilization ratio S of a zener circuit is defined as

$$\frac{\text{small change in output voltage}}{\text{small change in input voltage}} = \frac{\Delta V_o}{\Delta V_i}$$

for a constant value of load resistance, show that $S = R_z/(R_z + R_s)$. What approximations, if any, have you made?

(28)   The circuit of *Figure 13.28* shows a zener diode being used in a meter protection circuit. The meter has an f.s.d. of 500 μA and an internal resistance of 360 Ω, and is scaled 0–10 V. If the zener is an 8.2 V device, find values for R_1 and R_2 so that when $V = 10$ V the diode conducts and the meter is protected against overloading. Explain how this circuit actually works, and mention any possible disadvantage it might have in practical application.

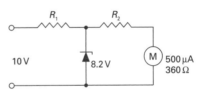

Figure 13.28

(29)   Sketch a model for a zener diode having the following characteristics:

Nominal voltage at 5 mA = 9.1 V

Maximum slope resistance at 5 mA = 10 Ω

A linear gradient above $I_z = 1$ mA

This zener stabilizes a load current of 15 mA at 9.1 V from a 24 V nominal d.c. source. If this source varies by 2 V, what will be the load voltage change?

(30)   The reverse characteristic of a certain zener diode is given by the equation $V = 9.1 + 5.01 I_z$ (I_z in amperes) for values of I_z greater than 10 mA. It is used in a simple regulator circuit for a load which can vary between 0 and 150 mA from a nominal 25 V d.c. source which may vary by +5 per cent.

What series resistor is required if the zener current does not fall below 10 mA for *all* input and output conditions? For this value of resistor calculate (a) the minimum power rating of the zener, (b) the maximum variation in the output voltage.

(31)   A load draws a current which may vary from 10 to 100 mA at 100 V. A zener regulator consists of a series resistor of 200 Ω and a zener diode represented by the model $V_z = 100$ V, $R_z = 20\ \Omega$. If the zener minimum current is 10 mA, then for a load current of 50 mA, find the permissible variation in V_i corresponding to a 1 per cent increase in V_o.

(32)   A voltage source with an internal resistance of 100 Ω is to supply a current of 0.1 A to a 310 Ω load. A 30 V zener with a slope resistance of 10 Ω and a maximum I_z of 0.3 A is to be used to reduce load voltage variations. (a) Draw an appropriate circuit diagram and calculate V_i and V_o. (b) If the source voltage falls by 10 per cent, what will be the percentage change in V_o? (c) For the same load, calculate the new I_z and V_o if the source voltage falls to 29 V.

(33)   What is the purpose of the diodes in the circuits of *Figure 13.20*?

(34)   A diode is often placed across the integrated regulators of *Figure 13.20*. What purpose does this serve? Indicate the polarity of the diode in each case to suit your explanation.

(35)   What would happen to the output voltage, if anything, in the circuit of *Figure 13.27*, if the zener diode went (a) open-circuited, (b) short-circuited?

14

Combinational logic

> *Basic gates*
> *Truth tables*
> *Gate combinations*
> *Integrated logic and characteristics*
> *Logical algebra*
> *TTL and CMOS systems*

Most computer and digital systems use binary code to represent all numbers and quantities. This is done because it allows the required number of digits to be represented by only two symbols, 0 and 1. For any circuit system based on decimal counting, ten such digits would need to be employed and this would lead to an unmanageable circuit arrangement. So the use of binary counting is a convenient philosophy since two very determinate states of current or voltage can be assigned to the two symbols we require: 'switch OFF' for zero and 'switch ON' for 1.

The basic building blocks of digital systems are therefore logic gates, so called because they manipulate the two binary states in a logical way. We briefly revise the basic examples of these gates at this point as a necessary introduction to the requirements of the syllabus which cover combinational and sequential logic systems.

BASIC GATES

AND gate

Figure 14.1(a) shows a simple circuit with two switches wired in series. The lamp in the circuit will be only ON if switch A is on and switch B is on. This arrangement illustrates the logical AND function, which is represented by the symbol shown on the left of the diagram. Calling the output F we can say

A and B = F

or, more usually

A.B = F

the dot standing here for AND. This is the Boolean expression for a two input AND gate. Such a gate can have any number of inputs; the expression for a three-input AND gate, for example, would be

A.B.C = F

We can now gather the information that the circuit symbol tells us into the form of a table, and such a table shows us the output state for each possible input combination. This is known as a truth table and *Figure 14.1(a)* shows the truth table for the two-input AND gate on the right. We now have the gate symbolization, the circuit representation and the truth table. Since there are two binary input variables, A and B, there are $2^2 = 4$ possible input combinations, these being given in the truth table. The four possible states of the output for each of these input combinations are indicated in the F column.

It is important whenever you set up a truth table always to show the output state for all possible input combinations. An easily remembered way to do this is to put down the input states in binary counting sequence, e.g. 00, 01, 10, 11 as for the two-input example above. For a three-input AND gate there will be $2^3 = 8$ possible combinations; these will be the binary sequence 000, 001, 010, 011, 100, 101, 110 and 111.

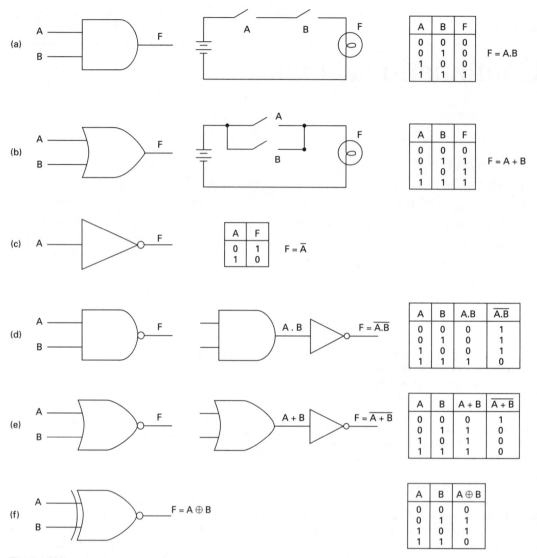

Figure 14.1

OR gate

Let the switches A and B now be connected in parallel as shown in *Figure 14.1(b)*. This time the lamp will light if either A or B is closed (either = 1) or both A and B are closed (both = 1). As before, there are four possible switch arrangements, and if either A or B = 1 or if A = B = 1, the circuit will be closed and the output F will be 1. Such a circuit represents the logical OR gate, or more strictly the *inclusive* OR gate, since its output is 1 when A = B = 1, additional to those conditions where A = 1 or B = 1. The symbol for the OR gate is seen on the left of the figure, with the truth table on the right. Again, the A and B input states have been set down in all four possible ways;

only if A or B (or both) are present will there be an output F. From the table, the Boolean expression can be deduced to be

A or B = F

or, more usually

A + B = F

The plus sign indicates the OR function; it must not be confused with its meaning for addition as it does for ordinary algebra. An OR gate can have any number of inputs in the same way as the AND gate may.

Follow the next worked examples carefully to check your revision of previous work on these two basic logical gates.

Example (1)   How many rows would there be in a truth table for a circuit containing n switches?

Solution   Since the variable quantities of logic have two states or values, a certain number of binary variables taken together will lead to a finite number of possible combinations. For example, if our variables are two switches each of which have two operating positions, both switches taken together provide four possible combinations. If we used three switches, each of which again have two operating positions, we would get eight possible combinations, since for each of the above cases two of the switches can be combined with the two positions of the third. Hence, four switches would result in 16 combinations and so on. The relationship is not difficult to deduce as it goes up in powers of 2; the number of possible combinations of the variables (or switch positions) is 2^n where n is the number of switches.

Example (2)   Draw circuit diagrams of switch arrangements and deduce the truth tables for each of the following logical expressions

(a) A + (B.C)   (b) A.(B + C)

Solution   (a) This expression reads as 'A *or* (B *and* C)'. So there is an output F when *either* A is operated or B *and* C are operated. We must have three switches in this circuit: the AND function represents a series arrangement of switches B and C, while the OR function puts switch A in parallel with them. The circuit is then assembled as shown in *Figure 14.2*. Satisfy yourself that this circuit will indeed perform the requirements stated in the given logical expression.

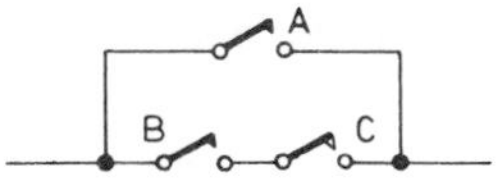

Figure 14.2

For the construction of the truth table, there are three switches and these will have eight possible combinations. So the table must have eight rows, and writing these in order under A, B and C headings gives us the first three columns of the table. To complete the fourth column, we must notice the conditions where the lamp is either lit and F = 1, or the lamp is unlit and F = 0. The lamp will be lit, of course, whenever A *or* (B *and* C) are switched on, hence we get the truth table of Table 14.1:

Table 14.1

A	B	C	F
0	0	0	0
0	0	1	0
0	1	0	0
0	1	1	1
1	0	0	1
1	0	1	1
1	1	0	1
1	1	1	1

Whenever B or C (or both) are open, together with A open, there can be no output, and these conditions provide the 0 output in the first three rows of the truth table. In the fourth row, B and C are both closed and the output is 1. After this, A is closed for each remaining cases, and so the output will be 1 irrespective of what B and C happen to be.

(b) This expression reads as 'A and (B or C) = F', so an output will be obtained when A together with B or C are closed. There are again three switches in the system: the OR function represents a parallel arrangement of switches B and C, while the AND function puts switch A in series with them. The circuit is then assembled as shown in *Figure 14.3*.

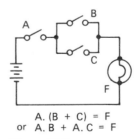

A. (B + C) = F
or A. B + A. C = F

Figure 14.3

For the truth table there will be eight rows identical with those set out in the previous example. This time, an output F is only obtained if A = 1 and either B or C (or both) = 1 (*Table 14.2*).

Table 14.2

A	B	C	F
0	0	0	0
0	0	1	0
0	1	0	0
0	1	1	0
1	0	0	0
1	0	1	1
1	1	0	1
1	1	1	1

As A = 0 in the first four rows, F must also be 0, irrespective of what B and C happen to be. The fifth row has A = 1 but both B and C are 0, hence F remains at 0. Only the last three rows satisfy the condition of A present with B or C (or both) present at the same time.

Try the next two problems for yourself.

(3) Draw the circuit diagram and set up the truth table applicable to the logical function A + B + C = F.

(4) How many switching operations are possible with the circuit of *Figure 14.4(a)*. Write down a logical expression for this circuit and draw up a truth table.

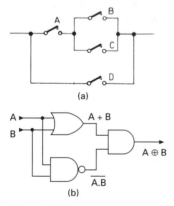

(a)

(b)

Figure 14.4

We return now to *Figure 14.1* and other logical gate systems.

NOT gate

The NOT gate (or inverter) simply provides an output F that is the opposite, inverse, negation or comple- ment, of the input. Hence for a logical 1 at the input, the output is logical 0, and conversely. The Boolean expression for this is

$$F = \overline{A}$$

the bar over the input indicating inversion, read as 'not A'.

There is clearly only one possible input to a NOT gate. The symbol this time is shown in *Figure 14.1(c)*, and you should have no difficulty in following the truth table for this circuit!

NAND gate

This is basically a combination of an AND gate fol- lowed by a NOT gate to form a NOT-AND or NAND gate. *Figure 14.1(d)* shows this arrangement and the symbol for a single combined gate. The small circle shown at the output of the NAND gate indicates that the output from this device is simply the inversion of those for the AND. The Boolean expression for the NAND gate is

$$\overline{A.B} = F$$

Hence the output F is 1 whenever *any* or *all* of the inputs are 0.

NOR gate

This gate, like the NAND, is a combination: an OR gate followed by a NOT gate to make up a NOT-OR or NOR gate. *Figure 14.1(e)* shows this arrangement together with the symbol for a single combined gate. The addition of the small circle at the output again indicates that the output from this device is simply the inversion of those for the OR gate. The Boolean expression for the NOR gate is

$$\overline{A + B} = F$$

Hence the output is 0 if any of the inputs are 1.

XOR gate

The last of the basic gates which are of interest to us is the XOR gate or the *exclusive* OR gate. We have noted above that for the inclusive OR gate the output is 1 whenever *any* or *all* of the inputs are 1.

In the exclusive OR, as the truth table of *Figure 14.1(f)* tells us, the output will be 1 only if the two inputs disagree; for both (or all) inputs at 1 the output is 0. We may write the Boolean expression for the XOR gate, therefore, as

$$F = A.\overline{B} + \overline{A}.B \text{ or } (A + B)\,\overline{A.B}$$

which is read as 'A or B but not A and B'. The gate is sometimes called an *inequality comparator* because it provides an output of 1 only if the two inputs disagree with each other.

The exclusive OR is represented by a special symbol to distinguish it from the inclusive case and the Boolean expression is reduced to this more compact form:

$$A \oplus B = A.\overline{B} + \overline{A}.B = (A + B) \overline{A.B}$$

An exclusive OR can be made up from a combination of other gates, one example of which is shown in *Figure 14.4(b)*. Here an AND, a NAND and an OR gate are used. The combination of the input states should be self-explanatory.

LOGICAL EQUIVALENCE

A constantly recurring problem in the design of logic circuits is how to assemble a combination of the basic gates to realize a desired function. There are two methods which will be of interest to us: by the comparison of truth table outputs or, more directly, by the application of Boolean algebra. The first method is more tedious than the second but it is a method worth knowing. The choice between the two methods can be likened to the solution of an a.c. problem by scaled diagram or by direct trigonometry. The scaled diagram is the more tedious of these and it has slight inaccuracies brought about by the quality (or lack of it) of the measurements and angles of the phasor lines. The trigonometric method, on the other hand, provides results to whatever accuracy we want.

In logical equivalence, however, the comparison of truth tables does give us a positive result: right or wrong, without half measures or inaccuracies, small or otherwise. For this reason we will go through a few illustrative examples of the method before looking at some not too difficult applications of direct Boolean algebra.

TRUTH TABLE COMPARISON

Suppose we have two boxes, one containing the switching circuit shown in *Figure 14.5(a)* and the other the circuit shown at *(b)*. The logical equation for circuit *(a)* should be no problem; it is

$$A = F$$

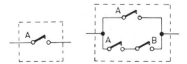

Figure 14.5

For circuit *(b)* you should be able to deduce that

$$A + (A.B) = F$$

As before, the F simply signifies that we get an output signal when the switches are operated in the manner indicated by the equations. If then we are to demonstrate that either equation represents identical circuit conditions, we should be able to say

$$A = (A.B) = A$$

With a little thought we can see that this equivalence is true: if switch A is operated, the circuit is completed in either of the boxes. The presence of switch B is irrelevant, so too is the other A switch in series with it. Only the upper A switch is strictly necessary, hence the circuit at *(b)* does exactly the same job as the more simple arrangement at *(a)*.

The equivalence of two logical equations is a matter of great importance in the design of logical systems, since the ability to reduce a complicated circuit with a simpler one doing exactly the same job enables the designer to achieve a needed result with a minimum number of circuit elements. The above, of course, was a very elementary example, but can we prove the equivalence of two equations without recourse to diagrams like those shown? Of course the answer is yes, and the comparison of truth tables is our first investigation.

The method is quite simple: a truth table is constructed for each of the expressions we are comparing for equivalence or otherwise. If the two expressions have the same output truth column, then the expressions are equivalent. If any single term in the output is different from the other, then the expressions are not equivalent. Worked examples will now illustrate this method.

Example (5) Show, using a truth table, the A + (A.B) = A.

Solution This is the equation we looked at above in terms of switching circuits. To verify the equality we draw up a truth table as follows:

A	B	A.B	A + (A.B)
0	0	0	0
0	1	0	0
1	0	0	1
1	1	1	1

The first two columns, as usual, represent all possible combinations of the quantities A and B. The third column is the logical AND function A.B and the fourth is constructed in accordance with the logical OR function, A + (A.B). Comparing this fourth column with the first, they are seen to be identical, hence the statement A + (A.B) = A is shown to be true.

Example (6)   Prove that $A + \overline{A}.B = A + B$, and verify the equivalence in terms of switching circuits.

Solution   As before, we draw up a truth table for both expressions and compare them.

A	B	$\overline{A}$	$\overline{A}.B$	$A + \overline{A}.B$	A + B
0	0	1	0	0	0
0	1	1	1	1	1
1	0	0	0	1	1
1	1	0	0	1	1

Here the third column ($\overline{A}$) is the negation of the first column (A). The fourth column is the logical AND function for $\overline{A}.B$ and has the value 1 only when both $\overline{A}$ and B are simultaneously 1, i.e. the second row condition. The fifth column is the logical OR function and has the value 1 when either the A column or the $\overline{A}.B$ column (or both) have the value 1. The last column is the OR function A + B.

Comparison of the fifth column with the sixth shows them to be identical, hence the statement $A + \overline{A}.B = A + B$ is true.

Figure 14.6(a) shows the circuit representation of $A + \overline{A}.B$ and *(b)* shows the representation of A + B. These should be identical in circuit function. In (a) the circuit is completed if A is closed (note that $\overline{A}$ then opens) or if A is left open (so that $\overline{A}$ remains closed) and B is closed, i.e. $A + \overline{A}.B$ is satisfied. All that is necessary for circuit completion then is the closure of either A or B; switch $\overline{A}$ is redundant to the operation. Circuit *(b)* representing A + B is then identical with circuit *(a)*.

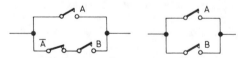

Figure 14.6

Example (7)   Use truth tables for (a) the NAND gate, (b) the NOR gate, to verify the equations

$$\overline{A.B} = \overline{A} + \overline{B} \quad \text{and} \quad \overline{A + B} = \overline{A}.\overline{B}$$

respectively.

Solution   We reproduce the NAND table from *Figure 14.1(d)* in the table below. We make additions to this so that the verification of the first of the equations can be made. These additions are columns for $\overline{A}$ and B and for $\overline{A} + \overline{B}$; we already have a column for A.B.

A	B	A.B	$\overline{A.B}$	$\overline{A}$	$\overline{B}$	$\overline{A} + \overline{B}$
0	0	0	1	1	1	1
0	1	0	1	1	0	1
1	0	0	1	0	1	1
1	1	1	0	0	0	0

The columns for $\overline{A.B}$ and $\overline{A} + \overline{B}$ are identical, the given equation is therefore verified. What this result tells us (and it is an important one in Boolean algebra) is that when a negation bar is drawn across both A and B the inversion involves a negation of A to $\overline{A}$, AND to OR, and B to $\overline{B}$. Hence $\overline{A.B} = \overline{A} + \overline{B}$.

The truth table for the NOR gate of *Figure 14.1(e)* has been reproduced below. By making additions to this as we did for the NAND gate above, we can verify the second equation. These additions are A and B, so providing us with A.B; this can then be compared with A + B which we have already.

A	B	A + B	$\overline{A + B}$	$\overline{A}$	$\overline{B}$	$\overline{A}.\overline{B}$
0	0	0	1	1	1	1
0	1	0	1	1	0	1
1	0	0	1	0	1	1
1	1	1	0	0	0	0

Comparing the columns for $\overline{A + B}$ and $\overline{A}.\overline{B}$ we find they are identical, hence the relationship is verified.

Keep these two equations in mind; they are what are known as De Morgan's rules and have many applications in logical algebra.

Try the next two examples for yourself.

---

(8)   Use truth tables to verify the following equations:

(a) $A + \overline{A}.B = A + B$
(b) $A + (B.C) = (A + B).(A + C)$

You will need eight rows for this last one.

(9)   Write down the logical expression representing the circuit shown in *Figure 14.7*. Verify

that your solution can be represented by the much simpler expression F = A.B.E.

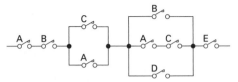

Figure 14.7

RULES OF LOGICAL ALGEBRA

Although the plus (+) and multiplication (.) signs used in ordinary algebra are not to be confused in any way with the OR (+) and the AND (.) signs associated with logical expressions, most rules for the manipulation of logical algebra follow those of ordinary algebra. There are a few surprises which only experience will overcome, but a convenient way of looking at logical algebra is to visualize the symbols and their interconnections in terms of the basic electrical circuits in the way we have been doing so far in this chapter. With experience, you will find that you can eventually dispense with this approach as the direct algebraic method becomes familiar. In the meanwhile, fix the following equivalents in your mind:

(a) 0 is equivalent to an *open* circuit;
(b) 1 is equivalent to a *closed* circuit;
(c) A (or B or C etc.) is equivalent to a switch which may be closed (=1) or open (=0);
(d) $\overline{A}$ is equivalent to a switch which may be closed (=0) or open (=1).

For example, A.1 is equivalent to a switch A in series with a closed circuit element, see *Figure 14.8(a)*. Since the closed element is always closed, the function of the circuit depends only on the state of A. Hence the 'product' A.1 = A.

Again, A + $\overline{A}$ is equivalent to two parallel switches,

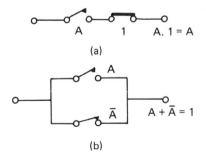

Figure 14.8

one closed (A) when the other is open ($\overline{A}$). Whatever the setting of the switches, the circuit is *always* closed by way of one of them. Hence A + $\overline{A}$ = 1. This is rather like saying that the result of a toss of a coin will result in either HEADS or NOT-HEADS; your answer is a dead certainty!

We can now make a table of what we might call a listing of product and sum rules for logical algebra; although some of the results may not seem very logical at a first glance, they are easily interpreted in terms of the circuit symbols appended to them. The symbols 1 and 0 are treated as 'constants' in the sense that the closed and open circuit which they respectively represent are *permanent*. The symbol A (or B or C etc.) is a variable in that it may adopt either of the two possible states of 0 or 1.

You may not be called on to manipulate problems in logical algebra in some parts of the syllabus, but here are some worked examples as a guide to the method.

Example (10)   Verify the following logical equations:

(a) $\overline{A}.\overline{B} + \overline{A}.B + A.\overline{B} = \overline{A} + A\overline{B}$
(b) $AB\overline{C} + ABC = AB$
(c) $\overline{A + B + C} = \overline{A}\,\overline{B}\,\overline{C}$
(d) $AB + A\overline{B} + \overline{A}B = A + B$
(e) $\overline{A}BC + AB\overline{C} + ABC = AB + BC$

Solutions   As we would do in ordinary algebra, we take the left-hand side of each equation and deduce that it is equal to the right hand side.

(a) $\overline{A}.\overline{B} + \overline{A}.B + A.\overline{B} = \overline{A} + A\overline{B}$

L.H.S.:

$\overline{A}.\overline{B} + \overline{A}.B + A.\overline{B}$

$= \overline{A}(\overline{B} + B) + A.\overline{B}$ by factorizing

$= \overline{A}.1 + A.\overline{B}$ since $(\overline{B} + B) = 1$

$= \overline{A} + A.\overline{B}$ since $\overline{A}.1 = \overline{A}$

$=$ R.H.S.

(b) $AB\overline{C} + ABC = AB$

L.H.S.:

$AB\overline{C} + ABC$

$= AB(\overline{C} + C)$ by factorizing

$= AB$ since $(\overline{C} + C) = 1$

$=$ R.H.S.

(c) $\overline{A + B + C} = \overline{A}\,\overline{B}\,\overline{C}$

Here we apply De Morgan's rules: on the L.H.S. the *individual* elements A, B, C are not negated but the whole term is; hence we can negate them separately and change all the OR signs to AND. This then gives us the R.H.S.

(d) $AB + A\overline{B} + \overline{A}B = A + B$

L.H.S.:

$AB + A\overline{B} + \overline{A}B$

$= A(\overline{B} + B) + \overline{A}B$ by factorizing

$= A + \overline{A}B$ since $(B + B) = 1$

But $A + \overline{A}B$

$= (A + \overline{A}).(A + B)$ by the distributive rule

$= A + B$ since $(\overline{A} + A) = 1$

$= R.H.S.$

(e) $\overline{A}BC + AB\overline{C} + ABC = AB + BC$

L.H.S.:

$\overline{A}BC + AB\overline{C} + ABC$

$= \overline{A}BC + AB(\overline{C} + C)$ by factorizing

$= \overline{A}BC + AB$ since $(\overline{C} + C) = 1$

$= B(\overline{A}C + A)$ factorizing for B

$= B(A + C)$ by the distributive rule

$= AB + BC$

$= R.H.S.$

(11)   Verify the logical equations:

(a) $(A + B).(A + B) = A$

(b) $A + (BC) = (A + B).(A + C)$

(c) $\overline{A}C + B\overline{C} + \overline{A}BC + ABC = B + \overline{A}C$
(Hint: factor out twice)

(d) $AB\overline{C} + ABC + \overline{A}BC = B(A + \overline{A}C)$

(e) $\overline{\overline{A}\overline{B} + C + D} = ABCD$ (Hint: use De Morgan twice and remember that $\overline{\overline{A}} = A$.)

ELECTRONIC SYSTEMS FOR LOGICAL OPERATIONS

All logic gates so far discussed have been represented as boxes of various diagrammatic shapes, with inputs consisting of 'signals' in the form of binary digits, 0 and 1. These signals which represent one or other of two possible states are, in practice, voltage levels.

There are two conventions about these levels: there is the 'positive logic' convention and the 'negative logic' convention. In positive logic the more positive voltage level is assigned the logic state 1 and the other voltage the logic state 0. In negative logic the converse applies. Since positive is the more commonly used of these conventions in practice, we will stick to this throughout the remainder of this chapter; and it does conform to our commonsense view of things. So we will take a high level voltage = 1 and a low level voltage = 0. 'High' and 'low' levels here are relative terms; we will see that specific and relatively tightly controlled levels are standardized in practical circuits.

There are a number of ways of converting the mechanical switching systems so far encountered into electronic circuits which will perform the same functions quickly and automatically; among these are diodes, transistors, silicon-controlled rectifiers and so on, used in a variety of combinations with passive components to secure the desired logical functions. Although some of these combinations have passed into obsolescence over the years, we will look at the general overall range briefly to provide an insight into, and a familiarity with, these basic techniques.

Primitive models

Consider a series combination of a diode and a load resistor, as shown in *Figure 14.9(a)*. If the input terminals A and B of this circuit are short-circuited so that the input is zero or logical 0, the diode will be forward biased and a current will flow down through the resistor. The output voltage will then be equal to that across the conducting diode which, for a germanium device, is about 0.25 V. We can treat this level as low or logical 0. The output is therefore low when the input is low. Suppose now that the input is raised to 5 V or more; the diode will switch off and the output will rise to approximately 5 V which is the supply voltage. The output is consequently high if the input is high. Notice that the input and output levels correspond closely with each other, that is, there is no inversion.

We have already seen that a transistor can perform the function of a switch, and *Figure 14.9(b)* shows one connected for this purpose. With zero input, the base-emitter junction is reverse biased by $-V_B$, the collector current is zero and the output is about 5 V. A positive input greater than V_B will switch the transistor on and a large collector current will flow through R_3. The collector potential, and hence the output, will

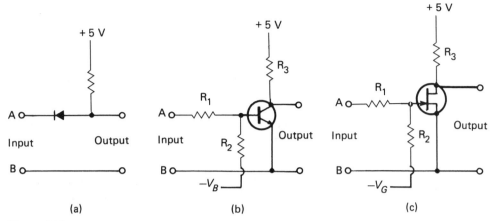

Figure 14.9

consequently fall to a low level, typically 0.1 V. Again the output levels are closely equal to the inputs, but this time inversion has taken place. The circuit here is doing the service of a NOT gate.

The junction FET will behave similarly when zero and 5 V levels are applied to its input terminals, see *Figure 14.9(c)*.

We can now apply the principles of these primitive switching models to the design of practical gates in general.

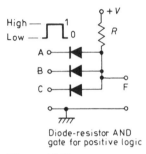

Diode-resistor AND gate for positive logic

Figure 14.10

Multiple input gates

We begin with a look at an elementary AND gate built on the diode resistor circuit of *Figure 14.9(a)*. We assume that the gate has three inputs A, B and C and that the magnitude of the supply voltage V is set to be a volt or so above that of logical level 1 of the input levels. The circuit is shown in *Figure 14.10*.

Suppose all input signals are simultaneously high; then all the diodes conduct because of the more positive anode voltage V, and the output is connected to the high level inputs directly through the diodes. Hence the output F is high. The switch analogy in this case is shown in *Figure 14.11(a)*. Assume now that any one of the inputs, A, for example, goes low (zero volts). The cathode of diode A is now low but the anode is high, hence the diode conducts and connects the output to the low level A input. This has the effect of bringing the anodes of diodes B and C also to a low level since all anodes are connected together. Hence the presence of high level conditions at the cathodes of B and C in conjunction with the low level anodes causes these two diodes to switch off, and no signals from inputs B and C can reach the output. The output is correspondingly low, as the switch analogy of *Figure 14.11(b)* shows. By exactly the same reason-

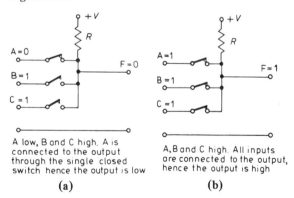

A low, B and C high. A is connected to the output through the single closed switch hence the output is low

(a)

A,B and C high. All inputs are connected to the output, hence the output is high

(b)

Figure 14.11

ing (which you should go through in detail for yourself), the output will be low whenever any two of the inputs are low and the third is high or all three are low. Only the simultaneous presence of high level inputs at all three input terminals produces a high at the output, hence this circuit performs as a logical AND for positive logic inputs.

Input pulses are not always of the same duration, so it is possible for the AND gate to act as a detector of coincidence among several inputs. A possible train of positive input pulses at points A, B and C is shown

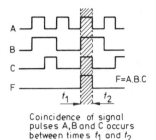

F=A.B.C

Coincidence of signal
pulses A,B and C occurs
between times t_1 and t_2

Figure 14.12

in *Figure 14.12*. Only when inputs A, B and C are simultaneously high will there be an output F. From the diagram, this state of affairs occurs only during the time interval t_1 to t_2.

(12)   Complete the following:

(a)   The AND gate is sometimes referred to as a . . . gate.
(b)   The AND gate behaves like . . . connected switches.
(c)   The output of an AND gate with one input held low will be at logic. . . .

Diode–resistor OR gate

Look now at the diode–resistor circuit of *Figure 14.13*. The diode cathodes are returned to a negative voltage point (this is sometimes simply zero level) and the input pulses conform to positive logic. This circuit will perform the inclusive OR function in which any single high input (or two or all three) will produce a high output. When all the inputs are low, all diodes

conduct, the cathodes being more negative than the anodes. The output is then connected directly to the input and so the output is low. This condition relates to the first row of the truth table in the figure. Suppose now that any input, say C, goes high. Diode C will conduct and the output will be high; diodes A and B then have high cathodes but low anodes and consequently switch off. This condition clearly applies for any other input, or combinations of the inputs, being high, hence the truth table can be completed as shown.

(13)   How would the truth table shown in *Figure 14.13* be modified if it applied to an exclusive OR gate?

(14)   *Figure 14.14* shows a train of input pulses which are applied in turn to three-input (a) positive AND gate, (b) inclusive OR gate, (c) exclusive OR gate. The time scale indicates milliseconds of duration. Between what times are there high level outputs from (a) the AND gate, (b) the OR, (c) the exclusive OR gate?

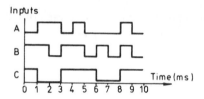

Figure 14.14

(15)   A diode–resistor OR gate provides a high output when one or more of its diodes are . . . biased.

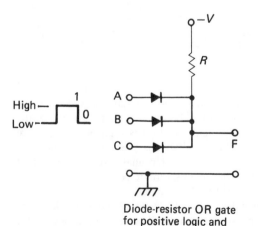

A	B	C	F
0	0	0	0
0	0	1	1
0	1	0	1
0	1	1	1
1	0	0	1
1	0	1	1
1	1	0	1
1	1	1	1

Diode-resistor OR gate
for positive logic and
truth table

Figure 14.13

DIODE–TRANSISTOR LOGIC

The logic gates available in modern integrated circuits are rather more sophisticated than the simple diode–resistor combinations we have discussed so far, and we now look at a few different families of logic systems to see how development has taken place.

The obvious step in possibly improving on the diode–resistor circuits is to follow the diodes with a transistor switch in a single system. This is a basic diode–transistor logic gate (DTL) which can perform either the AND plus NOT function or the OR plus NOT function, that is, either the NAND or the NOR function. A DTL NAND gate is shown in *Figure 14.15*. The value of resistor R_1 is chosen to suit the switching requirements of the transistor T_1. When all inputs A, B and C are high, the diodes are switched off and the transistor is held in saturation by the positive bias on its base. The output at its collector is then low. If one or more inputs are taken low, the relevant diode (or diodes) switch on and the base of the transistor goes low. If this base voltage falls below about 0.6 V, the transistor will turn off and its output will go high. We thus have the NAND function where the output is low only if all the inputs are high.

There is, however, a difficulty with this circuit as it stands. If silicon diodes are used, as they invariably are, the transistor will either not be turned completely off or it will only just be turned off, since the forward voltage drop of the diode will also be in the region of 0.6 V. This situation is much too risky for comfort. Some improvement will be possible by using germanium diodes where the forward drop will be about 0.2 to 0.3 V, but a better arrangement is to use a so-called level shifting diode as shown in *Figure 14.16*. As before, when all the inputs are high, all the diodes on the inputs are cut off and the current flowing through R_1 and D_4 (which is forward biased) makes the base of the transistor sufficiently positive to ensure saturation, the voltage level at point Z being $2 \times 0.6 = 1.2$ V, not 0.6 V as it was before D_4 was included. It is quite usual to have two diodes in series for D_4, so raising the potential at point Z to about 1.8 V.

When any one or more of the inputs go low, the voltage at point Z becomes 0.6 V but this time the presence of D_4 prevents sufficient current flowing into the base of the transistor to permit it to conduct, hence it switches off and the output goes high.

A DTL NOR gate is shown in *Figure 14.17* and you should have no difficulty in deducing how it operates. Look at it as an OR gate followed by a NOT gate. The transistor is switched hard on when one or more

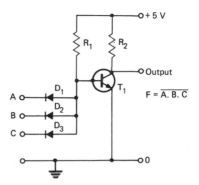

Figure 14.15

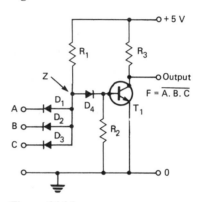

Figure 14.16

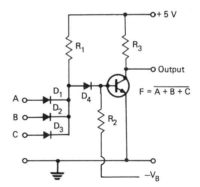

Figure 14.17

inputs are taken to logical 1, that is, the output is high only when *all* inputs are low.

TRANSISTOR–TRANSISTOR LOGIC

In integrated circuit technology it is just as simple to make a special transistor as it is a diode, and the next step is to replace the separate diodes used in the DTL circuits with a multiple emitter transistor as shown in *Figure 14.18*. Not only does this transistor replace the separate diodes, it replaces the level shifting diode D_4 of *Figure 14.16* as well by the presence of the base-collector junction. A clearer understanding of the way the special transistor does its job can be gained by interpreting the circuit in the form shown in *Figure 14.19* where its similarity to *Figure 14.16* will be apparent. Circuits of this sort are known as transistor–transistor logic or TTL.

When all input levels are high, none of the base-emitter junctions of T_1 is forward biased. The base-collector junction is, however, forward biased and current flows through this diode into the base of T_2 saturating it and causing the output to go low. So that T_2 can be kept saturated, it is necessary for point Z to be at about 1.2 V as we noted for the DTL circuit above. If one or more of the emitters are now taken low, those base-emitter junctions will switch on and the voltage at point Z will fall to about 0.6 V. As for the DTL circuit of *Figure 14.16*, this level will be insufficient to permit current to flow into the base of T_2 which will switch off and allow the output to go high.

Now although this circuit is an advance on the discrete diode systems, there is still a problem when it becomes necessary to connect the output of the gate to following gates. The ability of a logic gate to feed its output into a number of following gates without deterioration or erratic performance is known as its fan-out value. For a TTL form of gate element, this is typically 10, but to achieve this figure it is necessary to modify the output arrangements so far described in the circuits of *Figures 14.15* to *14.18*. We will look into problems associated with fan-out shortly; in the meanwhile, we look at a common form of output circuit found in most TTL integrated circuit gates.

The single transistor output stage so far considered is replaced by two transistors in what is known as a totem pole circuit, and this is illustrated in *Figure 14.20*. This diagram shows a complete integrated NAND gate which includes the multiple emitter input arrangement already discussed. For simplicity only two inputs, A and B, are considered. When A and B are both high, neither base-emitter junction of T_1 is forward biased but the base-collector junction is, so that current flows through it via R_1 into the base of T_2, switching it hard on together with T_4. With T_2 on the base-emitter potential of T_3 is too low for forward bias and T_3 is off. So if all inputs are high, T_2 and T_4 are switched on, T_3 is off, and the output is low. The diode D_1 is there to ensure that its forward voltage drop of 0.6 V will keep T_3 off when T_4 is on.

If now one or both of the inputs fall low, the emitter(s) of T_1 is/are effectively at zero potential and its collector voltage falls, turning off T_2. The collector

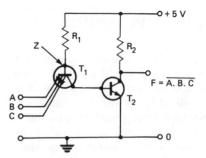

Figure 14.18

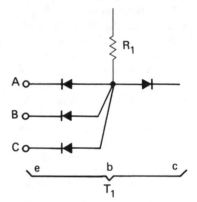

Figure 14.19

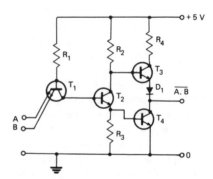

Figure 14.20

of T_2 then rises towards V_{CC} and the emitter falls to zero volts. T_3 then conducts (base high) while T_4 switches off (base low). The output then goes high. Notice from all this that when the output is high, T_3 behaves as a grounded collector (or emitter follower) stage which drives current into whatever load is connected to the output. This is known as current 'sourcing'. When the output is low, current flows from the load into T_4 which is saturated and has a very low resistance. This is known as current 'sinking'. So the totem pole arrangement presents a low resistance output in both logic states: either via T_4 when the output is low or via T_3 when the output is high. This ability to source and sink relatively large currents from and into low resistance paths gives the totem pole circuit a great advantage over the single transistor output stage. We will go into this in the next section.

Example (16) For the DTL NAND gate of *Figure 14.21*, the output transistor has $\beta_E = 50$ and a V_{CE} of 0.2 V when saturated. If input A is at 5 V and input B is at 0.4 V, what will be the output voltage at the collector of T_1? If input B now rises to 2 V, what will be the output voltage?

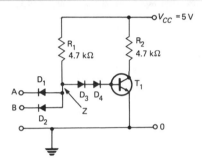

Figure 14.21

One output (A) is high, the other (B) is low. Since the state of a NAND gate is determined by the lowest input and B = 0.4 V, the potential at point Z (V_Z) cannot exceed 0.4 + 0.6 = 1 V. If current flows through the level-shifting diodes D_3 and D_4 the base-emitter voltage of T_1 is

$$V_{BE} = V_Z - 2(0.6) = 1 - 1.2 = -0.2 \text{ V}$$

This condition ensures that T_1 is completely cut off and the collector voltage stands at $V_{CC} = 5$ V.

When input B rises to 2 V, V_Z cannot exceed 2 + 0.6 = 2.6 V, but for T_1 to conduct, V_Z must exceed 3(0.6) = 1.8 V, hence base current will flow of a value

$$I_B = \frac{V_{CC} - 1.8}{R_1} = \frac{5 - 1.8}{4700} \text{ A} = 0.68 \text{ mA}$$

As the collector load R_2 is 4700 Ω, the greatest possible collector current would be

$$I_{C(max)} = \frac{V_{CC}}{R_2} = \frac{5}{4700} \text{ A} = 1.06 \text{ mA}$$

and a base current of $I_C/\beta = 1.06/50 = 0.021$ mA would be sufficient to produce this.

As the actual base current is 0.68 mA, the transistor is driven hard into saturation and the output voltage is now $V_{CE(sat)} = 0.2$ V.

This example shows that the input logic levels do not have to be equal or indeed up to the 5 V level for the state of the gate to be determined. What we refer to as 'high' or 'low' must clearly cover a range of voltage levels for a particular gate. We shall see that a 2 V level is right at the foot of the accepted range for which the term 'high' is applied.

(17)   Deduce that a NAND gate with all its inputs joined together behaves as a NOT gate.

(18)   Connect together two NOT gates and one NOR gate to perform the function of an AND gate.

(19)   Use two NOR gates and an AND gate to make an exclusive OR gate.

COMMERCIAL INTEGRATED LOGIC

We have now covered all the basic logic gates and a few of the possible combinations. Throughout we have assumed that these circuits are made up of discrete component parts, resistors, diodes and transistors, and indeed, gates can be made up in this way for many experimental investigations. For very complicated logical systems, however, even going no further than a simple hand-held calculator where several hundred gates may be necessary, such discrete component assemblies would become large and unwieldy, as well as consuming considerable power. Logic families have, as a consequence, appeared over the past 20 years or so in the form of integrated circuits and in a great variety of gate arrangements.

The earliest family was the 7400 series which used transistor–transistor logic and of which *Figure 14.20* is an example. This family is available to the present day at very low cost, each integrated package containing a number of identical gates. In the intervening period of evolution when a greater number of gates and functions were being called for, the basic TTL form was not wholly suitable for integration because the chip area necessary for each gate was too great and the heat dissipation excessive. The problem has been overcome by the introduction of large-scale integration (LSI) utilizing CMOS technology. Such circuits are found in the 4000 series, though this range in turn is being superseded by the 74HC series. We concentrate at this stage of the course on TTL and the 7400 series.

THE 7400 SERIES

TTL gates are found in everyday DIL (dual-in-line) packages having the general appearance shown in *Figure 14.22*, where pin number 1 is marked with a dot or indentation. The packages have usually 14 or 16 pin-outs and each contains a number of gates, the nominal 5 V operating voltage (V_{CC}) supply pins being common to all of them. The family is identified by a number of letters in the form $--74--$ or $--75LS--$. The LS series offers a superior performance over the standard range principally in respect of power consumption, but the internal gate arrangements are otherwise identical. The first letters are usually a code for the name of the manufacturer.

Table 14.3 shows five popular 7400 integrated circuits, together with the pin connections. The basic gate of the series is the 7400 itself which is a quad two-input NAND gate, each unit containing the circuit arrangement shown earlier in *Figure 14.20*. The 7402 is a quad two-input NOR gate, while the 7404 contains six inverters or NOT gates. The 7408 is a quad two-input AND gate and the 7486 is a quad two-input EXCLUSIVE OR gate. In a circuit system, one or all of the gates contained in the chip may be used, the V_{CC} supply being applied between pins 14

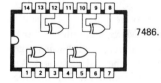

Figure 14.22

7408.   Quad two—input AND gate

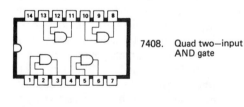

7486.   Quad two—input EXCLUSIVE OR gate

Table 14.3

(positive) and 7 (negative ground). All diodes, transistors and resistors are formed on the silicon chip as a compact integrated assembly, and the cost is little more in many cases than the price of a single discrete transistor.

(20)   TTL circuits are preferred to DTL because
(a)   they have a smaller fan-out;
(b)   they operate from a 5 V supply;
(c)   the multiple emitter transistor is easier to integrate than a number of diodes;
(d)   they are available in an extensive I.C. family.

(21)   The fan-out of a TTL gate is determined by
(a)   the output resistance of the gate;
(b)   the load current of the output;
(c)   the applied V_{CC} level;
(d)   the actual logic levels employed.

TTL CHARACTERISTICS

Manufacturers issue data sheets for each of their logic gates and a number of important operating parameters must be appreciated before any such gates are incorporated into a circuit system. You should obtain a few of these sheets and study them carefully. Here we shall consider some for their most important points.

1   Operating voltage. The 7400 series TTL gates operate from a nominal V_{CC} supply of 5 V, but there are limits to the variation which can be permitted from this figure (V_{CC} has an absolute maximum rating of 7 V), and if a variable power supply unit is being used in any experiment with TTL gates, great care must be taken that it is not set to a voltage greater than 7 V at switch-on or at any time subsequently. Any supply over this top limit will almost certainly destroy the gate. Using batteries is generally a safe way to avoid this possibility of damage; a 4.5 V battery, which when new will provide a terminal p.d. of some 4.75 V, is ideal. A 6 V battery may be used when nothing else is available, though the manufacturer's operating limit of V_{CC} is 4.75 to 5.25 V. Where it is important to keep the supply accurately at 5 V is in the matter of the speed of response of the gate. As V_{CC} is reduced, the time taken for a logical change to pass through the system increases. This may not be very important in simple bench experiments, but circuits such as high-frequency

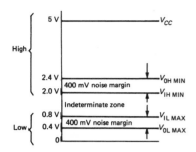

Figure 14.23

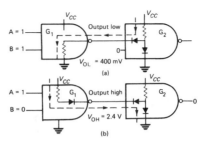

Figure 14.24

counters and dividers would not operate to their full specification under low V_{CC} conditions.

2   Logic levels and noise margin. Data sheets provide us with the voltage levels that a TTL gate will accept as a legitimate 1, or high, and the voltage levels it will accept as a logic 0, or low. We have, for convenience and simplicity, so far taken +5.0 V to be high and 0 V to be low, but as in the V_{CC} supply, there are limits to what the input levels may be in order to assure trouble-free operation. Generally, the minimum logic 1 level, V_{IHMIN}, is taken to be 2.0 V, and the gate is guaranteed to recognize an input of 2.0 V or more as a high and accept it as such. The maximum logic 0 level, V_{ILMAX}, is taken to be 0.8 V and any input signal equal to or less than 0.8 V is accepted by the gate as a legal low. *Figure 14.23* illustrates these level conditions. Notice that there is a region between 0.8 V and 2.0 V which comes into neither the high nor the low level category; this region is indeterminate and input signals must not settle at levels within this band.

Two other levels marked in *Figure 14.23* need our attention. These relate to the output voltage levels obtained from a gate. For the 7400 series, an output logic 1 is guaranteed to be equal to or greater than a minimum level V_{OHMIN} of 2.4 V. In actual practice, output highs are usually above 3.0 V. Make a note that the minimum output high (2.4 V) is 0.4 V or 400 mV higher than the minimum level required for an input high on a following gate. This 400 mV gap is known as the noise margin. Noise margin is a measure of the ability of a gate to reject a small noise impulse at its input and prevent it from erroneously changing state. A noise impulse has to be greater in magnitude than 400 mV before it can force the input to a following gate to go below the 2.0 V input high minimum. Similarly, a noise margin of 400 mV exists between the maximum logic 0 output level of 0.4 V and the maximum logic 0 input level of 0.8 V. Noise margins can be affected, and nullified, by excessive loading on a particular gate.

3   Sink and source currents. These terms were mentioned in passing earlier on; we now examine them in more detail. On the manufacturer's data sheets you will find output currents marked for logic 1 and logic 0 output voltage conditions. These are usually stated as 400 μV for logic 1 and 16 mA (often −16 mA) for logic 0 outputs, but we can divide these figures by a factor of 10 since they are given for a fan-out of ten, that is, when the gate output in question is connected to the inputs of ten others. We consider the case, then, of the output of one gate connected to the input of another, taking as our example the NAND gates of *Figure 14.24* shows the situation. Relate the 'internal' components shown in the gates to the totem pole output circuit in the one on the left, and part of the multi-emitter input circuit to the one on the right. Now in diagram *(a)* we assume that the output of gate G_1 is low; it then creates a current path (through the switched-on lower transistor of the totem pole) for the V_{CC} supply of gate G_2 to earth as indicated by the broken line. Gate G_1 is then sinking current as it pulls the input to gate G_2 low.

The manufacturer's data states that the maximum logic 0 input current for a 7400 TTL gate (I_{LMAX}) is −1.6 mA. The negative sign is simply a convention that indicates that the current (conventional) is flowing out of the relative input terminal; a positive sign indicates that the current is flowing into that terminal. So the figure of 1.6 mA means that for a TTL input to be pulled low, the output connected to it may have to sink a current of that magnitude to earth.

Figure 14.24(b) shows the situation when the output of gate G_1 goes high. A current now flows (conventionally) from the V_{CC} supply of gate G_1 (through the switched-on upper transistor of the totem pole) into the input of gate G_2, again indicated by the broken line. This current flows through gate G_2 to earth. So gate G_1 is said to source current to gate G_2 in order to pull the input to G_2 high. The maximum logic 1 input for a TTL gate is 40 μA, so gate G_1 may have to provide a current of that magnitude to earth.

We have noted that fan-out is a measure of the greatest number of gate inputs which may be connected to a single gate output without preventing the output from reaching the permissible high and low voltage levels. The fan-out for TTL gates is ten; the total available source current is guaranteed to be at least $(10 \times 40\ \mu A) = 400\ \mu A$, and the guaranteed sink current is $(10 \times 1.6\ mA) = 16\ mA$. You should now be able to appreciate the usefulness of the totem pole output circuit relative to the single transistor of the RTL circuits.

4   Open-collector outputs. In spite of what we have just said about the totem pole, the configuration has a disadvantage when two or more outputs from separate gates are to be connected together. You should be able to surmise for yourself what is likely to happen if you connect together the output terminals of two (or more) totem pole circuits. If, glancing back to *Figure 14.20*, the T_4 of one gate happens to be switched on when the T_3 of the other gate is switched on, a low resistance path to earth is provided through them and the relevant diode so effectively short-circuits one of the outputs. Resistor R_4 helps to reduce the current, but this resistor is not always included; make sure, then, that you never short the output of a TTL gate, either directly or by an attempt to connect two gates directly together.

One way out of the difficulty, of course, is to connect the outputs from, say, three NAND gates to a fourth gate as shown in *Figure 14.25*, this being followed by a NOT gate. The output is then clearly F = $\overline{AB}$. $\overline{CD}.\overline{EF}$. However, if the circuit of the totem pole is modified by omitting the upper transistor and its resistor (see *Figure 14.26*) we have what is known as an 'open-collector' output. When several outputs are now joined together, a single resistor can take the place of the original individual resistors; this is known as a pull-up resistor. When the transistor T_4 is off, this resistor brings the output to a high. The 7400 series has a family of gates with open-collector outputs: the 7401 and 7403, for example, are both quad two-input open-collector NAND gates, the 7405 is an open hex inverter and the 7409 is a quad two-input open-collector AND gate.

The open-collector output system tends to be noisier than the standard totem pole configuration and slower to respond to a change of state, and these are disadvantages. For this reason, the open-collector has given way in many cases to what is known as the three-state output logic where the output has the normal TTL low and TTL high, but also has a floating output

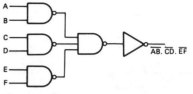

Figure 14.25

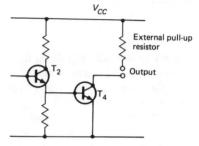

Figure 14.26

state created by turning off both output transistors with a separately applied signal.

THE 4000 SERIES

Most of your work in logic systems at this stage of your studies will be done with the 7400 TTL series of integrated circuits. It is necessary, however, to be aware of the 4000 series of logic gates which make use of the CMOS-based (or complementary metal oxide semiconductor) technology. These gates make use of the field effect transistor which gives a number of advantages over the TTL series that use the bipolar transistor. These advantages include a very small current consumption, a wider range of operating voltage, typically from 3 V up to 15 V, and a high fan-out capability brought about by the very high input resistance of CMOS gates.

Like the TTL series, the CMOS series is numbered from a basic 4000 upwards. The 4001, for instance, is a quad two-input NOR gate, the 4011 a quad two-input NAND gate and the 4049 a hex inverting gate.

The basic arrangement for a 4000 series inverter is shown in *Figure 14.27(a)*, with a NAND gate in *(b)*. The inverter is built up from one N-channel and one P-channel enhancement MOSFET as described in an earlier chapter. When the input is low (logic 0), the N-MOSFET is cut off and switched off and the P-MOSFET is switched on. Thus there is a low resistance path from the supply rail V_{DD} to the output; hence the output is high and closely equal to V_{DD}. When the input goes high (logic 1), the transistors change state and the output is taken low through the switched-on

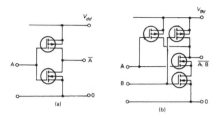

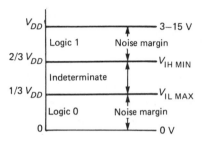

Figure 14.27

Figure 14.28

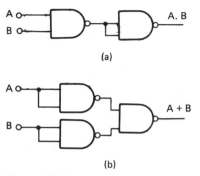

(a)

(b)

Figure 14.29

N-MOSFET. The circuit thus behaves as an inverter or NOT gate.

This gate can be extended into a NAND gate by the addition of two more devices, *Figure 14.27(b)*. It is left as an exercise for you to deduce that when the inputs A and B are both high the output is low, and when either A or B or both are low the output is high.

The logic levels for the 4000 series are shown in *Figure 14.28*. They are given as percentages of the applied V_{DD} which, as we have noted, can be anything from 3 V to 15 V. Notice that the noise margins here are one-third of the level of V_{DD}.

GATE ECONOMY

It is possible to make up any logical circuit combination to perform any logical function using only NAND or NOR units, and it is common practice for circuits to be designed in this way. The obvious advantages are (a) the economy obtained in bulk buying of gates, (b) the ease of replacement, hence ease of maintenance, and (c) the fact that logic levels remain nearly constant throughout the system.

We have noted earlier that a NAND gate with all its inputs connected together acts as a NOT gate. *Figures 14.29(a)* and *(b)* show, respectively, how NAND gates can be used to perform the AND and OR functions. The AND gate in diagram *(a)* uses two NAND elements, the first used as a NAND and the second as a NOT. The output is then the same as that of a single AND gate.

In diagram *(b)* three NAND elements are used to perform the function of an OR gate. The first two NANDs are employed as NOTs, followed by a normal NAND; the output is then a double-negated product which is equivalent, by De Morgan's theorem, to the OR gate A + B.

> (22) Using NAND gates only, make up circuits performing the functions of (a) NOR gate, (b) EX-OR gate.

It is an interesting point that though NAND and NOR gates may be used to make any other type of gate (and are often known as 'universal gates' for this reason), AND and OR gates can never be assembled in any manner to make NAND, NOR or NOT relationships.

> Example (23) Show that the circuit of *Figure 14.30*, which uses four different gate elements, can be replaced by a single NAND gate.

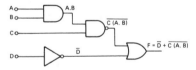

Figure 14.30

Solution This kind of question can be solved by the use of truth tables or by the application of logical algebra. We will illustrate the use of logical algebra in this case. On the diagram the various intermediate logical conditions have been indicated and the final output is F = $\overline{D}$ + $\overline{C(A.B)}$. This does not look like the output of a single NAND gate, but if we apply our algebraic rules we get:

$F = \overline{D} + \overline{C(A.B)} = \overline{D} + \overline{C} + (A.B)$
by De Morgan

Applying De Morgan a second time, we have

$\overline{D} + \overline{C} + \overline{(A.B.)} = \overline{D.C} + \overline{A.B}$

$= \overline{A.B.C.D}$

which is the output of a four-input NAND gate. Why use four gates when one will do?

SOME HINTS ABOUT USING TTL AND CMOS

We conclude with a few important hints about the way TTL and CMOS gates should be used and handled in bench experiments. CMOS devices are particularly sensitive to mishandling and careless work can easily lead to the destruction of the integrated package.

For TTL, any unused inputs should be either connected up to the V_{CC} line by way of a 1 kΩ resistor or tied directly to the earth (zero) line. Remember, also, that the maximum V_{CC} permitted is 7 V; try to keep the V_{CC} line as close to 5 V as you can.

The outputs of TTL gates must not be connected together unless you are dealing with open-collector devices. Any unused outputs should be left floating.

When experimenting with TTL gate circuits, particularly on protoboards, plug-in systems, S-Decs or Veroboard, always shunt the V_{CC} supply points (usually pins 7 and 14 on TTL packages) with a ceramic 0.1 μF capacitor, keeping this as close to the pins themselves as possible. Also, keep all the interconnecting leads as short as possible. Anything longer than about 25 cm can lead to instability problems.

CMOS handling is more critical than TTL. Like TTL, all unused CMOS inputs should be connected to V_{CC} or earth, but the inputs can be damaged by discharge of static electricity built up on the fingers or particularly nylon shirt cuffs. Always make sure that the soldering iron you are using has its case securely earthed. Keep all CMOS devices wrapped in foil (aluminium kitchen foil will do) or embedded on conducting plastic until they are ready for use. Do not mix up TTL and CMOS packages in one circuit system. (They *can* be intermixed, but not at this stage of the course!) Finally, keep all leads as short as you can, and fit bypass capacitors across the supply pins of each package.

SUMMARY

- Digital systems employ discrete instead of continuous signals.
- Basic logic operations are: AND, OR, NOT, NAND and NOR.
- Binary data values 0 and 1 correspond to OFF and ON switch positions.
- The validity of any logical statement can be demonstrated by a truth table.
- Boolean algebra is used to manipulate binary relationships.
- Integrated circuits contain a number of identical gates in a compact package.
- TTL and CMOS integrated families are used in most logical systems.

REVIEW QUESTIONS

1 Draw symbols for AND, OR, NOR and NAND operations.
2 Distinguish between electronic gates and switches.
3 What is a truth table and how is it used?
4 Why is the binary system of value in electronic switching circuits?
5 What is Boolean algebra?
6 Which of the Boolean operations do not accord with the rules of ordinary algebra?
7 What is the difference between an inclusive and an exclusive OR?
8 Construct a truth table for a NOR gate having three inputs.
9 How can a logic circuit be analysed into a logic statement?
10 The logic circuit of *Figure 14.31* carries out a single operation. What is it?

Figure 14.31

11 Define noise margin. What are the noise margins for TTL gates?
12 How would you recognize pin 1 on a dual-in-line integrated package? How are the remaining pins numbered relative to pin 1?

13   The 74LS09 is a quad two-input AND gate with open collector. Explain what you understand by this description.

EXERCISES AND PROBLEMS

(24)   Complete the following statements:

(a)   When logical 0 is at the input of an inverter, the output signal is logical. . . .
(b)   The output $A \oplus B$ is provided by a(an) . . . gate.
(c)   All logic functions can be constructed using only . . . or . . . gates.
(d)   It is not possible to construct NAND or NOR gates from any combination of . . . or . . . gates.

(25)   Write down the output expressions for the logic circuits shown in *Figure 14.32*.

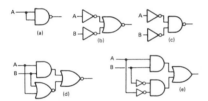

Figure 14.32

(26)   For the logic system shown in *Figure 14.33*, deduce that the output will be high if A = B and C = D.

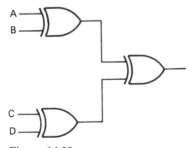

Figure 14.33

(27)   Deduce that the system of *Figure 14.34* will give a high output if an odd number of the inputs are high and a low output if an even number of the inputs are high.

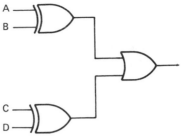

Figure 14.34

(28)   A student built up the circuit shown in *Figure 14.35(a)* as a design for an AND gate. Explain why this circuit failed to work. He then modified the circuit to that shown in diagram *(b)*. Did it work this time? Whatever you decide, explain your reasoning.

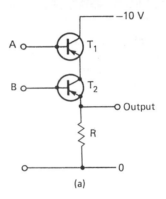

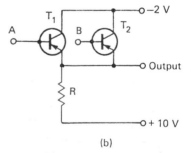

Figure 14.35

(29)   In the circuit of *Figure 14.36* the input signal levels are either 0 or +10 V. Answer the following:

(a)   Under what conditions will the potential at X be 10 V?
(b)   What then will be the potential at Y?
(c)   What logic function does this circuit perform?

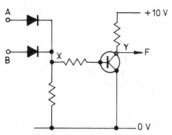

Figure 14.36

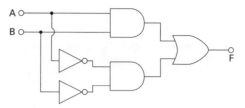

Figure 14.38

(30)   A logic system has inputs A and B to an OR gate whose output is AND-gated with an input C to give an output X. A second system has two AND gates with separate inputs A and B and which share input C. The outputs of these gates enter a two-input OP gate to give an output Y. Draw these systems and show by reasoning and a truth table that output S is identical to output Y.

(31)   The logic assembly shown in *Figure 14.37* exhibits a fault. A student takes voltage measurements at all the terminal points and records the values shown. Which part(s) of the circuit is/are giving trouble?

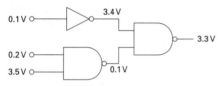

Figure 14.37

(32)   A machine operator controls red and green indicator lights by using four switches A, B, C and D. The sequence of operation is as follows: (a) red light is ON when switch A is ON and switch B is OFF or switch C is ON. (b) Green light is ON when switches A and B are ON and switches C and D are OFF. Write down logical expressions representing condition (a) and (b), and sketch logical circuits satisfying these conditions.

(33)   Using any gates you like, but keeping their numbers as small as you can, devise logic systems to satisfy the following conditions:

(a) $F = A\overline{B} + C$ (b) $F = A\overline{B} + BC$
(c) $F = (A + \overline{B})(\overline{C} + D)$ (d) $F = A\overline{B} + B\overline{C}$

(34)   Analyse the logical system of *Figure 14.38* and construct a truth table for the circuit.

Simplify the logical equation and check your answer against the truth table. What is the equation?

(35)   Using the 7400 quad two-input NAND package show how you would externally connect the appropriate pins on the package to make up the circuit of *Figure 14.39*. (There are a number of possible answers to this – any one will do.) What would you do (if anything) with the pins of the gate you did not use? What is the output equation?

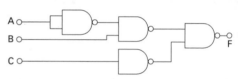

Figure 14.39

(36)   A person is entitled to apply for a certain job if they are:

(a)   a woman over the age of 40;
(b)   a married man.

Three switches are provided to test the entitlement:

Switch A puts logical 1 on the output if the person is over 40.
Switch B puts logical 1 on the output if the person is male.
Switch C puts logical 1 on the output if the person is married.

Write down a logical equation for an output F which indicates entitlement and draw a suitable arrangement of gates which would perform this function.

(37)   A security firm uses three keys to gain access to the vault; one is held by the manager, one by the assistant manager and one by the security officer. Devise a logical system which

will enable the door to be opened by using any two of the keys. In addition, the safe itself can only be opened by all three keys.

(38) When the transistor of *Figure 14.40* is switched on, $V_{BE} = 0.8$ V and $V_{CE} = 0.3$ V. If the forward drop in the diodes is 0.7 V and the levels at A go between 1 V and 2 V while B is at O, estimate the output voltages.

(39) *Figure 14.41* shows a two-input NAND gate. Explain how this circuit works. If the same transistor and diode types are used as for the previous example, estimate the output for inputs of 1 V and 2 V at A, assuming a 5 V input at B.

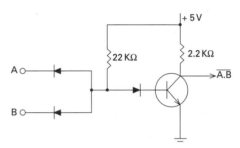

Figure 14.41

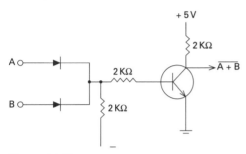

Figure 14.40

15

Sequential logic systems

The flip-flop memory unit
The D latch
The JK flip-flop
Shift registers and counters
Integrated counters
Numerical indicators

We come now to what are known as sequential logic systems in which the circuits, of which registers and counters are examples, work with timed sequences of clock (or enabling) pulses. This distinguishes these systems from the purely combinational gate arrangements we have discussed in earlier chapters. There the outputs of the gates were determined by the existing inputs; in sequential systems the outputs may be affected by past inputs as well as those actually present at that time. The basic circuit module from which most sequential logic is derived is the bistable multivibrator or flip-flop. In this chapter we will consider the operation of the flip-flop first in its discrete form and then in its integrated form, and some of the many applications it has in the design of digital systems.

A TRANSISTOR FLIP-FLOP

Digital systems are almost always built around registers which are groups of flip-flops. A flip-flop is a two-state element which can be set in either of the two states and remain in that state until some other input signal changes the state.

The operation of a flip-flop depends on the fact that a transistor is both a switch and an amplifier. *Figure 15.1* shows a single transistor set up as an elementary flip-flop. The base bias derived from the potential divider R_1 and R_2 is such that the transistor is normally cut off; the collector potential is then high. If a positive potential is applied to input A, the base-emitter

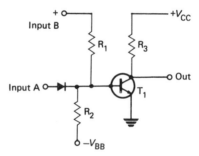

Figure 15.1

junction will be forward biased and the output will go low. This is simply a straightforward transistor switch or inverter.

However, there is an alternative input terminal we might make use of; if we make the potential at the positive end of R_1 more positive, the transistor could be switched on without anything being applied to input A. Hence we could get the transistor to switch if either input A or input B were made positive. In conjunction with the normal inversion (NOT) function of the transistor, the system is essentially a NOR gate. By using two such gates, a flip-flop can be created. The circuit for this is shown in *Figure 15.2*.

Here we have two transistors, each individually set up in the form of *Figure 15.1*; the resistors have been numbered to make this clear. The output of each transistor feeds back to the input of the other but in such a way that each output is effectively the input B point of *Figure 15.1* and the input points now marked R and S correspond to the input A point of *Figure 15.1*.

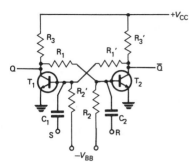

Figure 15.2

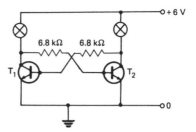

Figure 15.3

Suppose that both transistors are conducting and are drawing equal collector currents. Because the circuit appears symmetrical in its layout such a condition may seem possible, but actually it is impossible. No two transistors have precisely identical characteristics and no resistors have precisely the same ohmic values. Even if we select the components from close tolerance values, there are going to be electrical differences between one side of the circuit and the other. Hence, one collector current is going to be greater than the other. So assume that the collector current of T_1 is slightly greater than that of T_2 and increasing. The voltage drop across R_3 will increase and the collector potential will fall. This fall will be coupled through resistor R_1 to the base of T_2 and T_2 will experience a drop in its collector current and a consequent rise in its collector potential. This rise is in turn coupled back through resistor R_1 to the base of T_1 where it assists the increase in collector current already taking place. The effect is cumulative and the action continues until transistor T_1 is driven into saturation (fully conducting) and T_2 is cut off (non-conducting). Nothing further can then happen. Although this description has taken a minute or two to describe, the action takes place very rapidly after switch-on, usually in a fraction of a microsecond.

The circuit now rests in a stable state and will remain in this state indefinitely. Transistor T_1 is saturated and its collector is at (near) zero volts. This condition, in conjunction with the negative bias applied to its base via R_2, ensures that T_2 is firmly cut off. With T_2 cut off, its collector potential is large and positive and this is sufficient to overcome the effect of the negative bias via R_2' on the base of T_1. Hence T_1 remains conducting.

If we had started this analysis by assuming that the collector current of T_2 was greater than that of T_1, and increasing, the regenerative action would have swung in the opposite direction, with transistor T_2 being fully switched on and T_1 cut off in the stable state. The action would otherwise have been identical. There are

therefore two stable states possible with this circuit: either T_1 is ON and T_2 is OFF, or conversely. So the outputs at the collectors are complementary; if one of them is Q, the other is $\overline{Q}$.

Because it has two stable states, this circuit is known as the bistable multivibrator or, in digital applications, the R–S (RESET-SET) flip-flop.

AN EXPERIMENT

A simple experiment will demonstrate the operation of this basic flip-flop. Build the circuit shown in *Figure 15.3* using a pair of BC107 or BC108 transistors. The bulbs, which act as collector loads and which indicate which of the transistors is ON and which is OFF, should be low current types, typically 6 V at 60 mA. When you switch on, one or other of the bulbs will light. If you switch on and off a number of times you will most likely find that the same bulb lights up each time; this is because the circuit 'unbalance' favours that particular side.

If now you momentarily short out to the negative (earth) line the base of the transistor which is ON, the circuit will switch states and remain in this new condition until you repeat the shorting procedure. What happens is that when you short out the base of the ON transistor, its collector current falls to zero and its collector potential rises to 6 V. The OFF transistor now receives a base current flowing through its base resistor and switches ON. Its collector consequently falls to near zero volts, thus preventing the other transistor (now OFF) from receiving any base current even when the short circuit is removed. Make a note of the fact that the circuit will not switch states if the already OFF transistor is shorted.

TRIGGERING THE FLIP-FLOP

Go back to the circuit of *Figure 15.2* and assume that T_1 is switched ON and T_2 OFF, a stable state that gives the output at the collector of T_1 as low, or logical 0.

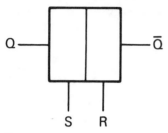

Figure 15.4

Now suppose a negative pulse to be applied to the base of T_1 by way of the S (or SET) terminal. This negative pulse will switch T_1 OFF and the circuit will change state in the manner already described. In this condition the circuit is again perfectly stable and will remain so even when the trigger pulse is removed. The output at the collector of T_1 is now high, or logical 1. Any further negative pulses applied to the SET terminal will have no effect since T_1 is already switched off. However, a negative pulse applied to the R (or RESET) terminal will switch T_2 OFF and a change of state will immediately follow. The conventional way of representing this elementary bistable flip-flop is shown in *Figure 15.4* where the outputs are designated Q and $\overline{Q}$.

Triggering the R–S flip-flop in this way would not be very convenient in practice because it would be necessary to interchange the S and R inputs for each required change of state. This difficulty can be

overcome by modifying the circuit to that shown in *Figure 15.5*. Here the setting and resetting operation is carried out automatically, each negative pulse applied to the single input terminal causing a changeover. Two diodes have been added to the basic circuit and these direct or steer the input pulses to the appropriate transistor base. For this reason they are known as steering diodes. Assuming that T_1 is switched ON, output Q is consequently at logic level 0 and output $\overline{Q}$ is at logic level 1.

Now looking at diodes D_1 and D_2, under our assumed conditions D_2 is reverse biased and D_1 is forward biased. The anode of D_2 connects via resistor R_1 to the collector of T_1 which is low (near 0 V) but its cathode connects via resistor R_6 to the collector of T_2 which is high (V_{CC}), hence D_2 is reverse biased. The anode of D_1, however, connects via resistor R_2 to the high collector of T_2 while its cathode is returned via resistor R_5 to the low collector of T_1, hence D_1 has a small forward bias. When a negative trigger pulse is fed to the diode cathodes through capacitor C_1 and C_2, the conducting diode D_1 passes the negative change on to the base of T_1 but diode D_2 remains non-conducting. As T_1 switches OFF its collector potential rises, T_2 switches ON and its collector potential falls, so reducing the base current of T_1. The circuit rapidly changes state and the output logic levels reverse, Q now becoming $\overline{Q}$, and $\overline{Q}$ becoming Q. The bias conditions on the steering diodes are now reversed also, so that the following negative trigger pulse cuts off T_2

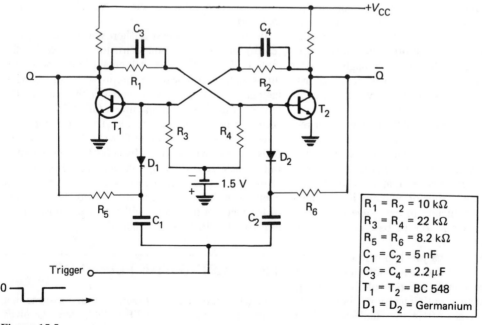

| $R_1 = R_2 = 10\ k\Omega$ |
| $R_3 = R_4 = 22\ k\Omega$ |
| $R_5 = R_6 = 8.2\ k\Omega$ |
| $C_1 = C_2 = 5\ nF$ |
| $C_3 = C_4 = 2.2\ \mu F$ |
| $T_1 = T_2 = BC\ 548$ |
| $D_1 = D_2 = Germanium$ |

Figure 15.5

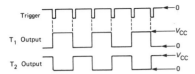

Figure 15.6

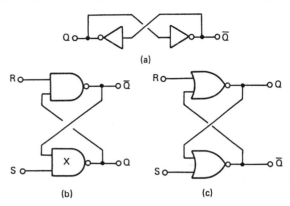

Figure 15.7

and the circuit reverts to its original state. The capacitors C_3 and C_4 shown in the diagram are normally included to assist in the rapidity of switch-over between the two states.

Figure 15.6 shows the waveforms of the input and output signals for a succession of trigger pulses. You should particularly notice that the bistable changes state on the negative-going edge of the trigger pulses. If the output is taken from one collector only, one output pulse is obtained for every two input pulses. Thus the circuit functions as a binary divider. In this way, any input frequency may be divided by any power of two (four, eight, 16, etc.) by combining the requisite number of bistables in cascade.

> (1) In the experimental circuit of *Figure 15.3* can you think of an alternative method of getting the bistable to switch state, again by shorting two points together?
>
> (2) Using the values given in the list alongside *Figure 15.5*, make up the circuit, replacing the collector resistors by low-current bulbs (6 V, 60 mA), as indicators. Feed in trigger pulses from a low-frequency square-wave oscillator, negative-going, with a frequency of a few hertz. Note that the circuit output pulses are half the frequency of the input pulses. (A V_{CC} supply of 9 V is suitable.)

INTEGRATED SYSTEMS

The basic flip-flop so far described is essentially nothing more than a pair of cross-coupled NOT gates: *Figure 15.7(a)*. If the gates are replaced by a pair of two-input NAND gates or a pair of two-input NOR gates, a controllable bistable becomes available in integrated circuit form: see *Figures 15.7(b)* and *(c)*, respectively.

Strictly these two circuits are what are known as latches, which will be explained as we go along, but they behave in much the same way as the R–S bistable already discussed and are commonly referred to as R–S bistables.

The SET and RESET inputs connect to one input on each gate, the other inputs being connected to the opposite outputs in cross-coupling. We will analyse the two-NAND system, and after that you should be able to work out the operation of the two-NOR circuit for yourself. Referring to *Figure 15.7(b)*, suppose R and S to both be held at logic 1, then the NAND gates simply act as inverters and the system holds its output state indefinitely, i.e. Q and $\overline{Q}$ are 'stored' in whatever condition they were in already and the circuit is said to be in its latched state. If S now falls to logic 0 while R remains at logic 1, output Q will switch to logic 1 because NAND gate X has an input at logic 0. In this condition the circuit is said to be SET. Returning S and R both to logic 1 will not reset the bistable and it will remain latched with Q = 1, $\overline{Q}$ = 0. Similarly, if R drops momentarily to logic 0 while S remains at logic 1, output Q will switch to logic 0, and hold there when R is returned to logic 1. The truth table for this R–S function is shown in *Table 15.1*. By convention, Q = 0 and $\overline{Q}$ = 1 in the RESET state.

Table 15.1

	S	R	Q	$\overline{Q}$
Latch	1	1	Q	Q
Set	0	1	1	0
Reset	1	0	0	1
Indeterminate	0	0	*	*

You will notice that the output state when both S and R are at logic 0 is described as indeterminate. This follows from the fact that any low input on a NAND gate makes the output high, hence for S = R = 0 both Q and $\overline{Q}$ would be high. But this is a contradiction of the definition that Q and $\overline{Q}$ are complementary; in other words, the output cannot be determined if R and S are both low.

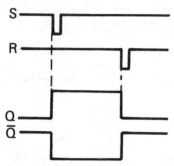

Figure 15.8

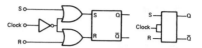

Figure 15.9

The waveforms shown in *Figure 15.8* illustrate how the circuit operates. Suppose both inputs to be initially high and Q low, then if S goes momentarily low, Q will go high. The pulse width, provided it exceeds a certain minimum, is not important. The high on Q is now latched and held there and the circuit is SET. The only way to RESET the circuit (get Q back to low) is to put the R input momentarily low after the S input has gone back high.

(3)   Draw up a truth table for the circuit of *Figure 15.5* covering a complete transition of the output cycle.

(4)   Analyse the bistable latch illustrated in *Figure 15.7(c)* and draw its truth table. (Hint: use the analysis above for the NAND circuit as a guide.)

THE CLOCKED R–S FLIP-FLOP

The addition of extra circuitry to the basic R–S flip-flop will not affect its fundamental operation but will give it added flexibility. The circuit of *Figure 15.7(b)* may be modified to bring it into line with the triggered bistable of *Figure 15.5* by adding a pair of OR gates and a NOT gate.

By this means, the facility of separate R and S inputs is retained while providing a common trigger input. In this regard, trigger pulses, known as clock or enabling pulses, are generated by (often) a high-accuracy pulse generator. This clock input allows the input conditions to be transferred to the output on one edge of the clock. The circuit is shown in *Figure 15.9*, together with its conventional symbol.

This circuit can only change state when the clock input is high. If the clock input is low, the output of the NOT gate is high, hence each OR gate has an input at logic 1 and their outputs must also be 1. Both

the S and R inputs to the basic flip-flop are therefore at 1 and, from the truth table, the Q and $\overline{Q}$ outputs are held. When the clock input goes high, however, the output from the NOT gate goes low and the OR gates each have an input at logic 0. The S and R on the flip-flop then depend only on the logic levels at the external S and R inputs; the state in which the flip-flop will set itself will then be in accordance with the truth table. The clocked R–S flip-flop therefore responds to the R and S inputs only when the clock input is high. For this reason, this logic level is referred to as the enabling pulse.

THE D-LATCH

The D (or data) flip-flop overcomes the problem of the basic R–S model having an indeterminate state when both the S and R inputs are at logic 0. The difficulty can be overcome by putting an inverter between the R and S inputs as shown in *Figure 15.10*. This circuit is available in quad format in the 7475 TTL or the 4042 CMOS. As the truth table of *Table 15.2* shows (try analysing the circuit for yourself), the Q output follows the D input as long as the clock logic is high. The logic state present on the D input just before the clock goes low will be latched on the output.

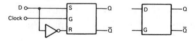

Figure 15.10

Table 15.2

D	Clock	Q	$\overline{Q}$
*	0	Q	$\overline{Q}$
0	1	0	1
1	1	1	0

One of the main uses for the D-latch is to hold or remember a certain binary number or a displayed output on a counter unit while the actual count sequence continues. For normal counting, the clock input is held high (connected to $+V_{CC}$); for the latched output it is connected to low (earth line). The symbol for the D-latch is shown alongside *Figure 15.10*. The clock input is usually designated G.

The D flip-flop is sometimes confused with the D-latch. Whereas in the D-latch the output Q follows the D input as long as the clock is high, in the flip-flop the Q output consists only of data which is present on the D input at the time of a positive transition on the clock input. This distinction is not particularly important to us at this stage because the terms 'flip-flop' and 'latch' are both used in a general and perhaps haphazard sense, but it is mentioned to show that such a distinction exists.

THE J–K FLIP-FLOP

This very versatile kind of flip-flop can be looked on as a combination of R–S and D-type flip-flops with additional control, these being preset and clear input points. These additions enable the flip-flop to be set up in a known state, say on switch-on, so that the initial conditions are established prior to full operation. When not required, these inputs are held low and do not influence the normal operation of the circuit.

A circuit arrangement (one of several possible) is shown in *Figure 15.11*. Two flip-flops are involved, one being the master and the other the slave. The master responds to data from the J and K inputs while the clock pulse is high (as for the clocked R–S flip-flop already described), but this is only transferred to the output (slave) flip-flop as the clock goes low. The slave is a D-type latch. The clock input feeds directly to the master but by way of an inverter to the slave. The slave consequently receives a high clock input when the true clock is low, and the Q output is waiting to respond to any change in the D input. There is no possibility of any such change, however, all the time the clock is low because the master flip-flop is in a stable state reached when the clock was last high.

When the clock does go high, the Q output of the master responds to the R and S inputs but the D input on the slave is unresponsive because of the clock input now being low via the inverter. When the clock returns to low, the Q output of the master is held in the new state and, simultaneously, the slave clock

input goes high and enables it to respond to this new state. Hence the J–K flip-flop changes its output state on the falling edge of the clock pulse, in that the master receives the information on the positive edge and transfers it to the slave on the falling edge. Since the clock does the triggering, the inputs themselves need not be in pulse form. To summarize: the operating sequence is that first the slave is isolated from the master, then the J and K inputs are entered into the master. The J and K inputs are then disabled and finally the information is transferred from the master to the slave.

The truth table for the J–K flip-flop is shown in *Table 15.3* where Q_0 indicates the present state of the output. With low inputs at J and K, the clock pulse has no effect and the existing state is maintained. With input J high and K low, any clock pulse sets the output high; with J low and K high, any clock pulse resets the output low. Notice that when both inputs go high, the output changes state. This is the divide-by-two action produced by the discrete circuit of *Figure 15.5* and is known as 'toggling'.

Table 15.3

	J	K	Q
Output stored	0	0	Q_0
Output set to 1	1	0	1
Output set to 0	0	1	0
Output changes state (toggles)	1	1	$\overline{Q_0}$

J–K flip-flops (as are R–S and D flip-flops) are available in integrated circuit packages, the 7474 being a dual D-type latch and the 7476 being a dual J–K flip-flop with preset and clear facilities. *Figures 15.12(a)* and *(b)* show the internal arrangements of these two packages. The preset input allows either a 1 or a 0 to be stored in the system; the clear input allows all old data to be cleared to 0, an example of which you will recognize in the 'Clear' button of a pocket calculator.

Example (5)   *Figure 15.13* shows a possible arrangement of a flip-flop that will store a single bit of information. Use logical algebra to show that the S input will be stored when a clock pulse is applied.

Solution   Let the input, output and intermediate levels be marked as shown in the diagram. Then when the clock pulse is not present the outputs X and Y must be high, since one input to each NAND gate is low. The inputs to the

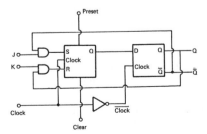

Figure 15.11

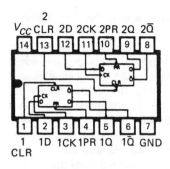

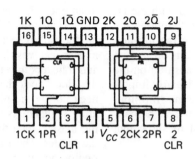

(a) 7474 dual D-type flip-flop

(b) 7476 dual J–K flip-flop

Figure 15.12

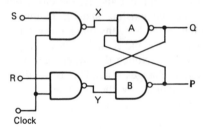

Figure 15.13

flip-flop itself are therefore Q and I on gate B, and P and 1 on gate A. Therefore

$$P = \overline{Q} \text{ and } Q = \overline{P}$$

This is a stable condition and the output will remain at Q indefinitely.

When the clock pulse is applied, C = 1, X = $\overline{S}$ and Y = R

$$\therefore \quad Q = \overline{S}.P$$

$$P = \overline{S.Q}$$

$$\therefore \quad Q = \overline{\overline{S}.(\overline{S.D})} = S + S.Q$$

$$= S(1 + Q) = S$$

Hence the output is S when the clock pulse is applied.

This example introduces us to the next section.

SHIFT REGISTERS AND COUNTERS

Shift registers are employed to store information for a great number of purposes: arithmetic calculations and operation, program control, buffers in the movement of data, to name only three. Essentially, since a register is a memory unit, it is based on the flip-flop, and a series of J–K flip-flops (using the 7476 TTL package, for example) can be used to store a series of binary digits. Digital signals are fed into the first stage of the series and are moved in sequence along the chain of flip-flops until a complete number (or word) is stored. Registers are used extensively in calculators and computers and in data processing for parallel-to-series (or conversely) conversion of information.

Figure 15.14 shows the connection of four J–K flip-flops to make a four-stage shift register: a register, that is, which will store a 4-bit binary number, anything from 0000 to 1111 or 0 to 15 in denary notation.

To illustrate the action we will enter the binary number 1010 (=10 denary) into the register. First of all, all the flip-flops are reset by 0 by momentarily connecting the CLEAR terminal to logic 1. The number 1010 is represented by the logic levels 1, 0, 1 and 0 and is fed one bit at a time into the J terminal of flip-flop A. The K terminal receives the complement of the J input by way of the inverter. At the first clock pulse, $J_A = 1$ and $K_A = 0$, so that Q_1 is set and the first digit has been stored in the A flip-flop. At the second clock pulse, J_B is found to be 1 so B is reset and this logic level is transferred to output Q_2. At the same instant, flip-flop A now has a logic 0 on J_A (and 1 on K_A) so that Q_1 switches to 0 on this second clock pulse. Two of the digits are now stored in the register. After four such clock pulses, the full number 1010 has been entered into the register. If LED indicators are connected to the Q output points, they will display the stored number by lighting where the output logic is high.

A shift register of this kind not only stores a piece of information but converts the input serial application of the bits into parallel output mode. Aside from

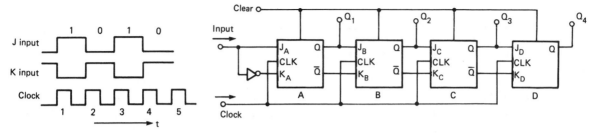

Figure 15.14

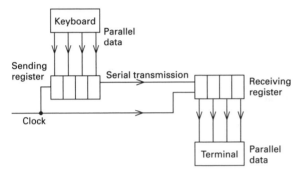

Figure 15.15

its storage properties, this facility makes the register a data-conversion device. The input sequence is clearly fed one digit after another; the output is available at Q_1, Q_2, Q_3 and Q_4 as simultaneously available digits. Of course the data can only be read out in this way when the register is not carrying out its internal shifting of the bits. Computers usually process their information in parallel form, but their inputs are in serial form, coming as a time sequence of 1s and 0s from the keyboard, electric typewriter, magnetic tape or other source. *Figure 15.15* shows how two shift registers might be employed to connect a keyboard to a data terminal over a single wire link (such as a telephone line). It would not be possible to transmit the bits in parallel form over such a link because each bit would require its own separate line. Hence it is necessary to convert any input to the link into a serial mode of transmission and reconvert it back to parallel mode at the output of the link. The system is kept in step (or synchronized) by using a common clock signal. As drawn in the Figure, the system allows 4-bit words to be transmitted from the sending register to the receiving register in serial form. The input from the keyboard and the output to the terminal, however, require to be in parallel form, so the registers are being used as parallel-in/serial-out (PISO) at the sending end and as serial-in/parallel-out (SIPO) at

the receiving end of the link. The 74164 is a serial/parallel data converter.

The parallel facility is obtained by using the PRESET input on each flip-flop (not shown in *Figure 15.14*). If, once the four digits are stored in the register, additional digits are introduced, those already in store are pushed out at the far end of the chain and lost. If data is entered at the preset inputs in parallel form, the clock train will feed out the digits in serial form, so making the register a PISO converter. The four bits, then, of *Figure 15.15* are first loaded from the keyboard into the sending register in parallel; then, on receipt of the clock signal, both sending and receiving registers shift four places. The information is pushed out of the sender and enters the receiver where it is read out in parallel by the same clock signal and transferred to the terminal.*

---

* The systems and protocols associated with serial transmissions over the telephone network are covered in a companion book, *Electrical and Electronic Principles 2*.

COUNTERS

We turn now to a sequential logic unit which derives immediately from the registers we have just discussed. This is the counter.

Counting is essentially a matter of being able to add up. If we can add one digit at a time, we are counting. By using the binary system of numbering, counting by electronic means is simplified: one binary counter need store only two digits, 0 or 1, in any column, rather than 10 digits, from 0 to 9, as would be necessary in a decimal (denary) counter. Counters, irrespective of what they might appear to be, basically can all be looked on as serial registers with a single input and a parallel output from each of the separate flip-flops making up the register. The function of any counter is to count the pulses as they arrive at the input. There are obviously many applications for such counting systems: batch counting on a production line, the measurement of time intervals and the measurement of frequency, again to name only three.

The basic element in most binary counters is the familiar J–K flip-flop, though a D-type will also serve. A J–K flip-flop with the J and K inputs tied high, or a D-type with the Q output connected to the D input will toggle: the output state will change for each input pulse. Since two input pulses produce one output pulse, the element is a divide-by-two unit as we have already seen. *Figure 15.16* shows how four J–K flip-flops might be assembled in cascade to form a 4-bit binary counter. Counters of this sort are called asynchronous, this term merely implying that the counting sequence is not synchronized to or by anything else, or compared with an external frequency to check its accuracy. Errors can occur, particularly when the counting rate is high because of the time delay experienced by the signal as it passes through the flip-flops, but this aspect need not concern us as this stage.

In the circuit, the J and K inputs are both taken high so that each flip-flop will toggle. As for the register, the outputs can be set to zero by momentarily connecting the CLEAR terminals high, after which the system is ready to count. Notice that the pulse inputs which are fed in at the CLOCK terminal of the first flip-flop have been arranged in the diagram to be on the right-hand side. This has been done to get the output information in the 'right order'. This is also applicable to shift registers. The count is available as a four-digit binary number on the (parallel) Q_1, Q_2, Q_3 and Q_4 outputs, and of these Q_1 is the least significant bit (LSB).* As the diagram is drawn, therefore, the digits are in their correct order of significance. The binary outputs for the input pulse numbers from 0 to

15 are shown in *Table 15.4*. From this we see that, because of the divide-by-two function of each flip-flop, each time any output goes from 0 to 1 the following flip-flop toggles to the opposite state. It consequently takes two 1 to 0 transitions at the input to cause output Q_1 to change once; four transitions to cause output Q_2 to change once; eight transitions to cause output Q_3 to change once, and sixteen transitions to cause output Q_4 to change once. Thus after each input pulse the circuit gives a binary representation of the number of pulses received up to that point; so the circuit is counting in binary code. After the sixteenth pulse (1111) the stages reset to zero and the count begins again. This is a scale-of-sixteen counter. If numbers greater than 1111 (decimal 15) have to be counted, more stages must be added. As we have already noted, n stages will give a scale of 2^n counter.

Table 15.4

Input pulse	Q_4	Q_3	Q_2	Q_1
0	0	0	0	0
1	0	0	0	1
2	0	0	1	0
3	0	0	1	1
4	0	1	0	0
5	0	1	0	1
6	0	1	1	0
7	0	1	1	1
8	1	0	0	0
9	1	0	0	1
10	1	0	1	0
11	1	0	1	1
12	1	1	0	0
13	1	1	0	1
14	1	1	0	0
15	1	1	1	1

* The least significant bit is the number 'on the right'. In the denary number 3675, for example, 5 is the LSB. It represent the 'units'. The 7 represents 70, the 6 represents 600 and the 3 represents 3000. This left-hand number is the most significant digit.

(6) It is desired to count up to a maximum of decimal 512 using a scale-of-sixteen binary counter. How many flip-flop stages are necessary?

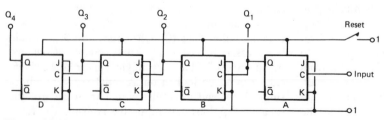

Figure 15.16

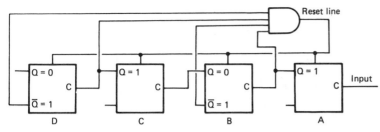

Figure 15.17

Now counting in blocks of powers of two (or modulo-2) is not a particularly convenient way from the point of view of reading out the count. Of course, it is perfectly possible to be able to calculate in binary and many people can do this with the same fluency as the rest of us count in the decimal or modulo-10 system. If the counter can be modified to count up only to ten before resetting to zero, the counting pattern will at least be a little more familiar, although the output total will still be seen in binary form. A modification of the basic counter is shown in *Figure 15.17*. This is only one of several alternatives but it is probably the easiest to follow. For clarity, the J and K inputs are omitted. The binary equivalent for decimal ten is 1010; this has four digits and can be handled by a four-input logic gate. In the Figure an AND gate has been selected. Notice that the four inputs to this gate are taken from the Q outputs on the A and C flip-flops and from the Q̄ outputs on the B and D flip-flops. When the count reaches 1010 at the four Q terminals respectively, the inputs to the AND gate will all be simultaneously at logic 1. The output of the AND will then change to 1. This output applied to all four reset (CLEAR) inputs on the flip-flops will cause the counter to reset to 0000 on the count of 10, after which the cycle will recommence. We now have a decade or modulo-10 counter.

The system can be easily adapted to other counting cycles by changing the inputs and the AND gate, noting that A is the most significant digit. For example, a scale-of-12 counter can be made by noting that the binary equivalent for decimal twelve is 1100. The inputs to the AND gate are therefore taken from the Q outputs on the A and B flip-flops and from the Q̄ outputs on the C and D flip-flops.

(7)   How would you adapt the circuit of *Figure 15.16* to count in (a) a scale of six, (b) a scale of eleven?

INTEGRATED COUNTERS

Binary counters are readily available in integrated circuit form and these usually include additional functions so that different counting sequences are available to the user. They are also available to drive seven-segment read-out display devices directly or by way of a suitable decoder so that the output is presented in decimal form. Any pocket calculator is an immediate example.

We will examine one or two of the more popular TTL and CMOS packages available for counting purposes.

The 7493 TTL integrated circuit is a self-contained 4-bit binary counter and is similar to the circuit shown in *Figure 15.16* with an additional two-input AND gate included to make provision for counting sequences other than scale of sixteen. *Figure 15.18* shows the connections and the circuit arrangement of the 7493. All four flip-flops have a common reset line fed from the output of the AND gate. Also flip-flop A has its output brought separately to pin 12 and not tied to the input of flip-flop B as the other three flip-flops are connected. Flip-flop A can therefore be used separately as a straightforward divide-by-two (using pins 14 and 12) while input B (pin 1) and outputs B, C and D will provide a two-, four- or eight-times division, respectively. For use as a 4-bit counter, output

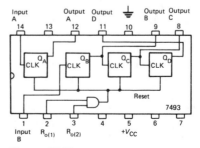

Figure 15.18

A is connected to input B. Notice here that output A is the least significant digit.

Resetting the outputs back to zero takes place through the included AND gate whose input is brought out on pins 3 and 4 and indicated as $R_{o(1)}$ and $R_{o(2)}$, respectively. This is an arrangement similar in form to that described at *Figure 15.7* earlier. To reset the counter, the reset line must be taken momentarily high; this will occur when both inputs to the AND gate are high. Counting can only take place when at least one of the AND inputs is low.

To count to a scale of ten (binary 1010), the inputs to the AND gate are taken from those outputs which are high on the count of ten, that is, from outputs B and D. Hence pin 2 goes to pin 9 and pin 3 goes to pin 11.

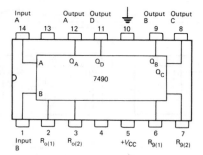

Figure 15.19

(8)   To which outputs would you connect the two reset pins on a 7493 counter so that you could count on a scale of (a) three, (b) four, (c) five, (d) six, (e) nine and (f) twelve? (Part (b) might just be tricky!)

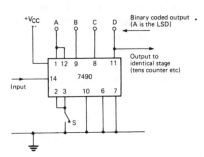

Figure 15.20

The 7490 TTL integrated circuit is a unit which gives a direct divide-by-ten facility, and so is often more convenient that the 7493 which, while it can be made to reset after binary 9, cannot then have its reset facilities available for other purposes – at least, not without the addition of further gates. The pin connections for the 7490 are shown in *Figure 15.19*. The A flip-flop in the 7490 is separate from the other three, as it is in the 7493, so that divide-by-two and divide-by-five functions are available.

By cascading a number of 7490s a counter can be made to count in powers of ten; units, tens, hundreds, etc. A basic circuit is given in *Figure 15.20* which is simply repeated for each decade required. The A input of the first counter is fed with the input pulses; the D output then feeds the A input of the following counter. As the units counter reaches 9 and returns to 0, the D output goes from high to low and the following (tens) counter responds to this transition, recording the 'carry' digit each time. The stages can be reset to zero by momentarily operating switch S which is common to all the connected reset pins and is normally closed. Pins 6 and 7 are not used in this application; they are a pair of reset-to-nine inputs which have a particular application of no interest to us at this point.

The divide-by-two and divide-by-five functions in the 7490 may be connected in either order and produce 'different' outputs accordingly. This difference is not in the counting, for although the output waveform at

the D terminal is necessarily one-tenth of the input frequency, it can take the form of what is known as either an *asymmetrical* or a *symmetrical* divide-by-ten.

The asymmetrical output occurs when the divide-by-two function is in front of the divide-by-five function; the output then has a mark-space ratio of 1:4, D being low on counts up to 7 and high only on 8 and 9. *Figure 15.21(a)* shows the timing diagram for this form of connection. If a ten-times division is required with a symmetrical mark-space ratio, the divide-by-five function has to be in front of the divide-by-two. This means that the D output must be connected to the A input and the input pulses fed to the B input. The symmetrical output is then available at pin A. The timing diagram for this arrangement is shown in *Figure 15.21(b)*.

By isolating the A flip-flop, division by five is possible with the 7490; the connections this time are indicated in *Figure 15.22*. This circuit, in conjunction with *Figure 15.20*, is useful for a total division by 50, so that if an input at 50 Hz mains frequency (suitably reduced!) is applied at the input, the output from the two dividers will be 1 Hz. This could then drive further counters to record second time intervals. Resetting of this first divider would not normally be used because of the very rapid counting taking place, but if it were required, a normally closed switch would be included in the earth connection of pins 2 and 3.

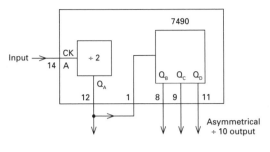

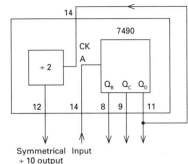

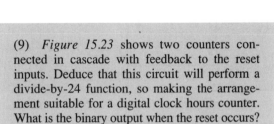

Figure 15.21

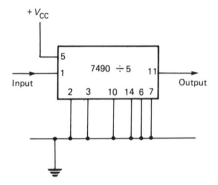

Figure 15.22

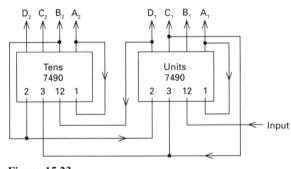

Figure 15.23

(9) *Figure 15.23* shows two counters connected in cascade with feedback to the reset inputs. Deduce that this circuit will perform a divide-by-24 function, so making the arrangement suitable for a digital clock hours counter. What is the binary output when the reset occurs?

NUMERICAL INDICATORS

Calculators, to take an everyday example, would not be very useful if their displays were indicated in binary notation. The insides of the calculators work, of course, in binary, but their outputs need to be in a form that is immediately comprehended: that is, decimal form. We know what 357 means straight away but 101100101 is a different matter!

The usual device available nowadays for the decimal representation of binary numbers (or of words, come to that) is the LED (light emitting diode) or the LCD (liquid crystal display). The LED types are best for general experimental work as they can be bought as single units and arranged as required. The seven-segment variety used for the numerals 0 to 9 is illustrated in *Figure 15.24*. These displays have horizontal

Figure 15.24

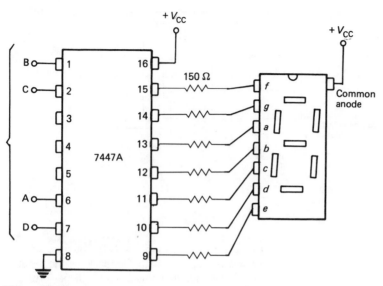

Figure 15.25

and vertical segments lettered a to g which may be lit up individually to produce the digit required as shown on the right of the figure. In some cases, the upper tail in the 6 and the lower tail in the 9 are omitted. The LEDs that the segments include are brought out to seven connecting pins, with a further pin being the common anode or common cathode connection. If the binary coded output is obtained from a 7490, a decoder is necessary between it and the display. One such decoder suitable for experiments is the 7447 A which contains all the logic for switching on the appropriate segments of the display when a binary number is entered.

Figure 15.25 shows the arrangement. The A, B, C and D binary coded inputs from the circuit of *Figure 15.20* are fed in as shown and the display unit is connected via current-limiting resistors to the output. One display unit is required for each 7490/7447 used.

For a common cathode display the 7448 is suitable in place of the 7447 A.

CMOS DIVIDERS

Dividers are available in the CMOS 4000 range of logic systems and a typical divide-by-ten circuit using the 4017 is shown in *Figure 15.26*.

A CMOS equivalent to the 7448 is the 4511 and these packages also have additional control inputs that energize the whole of the seven segments for testing purposes, blank the segments or blank leading zeros

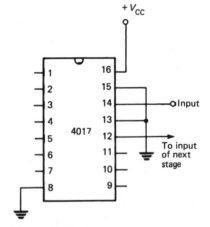

Figure 15.26

if these are not required, for example in a digital voltmeter.

A separate decoder such as the 7447 A is not always necessary as some CMOS counter packages contain an integral decoder and can be connected directly to a LED display. The CMOS 4026 is an example which contains five flip-flops for division purposes and a decoder which provides an output to operate a seven-segment display directly. The 4026 is mounted in a 16-pin DIL package and a suitable circuit is shown in *Figure 15.27*.

The usual handling precautions must be observed when working with CMOS devices. The V_{CC} supply may be anything from 3 V to 15 V; 9 V is usual for circuits driving LED indicators.

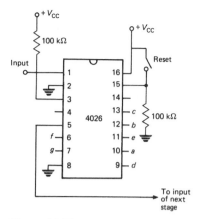

Figure 15.27

SUMMARY

- A flip-flop is the basic memory unit.
- Memory units store data processed by decision units.
- The basic R–S flip-flop is set by S = 1 and reset by R = 1.
- The J–K flip-flop operates on each clock pulse as determined by the inputs.
- The clocked R–S flip-flop can only change state when the clock input is high.
- Shift registers can store a series of binary digits.
- Counters are derived from modified shift registers.
- Integrated counters are available in TTL and CMOS forms.
- LED and LCD displays present binary numbers in decimal form.

REVIEW QUESTIONS

1. Distinguish between combinational and sequential logic systems.
2. What is a flip-flop? What is its function?
3. Distinguish between R–S, D-type and J–K flip-flops.
4. Explain how the J–K flip-flop counts. How many are needed to count to 10?
5. What keeps the outputs of a J–K flip-flop from switching when one input changes to 0 while the other is held at 0?

6. What is the purpose of clock signals in sequential systems?
7. A group of bits transmitted and received at the same time is called. . . .
8. What is the main difference between a D-latch and a D flip-flop?
9. What is the final output frequency of an eight-stage binary divider with an input of 10.24 kHz?
10. Draw a schematic diagram of a binary counter that can count up to 100.

EXERCISES AND PROBLEMS

(10)   Draw a block diagram of an R–S bistable using (a) NAND gates, (b) NOR gates. Explain the operation in both cases.

(11)   Two J–K flip-flops which respond to a falling edge transition are connected as shown in *Figure 15.28*. If a 1 kHz square wave is applied at the input, what will be the output?

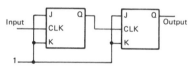

Figure 15.28

(12)   Explain the term 'modulo' in connection with binary counting.

(13)   Draw a circuit for a 7493 package used as a modulo-10 counter.

(14)   With the aid of a diagram, explain the operation of (a) a 3-bit shift register, (b) a 3-bit counter.

(15)   The 74164 is described as an 8-bit SIFO register. Explain, using diagrams, exactly what this description means.

(16)   Draw a block diagram of a 4-bit register demonstrating (a) serial input, (b) parallel input, (c) serial output, (d) parallel output.

(17)   A seven-segment display unit was shown earlier in *Figure 15.24*. Decimal 1 is displayed when segments b and c are lit; decimal 2 when segments a, b, g, e, d are lit, and so on. Draw

up a truth table showing the state of each segment (lit = 1) for the ten decimal symbols.

(18) Connect the numbered pins of the hypothetical integrated circuit shown in *Figure 15.29* to construct an OR gate. All internal gates must be used and the output is to come from pin 8.

(19) How would you describe the integrated package of the previous example?

(20) A father made his young son an elementary traffic control system, three lights, RED, GREEN and AMBER, being operated by three momentary action push-button switches. Draw a block diagram using bistable units showing how a workable controller was made up.

(21) An electric light is controlled by three switches A, B and C. The light is to be ON

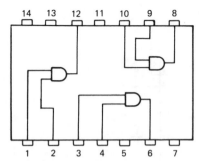

Figure 15.29

whenever A and B are in the same position; and when A and B are in different positions, the light is to be controlled by C. Draw up a truth table for this circuit, representing the light function F in terms of A, B and C. Simplify this function and design a practical circuit.

16

Generators and motors

Basic principles
Characteristics of d.c. machines
Operational features
Speed control and starting
Efficiency

There is a great variety of electrical machinery available these days, but the most important group, which turns up in the majority of electrical appliances, comprises the electric generator and the electric motor. These are rotating machines in which a conversion takes place between electrical energy and mechanical energy. If mechanical energy is supplied to the machine and electrical energy is taken from it, the machine is a *generator*, or *dynamo*; if the interchange is in the opposite direction, the machine is a *motor*.

In many cases the functions are completely interchangeable. Motors and generators operate on exactly the same physical principles, and can produce, or operate on, either alternating or direct current supplies. We shall be particularly interested in this chapter in direct current machines and their working characteristics, though we will get under way with a brief recapulation of the fundamental principles underlying all such devices.

THE GENERATOR PRINCIPLE

The simple a.c. generator, as we have seen in the introduction to alternating current principles earlier, consists in its most simple form of a coil of wire mounted on a shaft in such a way that it can be rotated between the poles of a permanent magnet. So that the generated current can be conveyed to and from the coil by way of an external circuit, the ends of the coils are terminated on brass or copper slip-rings which rotate with, but are insulated from, the shaft. Two carbon contacts, or brushes, bear against the slip-rings so enabling contact between the moving coil and the external circuit.

During one revolution of the coil, the generated e.m.f. follows a sinusoidal variation, the frequency being dependent on the speed of rotation of the coil in the magnetic field, as is the magnitude of the generated e.m.f. Such an elementary dynamo is illustrated in *Figure 16.1*. The polarity of the generated e.m.f. is determined by considering the force acting on positive charges in the moving conductor. Fleming's Right-Hand rule indicates that the force f_1 acting on positive charges in the upper conductor in *Figure 16.1* is into the paper, while the force f_2 acting on the positive charges in the lower conductor is out of the paper;

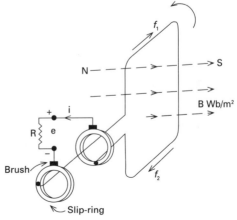

Figure 16.1

hence the polarity of the generated e.m.f. is as indicated at the slip-rings at the instant considered. In this way, the e.m.f.s in the two conductors will combine to drive a current around the circuit, and the total e.m.f. acting round the loop will be double that generated in either conductor.

The polarity reverses for each 180° rotation of the coil; during one complete revolution of the coil, therefore, the e.m.f. follows a variation in magnitude which, as we have determined earlier, can be expressed as a sinusoidal waveform.

A single turn of wire would have to be rotated at an impracticable speed in a very intense magnetic field in order to generate an e.m.f. that would be of any practical significance, so, although the principle is unchanged, the construction of a useful alternator is quite elaborate. As the principle of the device depends only on relative motion between field and conductor, it is immaterial whether the rotating part (the *rotor*) is the element which actually moves with respect to a fixed field, or the other way about. For this reason large alternators usually comprise a two-pole electromagnet as the rotor, and the alternating e.m.f. is generated in a surrounding assembly of fixed *stator* winding; that is, the coil is stationary and the field rotates.

Smaller alternators follow the pattern of our simple design above and use stationary magnetic field systems with e.m.f.s induced in an arrangement of rotor windings. The advantage of the former type is that the slip-rings and brushes carry only the excitation current to the rotating electromagnets and this current is relatively small compared with what would be carried if the rotor output was the generated current. The slip-rings and brushes for the rotor therefore need be only of light mechanical construction with small losses, no overheating or sparking. If the rotor coil carried the load current at high voltage the slip-rings, brushes and insulation requirements would be formidable and considerable energy would be wasted in friction and electrical resistance.

THE D.C. GENERATOR

The basic alternator described in the previous section becomes a d.c. generator by replacing the slip-rings by a split-ring automatic synchronous switch or *commutator*. The constructional form of a d.c. generator is, in almost every case, that of a fixed field magnet or magnets, with the e.m.f. induced in rotating conducting coils. A unidirectional current can then be obtained in an external load if the rotor windings are switched over at the correct times to reverse one or other of the alternate half cycles then being generated.

Considering again for the moment the single rotating coil of the alternator described; the ends of this coil, instead of going to slip-rings, are connected instead to the split-ring commutator as shown in *Figure 16.2(a)*. The e.m.f. is collected from two carbon brushes as before, but these are now positioned so that the contacts change from one half of the ring to the other at the instant the e.m.f. is about to change polarity, that is, when the coil is vertical as illustrated and the instantaneous e.m.f. is zero. The voltage across the brushes consequently pulsates as it follows each half cycle of induced e.m.f. but never reverses its direction as it did for the slip-ring arrangement. No practical d.c. generator is made up of a single rotating coil and a simple split-ring commutator. The output waveform obtained from such a system would be that of a full-wave rectifier as the lower curve of *Figure 16.2(b)* shows, which, as it was in the case of d.c. power supplies, requires quite a lot of 'smoothing'.

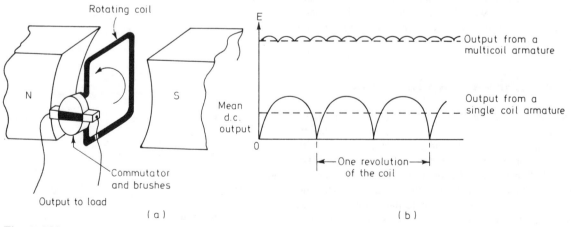

Figure 16.2

To obtain a smooth output voltage it is necessary to reduce the amplitude of the alternating ripple component; this is not done by connecting a capacitor across the output terminals but by increasing the number of segments on the commutator along with the number of conductors on the rotor. The output is then taken from each coil not over a 180° rotation period but only over a small angle when the position of that particular coil is such that the instantaneous induced e.m.f. is at its maximum. The form the 'smoothing' takes for an increasing number of coils and commutator segments is then as illustrated in *Figure 16.2(b)*, the uppermost line being the output from a multicoil rotor winding.

The armature

A practical rotor system, or *armature* as it is called, consists of a great many conductors wound in slots along the outer surface of a laminated iron cylinder, and are arranged so that they, together with the commutator segments to which they are connected, form a closed circuit, and each coil always comprises part of a circuit. *Figure 16.3* shows a single coil of typical form at *(a)* and the way these coils are assembled on a typical armature from, say, a power drill, is shown at *(b)*. Each commutator segment is insulated from its neighbour by a strip of mica and the whole assembly forms a cylinder concentric with the armature and the central shaft. The brushes which bear on the commutator are usually of graphite carbon which has a relatively high resistance to keep unwanted sparking to a minimum, keeps the commutator clean and by being

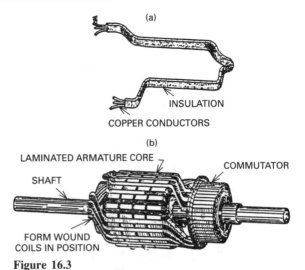

(a)

INSULATION

COPPER CONDUCTORS

(b)

LAMINATED ARMATURE CORE

SHAFT

COMMUTATOR

FORM WOUND COILS IN POSITION

Figure 16.3

comparatively soft does not wear or groove the commutator surface but adapts itself readily to its shape.

The number of slots and commutator segments are always equal, but the number of brushes is determined by the electrical characteristics required of the machine. It is always an even number, brushes of the same polarity being connected in parallel. In most cases, a practical d.c. generator also has more than a single pair of field poles. Such poles are arranged N, S, N, S alternately around the stator, each pole piece carrying a field coil for magnetic excitation and a shaped extension to ensure a radial field in the air gap between the field coils and the armature surface. *Figure 16.4* shows the general constructional features of a four-pole d.c. generator. The stator consists of a

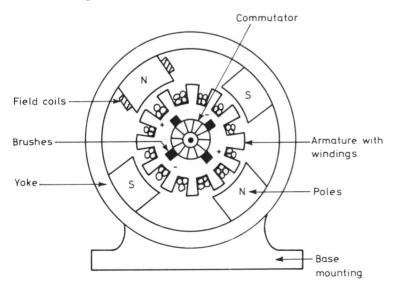

Commutator

Field coils

Brushes

Yoke

Armature with windings

Poles

Base mounting

Figure 16.4

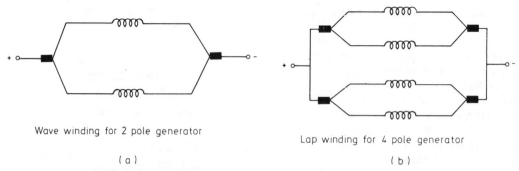

Wave winding for 2 pole generator

(a)

Lap winding for 4 pole generator

(b)

Figure 16.5

yoke, usually of cast steel, to which are bolted the field magnets, each with its field coil. These field coils are usually wired in series but in such a direction that adjacent poles exhibit opposite polarities as mentioned above. The slotted armature with its conductors and commutator segments then rotates within the magnetic fields of the stator coils.

There are two ways of arranging the conductors on an armature and these two ways determine the magnitude of the generated e.m.f. However the armature is wound there are always a number of conductors in series between the brushes, each conductor generating its own e.m.f. in the same direction. The e.m.f. between the brushes at any instant is the sum of the instantaneous e.m.f.s in these conductors just as the e.m.f. of a number of cells in series is the sum of their individual e.m.f.s. There will also be a number of such series systems of conductors in parallel between the brushes. Each series system produces the same e.m.f. and is unaffected by the number of parallel paths; but as in a series–parallel arrangement of cells, the internal resistance of the armature winding is decreased and its current-carrying capacity is increased, by increasing the number of parallel paths.

The winding methods concerned are known as (a) *wave* or two-circuit winding and (b) *lap* or multiple-circuit winding. In wave winding, see *Figure 16.5*, there are two paths in parallel *irrespective of the number of poles*, each path supplying half the total current output. Two sets of brushes only are necessary, but it is usual to fit as many sets of brushes as the machine has poles. Wave wound generators produce high-voltage, low-current outputs. *Figure 16.6* shows a 'rolled-out flat' view of a wave wound armature.

In lap winding there are as many paths in parallel as the machine has poles, and the total current output divides equally between them. There are as many sets of brushes here as the machine has poles; thus if there are 20 conductors arranged in four parallel paths, there will be a generated e.m.f. equal to the sum of 20/4 =

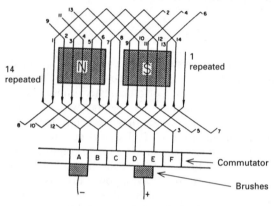

Development winding of a 2-pole
wave-wound armature

Figure 16.6

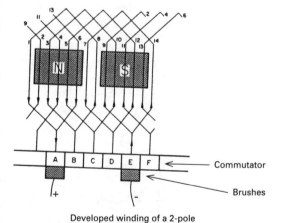

Developed winding of a 2-pole
lap-wound armature

Figure 16.7

5 of them, just as a battery of 20 cells arranged in four parallel groups of five cells each will have an e.m.f. equal to the sum of five cells. *Figure 16.7* shows the rolled-out arrangement of a lap winding, as well as the general form as described.

THE DYNAMO EQUATION

The derivation of an expression for the generated e.m.f. from the various winding arrangements can be found by obtaining the e.m.f. induced in a single conductor in its passage from brush to brush, and then multiplying this by the number of conductors between the brushes.

Let Z be the total number of conductors on the armature, Φ be the flux per pole, p the number of pole pairs, and N the armature speed in revs/min or $N/60$ revs/sec. Then the e.m.f. generated by the armature is equal to the e.m.f. generated by one of the parallel paths. Each conductor passes $2p$ poles per revolution and so cuts $2p\Phi$ Wb of magnetic flux per revolution. Hence the flux cut by each conductor per second is

$$2p\Phi \times \frac{N}{60} \text{ Wb}$$

and so the generated e.m.f. per conductor is

$$E = \frac{2p\Phi N}{60} \text{ volts}$$

For Z conductors and a parallel paths, the effective number of conductors generating an e.m.f. is Z/a and hence

$$E = \frac{2p\Phi N}{60} \times \frac{Z}{a} = \frac{2p\Phi ZN}{60a}$$

But angular velocity $\omega = 2\pi N/60$ rad/sec

$$\therefore \quad E = \frac{pZ}{\pi a} \cdot \omega \Phi \text{ volts} \qquad (16.1)$$

This equation is known as the generator e.m.f. equation. For a wave wound armature the number of conductors in series per path is $Z/2$, and so

$$E = \frac{Z}{2} \cdot \frac{2p\Phi N}{60} = \frac{p\Phi ZN}{60} \text{ volts}$$

Again, substituting for angular velocity ω, this becomes

$$E = \frac{pZ}{2\pi} \omega \Phi \text{ volts} \qquad (16.2)$$

For a lap wound armature the number of conductors in series per path is $Z/2p$ and this time

$$E = \frac{Z}{2p} \frac{2p\Phi N}{60} = \frac{\Phi ZN}{60} \text{ volts}$$

$$= \frac{Z}{2\pi} \Phi \omega \text{ volts} \qquad (16.3)$$

Notice that (16.1) and (16.2) above derive from (16.3) directly, since for wave winding there are just two parallel paths, hence $a = 2$, and for lap winding there are as many parallel paths as there are poles, hence $a = 2p$. Examples can be worked either in terms of angular velocity ω rad/s or in terms of revolutions per minute (or per second). Keep in mind that $\omega = 2\pi N/60$ rad/s where N is the speed in r.p.m.

Now follow through the next two examples carefully:

Example (1)   A four-pole wave wound armature has 480 conductors and a flux per pole of 25 mWb. Calculate the e.m.f. generated when the machine is running at an angular velocity of 105 rad/s.

Solution   For four poles, $p = 2$; for wave winding, $a = 2$; $Z = 480$ and $\omega = 105$. Then from equation (16.2)

$$\text{e.m.f } E = \frac{pZ}{2\pi} \Phi \omega$$

$$= \frac{2 \times 450 \times 10^{-3} \times 105 \times 25}{2\pi} \text{ volts}$$

$$= 376\text{V}$$

Example (2)   A four-pole lap wound armature has 54 slots with eight conductors in each slot, and for a speed of 1400 r.p.m. generates 420 V. Calculate the flux per pole.

Solution   For a lap winding $E = \Phi ZN/60$ volts.

Rearranging $\Phi = \dfrac{60E}{ZN} = \dfrac{60 \times 420}{432 \times 1400}$

$$\Phi = 0.041 \text{ Wb}$$

ARMATURE REACTION

So far we have been discussing the d.c. generator purely as a source of e.m.f. We now have to consider what happens when the e.m.f. is applied to an external circuit and current flows around that circuit. When the generator is on open-circuit no current flows in the armature conductors; there is therefore no magnetic field associated with any current flow in the armature conductors themselves, and the flux distribution is as shown in *Figure 16.8(a)*. When the machine is loaded, this is no longer true.

On a completed external circuit, a current flows in the armature windings and so these are now current-carrying conductors moving through a magnetic field.

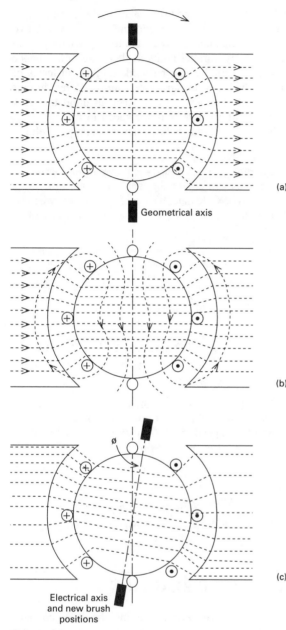

Geometrical axis

Electrical axis
and new brush
positions

Figure 16.8

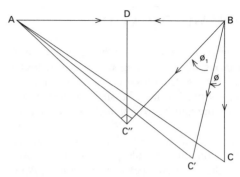

Figure 16.9

another example of nature making sure that we cannot get something for nothing; electrical power output *must* be matched by a corresponding mechanical
power input. It can be shown that the additional torque
is independent of the armature speed but is directly
proportional to the field flux and the armature current.

Now the current flowing in the armature conductors produces its own magnetic field which, along
with the original field, produces a resultant field; the
result of this *armature reaction* as it is called is a
distortion of the flux distribution (see *Figure 16.8(b)*),
leading to two practical results. Firstly, as *Figure
16.8(c)* shows, the compounded flux is no longer uniform over the pole faces, but tends to strengthen into
the forward pole tips in the direction of rotation, with
a corresponding weakening in the rearward tips. The
line along which any coil is enclosing the minimum
flux and generating no e.m.f. which coincided in diagram *(a)* above with the geometrical axis and the line
on which the brushes were assumed to be, the electrical
neutral axis in the open-circuit condition, has now
advanced to the new axis of symmetry between the
poles through an angle ϕ in the direction of rotation.

This now invalidates the original position of the
brushes; the brushes have now, therefore, to be advanced through angle ϕ to coincide with the new
electrical axis. In practice such a movement alters the
distribution of the field lines again; hence the brushes
have to be advanced over a further angle until the
field due to the armature current is perpendicular to
the compounded field.

The field phasor diagram shown in *Figure 16.9*
illustrates the effects of armature reaction. As *Figure
16.8(b)* showed, the armature current sets up a strong
magnetic field directed at right angles with respect to
the main field. Here, then, AB is the flux due to the
field magnets and BC is the flux due to the armature
current. The resultant field is therefore represented by
AC. The brushes are now advanced through the angle
ϕ, but the armature field retains its original strength,

Hence each conductor is acted on by a force which by
the left-hand rule will be acting in opposition to the
direction of rotation; Lenz's law is in action again.
This torque (which is now additional to the resistance
to motion offered only by the friction on open-circuited
conditions) has to be overcome by the mechanical
power supplied to the armature shaft. So additional
power must be supplied from the mechanical driving
source to maintain the armature rotation. This is

i.e. BC is constant and is now represented by BC′. The resultant field now adjusts its position to AC′ which is not perpendicular to BC′, hence a further correction is necessary. This is achieved by advancing the brushes through a further angle ϕ_1 until the armature field is finally perpendicular to the resultant field AC″.

From the diagram we see that the armature field BC″ can be resolved into two components: a cross-magnetizing component which accounts for the flux distortion, and a demagnetizing component BD which opposes the original field AB. Hence, and this is the second effect referred to above, the field flux is weakened as the armature current increases and the generated e.m.f. falls. This effect actually follows from the non-linearity of the magnetic circuit which *Figure 16.8(c)* showed; the flux density in high magnetic force regions is smaller than the decrease in low magnetic force regions, and so the total pole flux is reduced.

GENERATOR CHARACTERISTICS

When the condition is no load across the armature terminals the generator is said to be on open-circuit. The e.m.f. (*E*) generated in the armature conductors will be present at the output terminals as the open-circuit voltage. The open-circuit characteristic (O.C.C.) of a generator is a graph showing the relationship between the generated e.m.f. and the field current at a given speed. This graph can be plotted by setting up the machine as shown in *Figure 16.10*, the field coils being energized from a separate and adjustable supply.

From the generator equation (16.2) above

$$E = \frac{pZ}{\pi a} \Phi\omega$$

and since $pZ/\pi a$ is constant for a given machine, $E \propto \Phi\omega$. This is the e.m.f. expressed in terms of the variables Φ and ω.

The flux per pole Φ is controlled by the field current I_1 and when the field current and flux are constant, $E \propto \omega$. Hence a plot of e.m.f. E against angular velocity ω is a straight line passing through the origin, as shown in *Figure 16.11*. If the field current is reduced from, say, I_1 and I_2 another straight line characteristic is obtained with a reduced slope and hence smaller values of E at each speed setting.

Now with the speed constant, suppose the e.m.f. to be varied by adjustment of the field current and hence pole flux Φ. Then $E \propto \Phi$ and since $\Phi = BA$ and the pole area A is constant, $E \propto B$. But the flux density B is produced by a magnetizing force H At/m due to the field current I_F, hence $E \propto I_F$ provided that the magnetic circuit does not saturate.

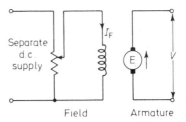

Figure 16.10

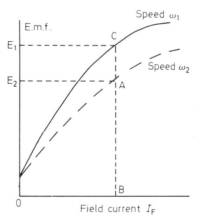

Figure 16.11

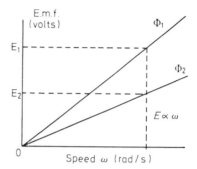

Figure 16.12

A graph of e.m.f. E plotted against I_F is seen to be similar to the B-H curve for the magnetic circuit of the machine. This open-circuit characteristic is sketched in *Figure 16.12*. You will notice that a small e.m.f. is generated at zero field current. There is always some residual flux in the magnetic circuit due to previous magnetization and when the generator is driven at speed the armature windings cut this flux and generate a small e.m.f.

(3) Why do the characteristics flatten out at high values of I_F?

If the generator armature velocity is reduced from ω_1 to ω_2 a similar magnetization curve is obtained with reduced values of e.m.f. at each setting of field current. Since $E \propto \omega$ when Φ is constant

$$\frac{E_2}{E_1} = \frac{\omega_2}{\omega_1}$$

hence, as shown in the diagram

$$\frac{AB}{BC} = \frac{\omega_2}{\omega_1}$$

$$\therefore \quad AB = BC \times \frac{\omega_2}{\omega_1}$$

and the O.C.C. at any other speed may be drawn in relation to the O.C.C. at speed.

THE GENERATOR ON LOAD

When a load is connected across the armature terminals, a load current I_L A will flow and the terminal voltage will fall from its open-circuit value E to some lower value, due to the voltage drop which now occurs in the armature winding resistance. We ignore here the drop in the e.m.f. due to the armature reaction effect on the main field.

If the armature resistance is R_A the voltage drop across it will be $I_A R_A$ (since $I_A = I_L$) and this is subtracted from the generated e.m.f. Hence the terminal voltage V is given by

$$V = E - I_A R_A \qquad (16.4)$$

Example (4)   Calculate the terminal voltage of a generator which develops an e.m.f. of 100 V and has an armature current of 20 A on load. The armature resistance is 0.28 Ω.

Solution

$$V = E - I_A R_A$$

$$= 100 - (20 \times 0.28)$$

$$= 100 - 5.6 = 94.4 \text{ V}$$

Example (5)   A generator has an armature resistance of 0.8 Ω and when connected to a load of 50 Ω passes a current of 5 A. What is the terminal voltage and the generated e.m.f.? A current of 5 A in a 50 Ω load represents a p.d. of $5 \times 50 = 250$ V. This is the terminal voltage of the machine.

Solution   Transposing equation (16.4) to make E the subject, we get

$$E = V + I_A R_A$$

$$= 250 + (5 \times 0.8)$$

$$= 254 \text{ V}$$

TYPES OF GENERATOR

The d.c. generator we have discussed above is a separately excited machine having field magnets whose windings are energized from an external and entirely separate d.c. supply. The provision of such a separate supply is obviously a disadvantage and such machines are used only in special applications. There is no reason why the field magnets should not be energized from the generator output itself, and machines in which this takes place are known as self-excited generators. Such generators may take one of three types:

(a)   Shunt-wound generators in which the field windings are connected in parallel with the armature output;

(b)   Series-wound generators in which the field windings are connected in series with the armature output;

(c)   Compound-wound generators which have a mixture of shunt and series windings designed to combine the advantages of each.

The shunt generator

This type of machine is shown in *Figure 16.13*. Because of the residual flux mentioned earlier, a small current flows in the parallel field windings as soon as the machine is started up. This small current increases the pole flux and there follows a rapid build up in both the field and the generated e.m.f. It is advantageous

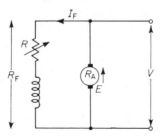

Figure 16.13

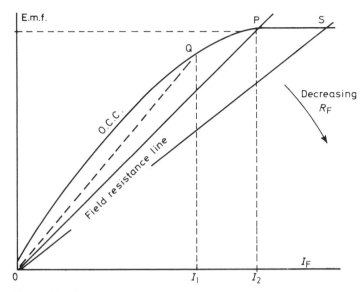

Figure 16.14

to keep the shunt current (I_F) as small as possible, getting the ampere-turns for the required flux from a large number of turns. Hence the field windings have a relatively high resistance in these machines, being wound with many turns of relatively fine wire.

Now the voltage across the field coils for different values of field current is given by the straight line graph connecting V and I_F for the particular value of winding resistance R_F. Let this field resistance line as it is called, cut the O.C.C. at a point P, see *Figure 16.14*. Then the open-circuit e.m.f. of the generator will build up to this value indicated by the intersection of the magnetizing curve and the field resistance line at P, and at this point (neglecting the armature resistance) the voltage required across the field is equal to the armature e.m.f. Conditions are then stable.

You will have noticed that a series regulator resistance R is shown in series with the field windings. If this resistance is increased, the effective value of R_F increases, hence the slope of the field line becomes steeper, point P moves down the curve to Q and the generated e.m.f. becomes less. If the resistance is decreased, R_F decreases, the slope of the field line becomes less and point P moves further up the curve to S, so increasing the generated e.m.f. subject to the saturation conditions of the magnetic circuit materials.

Example (6) A shunt generator supplies a load of 10 kW at 250 V through cables of resistance 0.1 Ω. If the resistance of the armature is 0.03 Ω

and of the field coils 75 Ω, calculate the terminal voltage and the generated e.m.f.

Solution

$$\text{The current in the load} = \frac{\text{power}}{\text{voltage}} = \frac{10\ 000}{250}$$

$$\therefore \qquad\qquad I_L = 40 \text{ A}$$

Volts drop in the cables = 40 × 0.1 = 4 V

Hence the terminal voltage = 250 + 4 = 254 V

$$\text{The field current } I_F = \frac{254}{75} = 3.387 \text{ A}$$

$$\therefore \text{ armature current } I_A = I_F + I_L = 3.387 + 40$$

$$= 43.387 \text{ A}$$

Volts drop in the armature = 43.387 × 0.03

$$= 1.3 \text{ V}$$

$$\therefore \quad \text{generated e.m.f.} = 254 + 1.3 = 255.3 \text{ V}$$

(7) *Figure 16.15* shows the O.C.C. for a certain shunt generator. Sketch lightly in pencil two field resistance lines for values of 150 Ω and 200 Ω. What is the open-circuit e.m.f. of this generator for each of these values of field resistance?

(8) In *Figure 16.15* the dotted line represents the field resistance line for 300 Ω. Check that

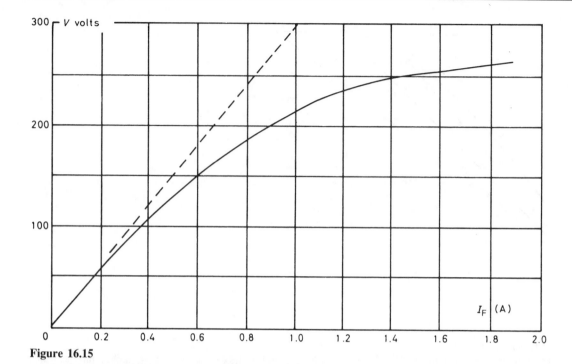

Figure 16.15

this is so and deduce what would be the result of running the machine up in this situation.

(9) A shunt generator has an armature resistance of 0.3 Ω and a field resistance of 150 Ω. What is the generated e.m.f. if the machine supplies a load with 10 A at a terminal p.d. 200 V.

It is important when solving problems of this sort not to try and remember formulae. You require only applications of Ohm's law.

The load characteristic of a shunt generator is the plot of terminal p.d. V against load current I_L for a given value of armature velocity ω. A typical curve is shown in *Figure 16.16*.

The terminal p.d. is greatest when the load current is zero as only the field windings are drawing current from the armature. As the load current increases, the armature volts-drop increases and the terminal p.d. (along with the voltage across the field coils) decreases. The net effect is a gradual and continuing fall in the output voltage. If the external current becomes so large that the machine is overloaded, the terminal p.d. falls off rapidly and difficulty is experienced in getting a field current large enough to produce an e.m.f. which will maintain the terminal voltage. When this adjustment becomes impossible, at current I_o in *Figure 16.16*, the terminal p.d. and the load current fall to zero and the machine shuts down. This situation should never be allowed to happen in practice.

(10)   How do you think the load characteristic of a separately excited generator would compare with the self-excited case shown in *Figure 16.16*?

The series generator

In this form of generator, shown in *Figure 16.17(a)*, the field coils are wired in series with the armature and so carry the full load current when the machine is operating. The resistance of the coils must, as a consequence, be small and relatively few turns of heavy gauge wire are used to provide the required

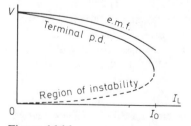

Figure 16.16

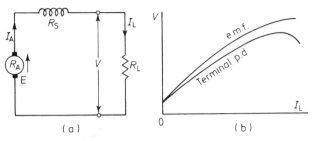

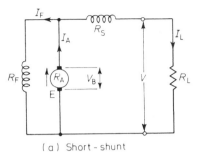

(a) Short - shunt

Figure 16.17

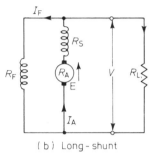

(b) Long - shunt

Figure 16.18

ampere-turns. If the output terminals are on open-circuit the load current is zero and so the field coils carry no current. Hence the field magnets will not excite and the terminal p.d. is very small, only the residual flux generating an e.m.f. in the armature.

The load characteristic of a series generator is shown in *Figure 16.17(b)*. Unlike the shunt generator characteristic which falls, the field flux and hence the terminal p.d. of the series machine rises along a smooth curve as the armature (and load) current increases, but bends over at high values of load current as the poles go into magnetic saturation.

Series generators are little used in practice.

The compound generator

This type of generator has field magnets excited partly by high resistance shunts coils and partly by low resistance series coils, the connections being made so that the fields produced are additive.

The two possible methods of connection are shown in *Figure 16.18* diagram *(a)* being the short-shunt and *(b)* being the long-shunt form of connection. The object of this kind of mixed winding is to give a substantially constant terminal p.d. irrespective of the load current being drawn.

When the load current increases, the armature volts-drop increases; this would normally result in a drop in terminal voltage as we noticed for the shunt generator above. The inclusion of the series winding, however, gives an additional field flux in proportion to the current being drawn and so provides a boost to the output voltage which compensates for the increased armature volts-drop. We are, in fact, combining the shunt and series generator characteristics to give either a practically level characteristic or a very slowly rising one. The former is called a level-compounded machine and the latter an over-compounded machine. These two forms of load characteristic are shown in *Figure 16.19*.

Any calculations for problems involving compound generators are, as usual, simply based on Ohm's law. For the short-shunt generator of *Figure 16.18(a)*

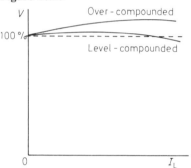

Figure 16.19

Volts drop in the series coil $= I_L R_A$

Voltage at the brushes $V_B = V + I_L R_A$

But $\qquad\qquad I_A = I_L + I_F$

$\therefore$ volts drop in the armature $= R_A(I_L + I_F)$

and the generated e.m.f. $E = V_B + R_A(I_L + I_F)$

(11) Now derive a similar expression for the generated e.m.f. of the long-shunt generator shown in *Figure 16.18(b)*.

THE D.C. MOTOR

In their basic form, d.c. motors and generators are identical and it is only in the direction of energy flow that the essential difference is found. A motor rotates and produces mechanical energy (torque) when given a certain terminal polarity; a generator gives the same electrical polarity when mechanically driven in the opposite direction.

The reason for this reversal of direction is that the electric motor depends for its action on the principle that when a current is passed through a conductor which lies in a magnetic field, the conductor is acted on by a force $F = BI\ell$ newtons which tends to move it in a direction perpendicular to itself and to the direction of the field: Fleming's Left-Hand Rule. This movement in the field in turn generates an e.m.f. which, in accordance with Lenz's law and the generator e.m.f. equation, opposes the change producing it, i.e. it is a back-e.m.f. acting in opposition to the applied voltage and therefore in accordance with Fleming's Right-Hand Rule. This back-e.m.f. consequently has the same values as if the machine were running as a generator, i.e. $E \propto \Phi\omega$ and so is proportional to the angular velocity ω. As regards type of armature and field windings, we have exactly the same classification for motors as for generators.

The current I_A which flows in the armature is due to the resultant of these two opposing voltages. If V = applied voltage and E_b = the back-e.m.f. then:

$$I_A = \frac{V - E_b}{R_A}$$

$$\therefore \quad I_A R_A = V - E_b$$

$$E_b = V - I_A R_A \qquad (16.5)$$

The back-e.m.f. determines the armature current and makes the d.c. motor a self-regulating machine. The speed of the motor automatically adjusts itself so that the electrical power required to drive the current through the armature is equal to the mechanical power given out through the shaft to the load. When a load is put on a motor a certain torque is demanded from the motor. This torque is controlled by the armature current and so the motor increases its armature current until it can provide the torque asked for. For any given load condition the motor adjusts its speed so that the back-e.m.f. induced in the armature is equal to the supply voltage minus the d.c. volts-drop $I_A R_A$ in the armature coils.

As we have noted above, $E_b \propto \Phi\omega$, where Φ is the pole flux and ω the angular velocity of rotation. Combining this relationship with (16.5) above we get

$$V - I_A R_A \propto \Phi\omega \qquad (16.6)$$

and this is a fundamental relationship from which the behaviour of any type of d.c. motor can be determined.

As the armature volts-drop $I_A R_A$ is usually very small compared with V, equation (16.6) can be simplified to

$$V \propto \Phi\omega \quad \text{or} \quad E_b \propto \Phi\omega$$

so if there is a change in operating conditions the new values may be calculated by proportion, that is:

$$E_{b1} \propto \Phi_1\omega_1 \quad \text{and so} \quad E_{b2} \propto \Phi_2\omega_2$$

Dividing we have

$$\frac{\omega_1}{\omega_2} = \frac{E_{b1}}{E_{b2}} \times \frac{\Phi_2}{\Phi_1}$$

Try this next example to see whether you have grasped the principle of this proportional relationship.

(12)   A motor runs at an armature velocity of 150 rad/s when the armature e.m.f. is 240 V. Find its speed when the flux per pole is reduced 10 per cent if at the new speed the e.m.f. is 235 V.

Torque

The torque T newton-metres (N-m) which a motor can exert is clearly related to the motor power. If the armature radius is R metres and the tangential force $F (= BI\ell)$ newtons causes the armature to rotate at angular velocity ω rad/s, then the torque $T = FR$ N-m and the work done by the force in one revolution $= 2\pi FR$ N-m. Therefore the work done by the force in 1 s $= 2\pi FR (\omega/2\pi) = FR\omega = \omega T$ joules, and the power developed

$$P = \omega T \text{ W} \qquad (16.7)$$

since 1 W = 1 J/s.

Considering the armature input power as VI_A W, we may multiply (16.5) above by I_A and then

$$E_b I_A = VI_A - I_A^2 R_A$$

Looking at each of these terms in turn we deduce

(a) VI_A is the power available for producing motion, the input power.
(b) $I_A^2 R_A$ is the power wasted in the armature (and brush) resistance.
(c) $E_b I_A$ is the armature power.

$I_A^2 R_A$ is the copper loss of the motor; $E_b I_A$ is known as the gross power of the motor, but the whole of this power is not available for doing useful work, some of it being required to overcome iron and friction losses. Neglecting these other losses for the moment

Gross power $= E_b I_A = \omega T$ W

Torque $T = \dfrac{E_b I_A}{\omega}$

But

$E_b \propto \Phi\omega$

$T \propto \dfrac{\Phi\omega I_A}{\omega} = k\Phi I_A$

where k is a constant.

> Example (13)   A d.c. motor takes 10 A from a 240 V supply. The machine losses amount altogether to 300 W. If the machine runs at a speed of 100 rad/s, what is the output torque?
>
> *Solution*   Input power $= 240 \times 10 = 2400$ W
>
> Output power $= 2400 - 300$
>
> $\qquad\qquad = 2100$ W
>
> Power $= \omega T = 100T$ W
>
> $\therefore\quad 100T = 2100$
>
> $T = 21$ N-m

MOTOR CHARACTERISTICS

Like generators, motors are classified according to their method of excitation and may be of the shunt-, series- or compound-wound variety. The shunt and series types are those which concern us here.

The shunt-wound motor

For a given terminal voltage this type of motor maintains an almost constant speed over a wide range of load torques. Like the shunt generator, the field windings are connected in parallel with the armature (and the input terminals) and are wound with many turns of fine gauge wire. Now $E_b \propto \Phi\omega$ and as Φ is constant the speed is directly related to E_b, which, from equation (16.4), is $V - I_A R_A$. Since V is constant, as I_A increases so E_b must decrease, hence ω must decrease, and the motor slows down. The plot of speed against armature current is drawn in *Figure 16.20(a)*.

The torque-current characteristic can be deduced by a consideration of the proportionality $T \propto \Phi I_A$. As more torque is demanded, the armature current increases to provide it and this can continue up to the

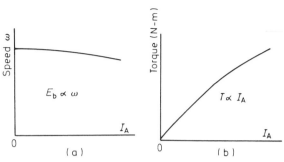

Figure 16.20

safe working limit of the armature capacity. For a constant Φ, torque is directly proportional to I_A, hence the characteristic of T against I_A is a straight line passing through the origin. In practice the line bends slightly as *Figure 16.20(b)* illustrates.

> (14)   Explain why the curve departs from a straight line in this way.

The shunt motor is used on steadily running constant-speed loads, such as printing or weaving machine drives.

The series-wound motor

The d.c. series-wound motor has a wide speed range and a large starting torque and finds considerable employment. The speed varies substantially with load variations. As the field coils are in series with the armature winding they are wound with a few turns of heavy gauge wire which carry the full armature current. The field flux depends on the armature current, and provided that the iron of the pole pieces is unsaturated, $\Phi \propto I_A$ as we have already noted. The gross armature torque is proportional to ΦI_A and $\Phi \propto I_A$. Hence $T \propto I_A^2$. This means that if the armature current is forced to increase by an excessive load slowing up the motor, a proportionally larger increase in torque is obtained to establish equilibrium again.

A typical plot of torque against armature current is shown in *Figure 16.21(a)*. The current does not follow the square-law shape at high current values as the poles gradually approach saturation and the field flux cannot then increase rapidly as I_A^2, hence the torque becomes more nearly proportional to I_A.

The speed characteristic of a series motor can be deduced by considering the fact that $E_b \propto \Phi\omega$. Assuming that E_b is constant then, if the armature and field drops are small, $\omega \propto 1/I_A$ as long as the poles

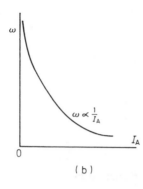

Figure 16.21

are unsaturated. The curve of ω against I_A is shown in *Figure 16.21(b)* and approximates to a rectangular hyperbola. The characteristic of high torque and low speed at high values of armature current is ideal in traction systems where considerable friction has to be overcome to get a vehicle moving from rest. Care has to be taken, however, that the load is not suddenly removed as the speed will immediately rise to a possibly dangerous level in an attempt to reduce the armature current by raising the back-e.m.f.

Example (15)   The armature of a d.c. motor has 300 uniformly spaced conductors. The armature diameter is 20 cm and the conductors are each 20 cm long. If the starting current taken by the motor is 15 A, giving a flux density if 0.5 T in the gap between pole and rotor, calculate the starting torque provided by the motor.

Solution

Force on each conductor

$= BI\ell$ N

$= 0.5 \times 15 \times 20 \times 10^{-2} = 1.5$ N

Total force on 300 conductors $= 300 \times 1.5$

$= 450$ N

Torque = force × arm of torque

$= 450 \times 10 \times 10^{-2}$

$= 45$ N-m

Speed control and starting

From the proportionality $E_b \propto \Phi\omega$, since $E_b = V - I_A R_A$ then

$$\omega \propto \frac{V - I_A R_A}{\Phi}$$

From this we see that the speed may be controlled by varying

(a)   the pole flux Φ;
(b)   the resistance in the armature circuit R_A;
(c)   the applied voltage V.

We shall be interested in the effect of adding resistance to the armature and field circuits to modify I_A and Φ, respectively.

In *Figure 16.22(a)* a resistor R has been added in series with the armature of a shunt-wound motor. At any particular load and armature current value I_A the extra resistance will cause a greater armature volts-drop and consequently a smaller value for $V - I(r + R)$. Hence the motor speed will be reduced when Φ is constant. A disadvantage of this method is that for a given value of R, a change in load torque will lead to a change in speed. A further disadvantage is the relatively large power loss in resistor R.

In *Figure 16.22(b)*, a resistor R has been added in series with the field windings. The effect of this is to reduce the pole flux Φ and so lead to an increase in speed. However, if the load torque is constant, then ΦI_A must be constant so that a decrease in Φ results in an increase in I_A. As a result the motor must be made larger and consequently costs more than it would if no speed variation were required. The method is, however, very simple and is an efficient way of

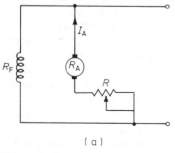

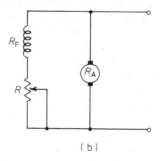

Figure 16.22

controlling speed above a minimum value obtained with full field current.

The next example illustrates this.

Example (16)   A shunt motor connected to a 240 V d.c. supply, runs at 1600 rad/s on no-load at an armature current of 1.0 A. If the armature resistance is 2 Ω, calculate the speed of the motor on load when the armature current is 8 A. By what percentage would the field have to be reduced in order to restore the speed to its original value?

Solution   We have the proportionality $\Phi\omega \propto (V - I_A R_A)$

∴ since Φ is constant, the ratio of the speeds will be

$$\frac{\omega_2}{\omega_1} = \frac{V - I_{A2}R_{A2}}{V - I_{A1}R_{A1}}$$

∴ for $\omega_1 = 1600$ rad/s

$$\frac{\omega_2}{1600} = \frac{240(-8 \times 2)}{240 - (1 \times 2)} = \frac{224}{238}$$

∴ $\omega_2 = \dfrac{1600 \times 224}{238} = 1506$ rad/s

Now as we have noted in the text $E_b \propto \Phi\omega$ or $\omega \propto E_b/\Phi$. A reduction in flux Φ will consequently increase the speed, so

$$\frac{\omega_2}{\omega_1} = \frac{\Phi_2}{\Phi_1} = \frac{1506}{1600} = 0.94$$

Hence the percentage decrease in Φ will be about 6 per cent.

Speed control of series-wound motors can be brought about by the methods illustrated in *Figure 16.23(a)* and *(b)* which depend on a variation in Φ and V, respectively.

In diagram *(a)* a variable resistor known as a diverter is connected in parallel with the series field winding; in this way any part of the main current may be passed through the field coils. Increasing the diverter resistance strengthens the field and the motor slows down; and conversely.

In diagram *(b)* the diverter is placed across the armature. For a given load torque, a reduction in I_A because of a reduction in the diverter resistance leads to an increase in Φ and a reduction in speed.

A method commonly used in fan and cookery mixer motors is shown in *Figure 16.23(c)*. The field coils are regrouped to provide a series of fixed speeds. Alternatively, the field coils are sometimes tapped so that the number of turns in circuit can be varied.

Variation of the applied voltage by a series resistor R is shown in *Figure 16.23(d)*. Increasing R reduces the applied voltage and hence the speed falls. The method leads to a considerable waste of power in the series resistor.

Starting

At the moment of switching on a motor there is no back-e.m.f. and the current that results is given by $I_A = V/R_A$. Once running, the back-e.m.f. limits the

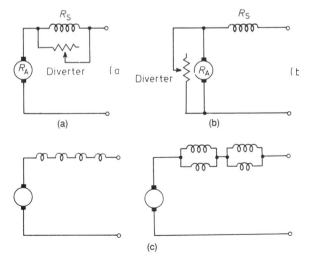

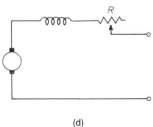

Figure 16.23

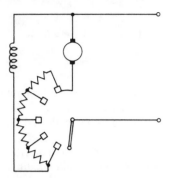

Figure 16.24

current to $(V - E_b)/R_A$. To reduce the starting current to a safe value as the back-e.m.f. builds up, a starter resistor is included in the armature circuit and this limits the current to about 1.5 times the normal full load value. This current, with a full pole strength of field, gives a good starting torque. *Figure 16.24* shows the basic shunt motor starter connections. As the motor e.m.f. rises the resistance is progressively cut out by moving the starter arm across the contacts.

LOSSES AND EFFICIENCY

The losses which occur in generators and motors are of the same kind, that is (a) copper losses, (b) iron losses, (c) friction losses.

Copper losses are the armature copper loss, the field windings copper loss and the loss due to the brush contact resistance at the surface of the commutator. There are all known as I^2R losses.

Iron losses mainly involve eddy-current loss in the armature and pole pieces, and the hysteresis loss in the armature.

Friction losses include the brush friction, bearing friction and air resistance against moving parts (windage), particularly when cooling fans are incorporated.

The energy changes and losses which occur in the transformation of electrical energy into mechanical energy are shown in *Figure 16.25*. This is the motor

case. Changing this diagram round end to end will provide the appropriate representation of the generator case. The copper losses are represented between sections A and B and the iron and frictional losses between sections B and C. We define

$$\text{Electrical efficiency } \eta_e = \frac{B}{A} \times 100\%$$

$$\text{Mechanical efficiency } \eta_m = \frac{C}{B} \times 100\%$$

$$\text{Overall efficiency } \eta_c = \frac{C}{A} \times 100\%$$

In the absence of any other indication, the efficiency stated is taken to be the overall or commercial efficiency η_c. Notice that

$$\eta_c = \frac{C}{A} = \frac{\text{output power}}{\text{input power}} = \frac{\text{output}}{\text{output + losses}}$$

Example (17)   A 240 V series motor draws a current of 35 A. The armature resistance is 0.1 Ω and the field resistance is 0.075 Ω. If the iron and friction losses are equal to the copper losses at this load, calculate the commercial efficiency.

Solution   Total resistance of motor = 0.175 Ω

$$E_b = V - I_A R_A = 240 - (35 \times 0.175)$$
$$= 233.88 \text{ V}$$

Input power = $240 \times 35 = 8400$ W

Armature power $E_b I_A = 233.8 \times 35$
$$= 8185.8 \text{ W}$$

Copper losses = $8400 - 8185.8 = 214.2$ W

Iron losses = 214. 2 W

Total losses = $2 \times 214.2 = 428.4$ W

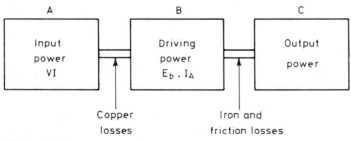

Figure 16.25

Output power = input power − losses

$$= 8400 - 428.4 = 7971.6 \text{ W}$$

$$\eta_c = \frac{7971.6}{8400} \times 100 = 95\%$$

Example (18) A shunt generator supplies 100 A at a terminal voltage of 200 V when the driving motor is developing 23 kW. If the armature resistance of the generator is 0.2 Ω, and the shunt field resistance is 50 Ω, calculate the copper losses, the iron and friction losses, and the overall efficiency.

Solution Shunt current $I = \dfrac{200}{100} = 2 \text{ A}$

∴ armature current $I_A = 100 + 2 = 102 \text{ A}$

Volts drop in armature $= I_A R_A = 102 \times 0.2$

$$= 20.4 \text{ V}$$

∴ e.m.f. generated $= 200 + 20.4 = 220.4 \text{ V}$

The mechanical power input is given as 23 kW

Electrical power produced in the armature $= EI_A$
$= 220.4 \times 102 = 22481 \text{ W}$

∴ iron and friction losses $= 23\,000 - 22\,481$

$$= 519 \text{ W}$$

Electrical power output $= VI = 200 \times 100$

$$= 20\,000 \text{ W}$$

∴ copper losses $= 22\,481 - 20\,000$

$$= 2481 \text{ W}$$

Overall efficiency $= \dfrac{\text{output}}{\text{input}} = \dfrac{20\,000}{23\,000}$

$$= 0.87 \text{ or } 87\%$$

Characteristic review

We might at this point conveniently review the characteristics of shunt and series motors:

Shunt motor	*Series motor*
Flux: Constant but adjustable	Varies with armature current
e.m.f.: $E = kN\Phi$	$E = k_1 N I_A$
Torque: $T = k\Phi I_A$	$T = k_2 I_A^2$
Speed: Nearly constant	Varies widely
Speed control: Efficient over a wide range by varying V_A or I_F	Some control by V or inserting R_{ser}
Starting torque: Fair	Good

STEPPING MOTORS

A stepping motor does not rotate smoothly and continuously in the manner of those motors already discussed, but 'steps' from one fixed position to the next in a sequence of discrete movements. The angular extent of the steps can be within a range covering up to 30°, so in a sense such a motor can be thought of as a 'digital device'. It is, indeed, usually operated under digital control, that is, with pulsed and not continuous input signals. Such motors are used in robotics, as head positioners in floppy disk drives, in computer controlled X−Y plotters and in teletype printers, to name only four common applications: anywhere, that is, where incremental and not continuous movement is called for.

An elementary form of stepping motor can be made using an escapement mechanism to advance the movement in discrete steps and many of this form are in use, but modern stepping motors work on the principle that various combinations and phasings of the fields and the interaction between these fields and the rotor field within the motor will cause the rotor to move forwards (or backwards) a defined number of degrees. The number of steps and the speed of rotation are respectively determined by the number of pulses and the frequency of the control signal. A 1.8° step angle is common, so giving 200 steps per revolution, and up to 800 steps per second can be achieved.

Figure 16.26 shows a possible arrangement of a stepping motor connected to an external control system. Suppose the coils to be energized in a particular way with switches Q_1 and Q_2 closed (ON), and the rotor held in a certain position: then changing to switches Q_1 and Q_4 causes the motor to move through a certain angle, say clockwise. A further change to Q_3 and Q_4 will move the rotor through a further clockwise step, after which, with Q_2 and Q_3 closed, the next step occurs. After this the sequence repeats and the motor will continue to step in a clockwise direction as long as the switching cycle continues. To get the motor to step in an anticlockwise direction, the switching sequence is reversed. *Figure 16.27(a)* shows the sequence we have discussed written in tabular form.

It might appear at first that this sequence is haphazard and arbitrary, but this is not so. The control must be precisely programmed. In the present example, for instance, if the switching sequence is interpreted in

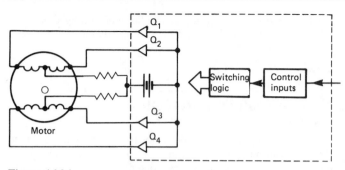

Figure 16.26

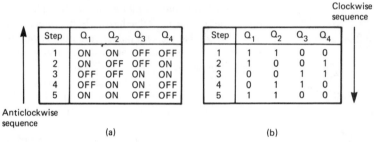

Clockwise sequence

Step	Q_1	Q_2	Q_3	Q_4
1	ON	ON	OFF	OFF
2	ON	OFF	OFF	ON
3	OFF	OFF	ON	ON
4	OFF	ON	ON	OFF
5	ON	ON	OFF	OFF

Step	Q_1	Q_2	Q_3	Q_4
1	1	1	0	0
2	1	0	0	1
3	0	0	1	1
4	0	1	1	0
5	1	1	0	0

Anticlockwise sequence

(a)   (b)

Figure 16.27

terms of control logic, where ON = 1, OFF = 0, a pattern becomes evident, see *Figure 16.27(b)*. To move from step 1 to step 2, the initial pattern 1100 is moved one place to the left, and thereafter each step is accomplished by rotating the pattern one more place to the left. For a reversal of the motor, the pattern is shifted sequentially to the right.

The sequence can be controlled by wiring the switches (which are, of course, electronic switches using bipolar or FET devices) to the output of a ring counter initially loaded with the binary sequence 1100. Each step is then controlled by clocking the counter. Clockwise or anticlockwise control can also be introduced by the counter. Alternatively, the switches may be controlled by a microcomputer which is programmed to operate the required step sequence. A suitable control chip for experimental purposes is the SAA1027.

SUMMARY

- A conductor moving in a magnetic field generates an e.m.f. and develops a torque.
- The e.m.f. can be evaluated in terms of the field strength, the velocity of the moving conductor and the length of the conductor.

- The generator e.m.f. equation is $E = \dfrac{pZ}{a}$ volts.
- An alternator generates a sinusoidal e.m.f.
- A commutator is an automatic switch that permits the generation of direct current and the production of a unidirectional torque.
- A wave winding has two paths in parallel between brushes irrespective of the number of poles and each path provides half the total output current.
- A lap winding has as many parallel paths between brushes as there are poles and the total output current is shared between them.

REVIEW QUESTIONS

1 A commutator may be called 'a synchronous full-wave rectifier'. Is this description justified?

2 Explain the operation of a commutator in a d.c. generator.

3 Outline the steps in deriving an expression

for the e.m.f. generated in a conductor moving in a magnetic field.

4 What are the two major components of the iron losses in a generator or motor?

6 In a d.c. motor, would the brushes need to be advanced or retarded relative to the direction of rotation, and why would this be necessary?

7 Make a diagram showing the energy changes and losses which occur in the transformation process of electrical energy into mechanical energy. Are these identical in the reverse process?

8 What factors determine the starting torque of a d.c. motor? Why is this torque high in a series-connected motor?

9 What is the O.C.C. characteristic of a generator or motor? What does it tell us?

10 Tabulate methods of speed control in a series-wound d.c. motor.

11 Explain, with diagrams, the essential differences in the construction and performance of series-connected and shunt-connected d.c. motors rated at the same full load power, supply voltage and speed.

12 Would an alternator work as a motor if the appropriate voltage input was applied to its terminals? If not, why not? If it would not, can you suggest a way in which it might be made to do so?

EXERCISES AND PROBLEMS

(19) Fill in the missing words or quantities:

(a) Motors convert . . . energy into . . . energy.
(b) Efficiency of a machine is defined as. . . .
(c) In a lap winding there are as many parallel paths as there are. . . .
(d) In a wave winding there are . . . parallel paths.
(e) At a constant speed, generated e.m.f. $E \propto$ . . . .

(20) A four-pole wave-wound machine has 420 conductors and is driven at 800 r.p.m. to generate an e.m.f. of 200 V. What is the flux per pole of the machine?

(21) A six-pole generator has a wave-wound armature and is driven at an angular velocity of 42 rad/s. Calculate the number of conductors if the machine produces an e.m.f. of 240 V with a pole flux of 0.04 Wb.

(22) A generator runs on load with an armature current of 75 A and a generated e.m.f. of 400 V. Determine the terminal p.d. given that the armature resistance is 0.2 Ω.

(23) A 50 kW, 250 V d.c generator has an armature circuit resistance of 0.15 Ω. What is the generated e.m.f. on full load?

(24) Find the value of the back-e.m.f. of a motor working from 220 V d.c. mains when the armature current is 10 A. Take the armature resistance to be 0.5 Ω.

(25) A d.c. shunt motor runs at an angular velocity of 188 rad/s when connected to a 100 V supply and in the unloaded condition the armature current is 0.5 A. If the armature resistance is 3.5 Ω find the speed of the motor when it is loaded and the armature current is 2 A.

(26) A 460 V motor runs at an angular velocity of 157 rad/s with an armature current of 120 A. If the armature resistance is 0.4 Ω, calculate the percentage change in the pole flux needed to obtain a velocity of 262 rad/s when the current is 125 A.

(27) What power is needed to drive a 150 kW generator when it is delivering its full rated output, if the machine has a full load efficiency of 91.5 per cent.

(28) Fill in the missing words or quantities:

(a) Motor torque $T \propto I_A \times$ . . . .
(b) Angular velocity $\omega \propto E$ . . . .
(c) The e.m.f. induced in the armature of a motor . . . the supply voltage.
(d) The armature power EI_A is called the . . . power.

(29) The open-circuit characteristic of a d.c. machine is given in the following table:

Generated e.m.f. (V)	100	176	218	240	256	266
Field current (A)	0.2	0.4	0.6	0.8	1.0	1.2

If the machine is used as a shunt generator, find (a) the field circuit resistance to give an open-circuit e.m.f. of 245 V, (b) the open-circuit e.m.f. when the field circuit resistance is 310 Ω.

(30)   Explain why a d.c. series-connected motor has a high starting torque. The armature of a d.c. series motor has 400 uniformly spaced conductors. The radius of the armature is 12.5 cm and the conductors are each 25 cm long. If the motor draws a starting current of 10 A, giving a flux density in the gap of $0.5T$, calculate the starting torque.

(31)   A 240 V shunt motor runs at 600 r.p.m. when the armature current is 10 A. If the armature resistance is 2 Ω find the speed when a 4 Ω resistor is placed in series with the armature. Assume that Φ and I_F remain unchanged. Why is this method of control inefficient?

(32)   *Figure 16.28* shows a shunt motor connected to a 100 V supply. Calculate the speed of this motor when a 2.5 Ω resistor is connected in series with the field winding, assuming that the load remains unchanged and that the flux is proportional to the field current.

(33)   A d.c. generator with a 120 Ω shunt field resistance is feeding a load of 20 Ω. The armature resistance is 2.5 Ω and the frictional

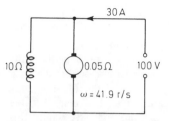

Figure 16.28

losses in the generator are equivalent to an additional shunt load of 100 Ω across the output terminals. What is the overall efficiency of the generator?

(34)   A d.c. motor operating from a 400 V supply is driving a d.c. shunt-wound generator having a field resistance of 125 Ω. This generator is supplying a current of 12 A at 250 V to a heater system. Calculate the fraction of the total power available at the brushes that is dissipated in the field winding. If the overall efficiency of the generator is 75 per cent and the efficiency of the motor alone is 70 per cent, calculate the current and power taken by the motor from the 400 V supply.

The induction motor

Three-phase alternators
The rotating polyphase field
Induction motor operation
Rotation and slip
Cage and wound Armatures
Single phase induction motors

All the machines described in the previous chapter were direct current machines. They have also been the type of machine in which the field has been stationary and the armature has rotated. From a consideration of the basic laws governing the operation of machines, however, there is no reason why the field should not rotate, and if a stationary system of coils, suitably wound and disposed, is supplied with alternating current, a uniformly rotating field of constant intensity is indeed produced. This principle is used in the design of several forms of a.c. motors.

If we go back to *Figure 8.5*, on page 123, we recall that the three-phase a.c. generator was considered as an arrangement of three identical coils, fixed 120° apart, rotating in a uniform magnetic field; and the observation was made that although the system was theoretically sound, a number of practical limitations prevented its use in that particular form.

One of these limitations is the problem of getting the generated power out of the rotating coils and into the external cables. The use of slip-rings is impracticable where very high currents are concerned and armature insulation is difficult to achieve at high voltage levels. By changing over the field and armature, this problem can be overcome.

In *Figure 17.1* a rotating magnet (the rotor) is turning between fixed coils (the stator coils). As the magnet rotates at a uniform angular velocity, each pole passes a coil in turn. This relative motion of field and coils is exactly similar in its effect with the case where the field is fixed and the coils move past it, hence first a

positive e.m.f. and then a negative e.m.f. is induced in each coil. One cycle of e.m.f. is produced for each complete revolution of the magnet and hence the generated frequency is a function of the speed of rotation.

In a practical generator, an electromagnet would be used as the rotor instead of the permanent magnet shown and this would be energized by direct current from an outside source by way of slip-rings. Such rings could be used here where the power requirement of the rotor would be small and constant. The stationary armature or stator coils on the other hand, in which the required e.m.f. is generated, can have direct connections to the external circuit.

A practical form of three-phase alternator is shown in *Figure 17.2*. This has six poles (but a much greater number are commonly used) and these are wound

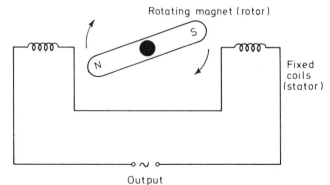

Figure 17.1

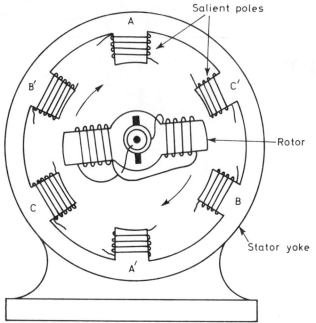

Figure 17.2

alternately right and left handed, similarly wound poles being placed 120° apart. The pole pieces are usually built up from steel stampings and either bolted or dovetailed to the outer frame. This form of construction is known as a salient pole, the pole pieces projecting as shown and shaped to produce a sinusoidal distribution of the flux in the air gap; the induced e.m.f. is then also sinusoidal. The interconnections between the pole windings are not shown in the diagram for the sake of clarity, but these may be arranged in either star or delta form.

GENERATED FREQUENCY

The direction of the induced e.m.f. as the field sweeps round will depend on the direction in which each pole is wound, and the e.m.f. generated in each coil will complete one cycle as the field moves past one pair of poles.

If ω is the speed in rad/s and p is the number of pole pairs, then $f = p\omega/2\pi$ Hz. Alternately, if N is the speed in rev/min, then $f = pN/60$ Hz. N is the synchronous speed of the alternator and is the speed at which the machine must be driven if the required frequency is to be generated. For power generation, 50 Hz is the standard frequency in the UK, although 60 Hz is common in the USA and some other countries. The problem of distribution is eased when the

frequency is low, but it must not be so low that flicker is noticed in any lighting.

> (1) Can you think of another disadvantage of a very low frequency?
> (2) Calculate the required speed (in rev/min and in rad/s) of an a.c. generator having (a) two poles, (b) four poles, (c) six poles, for a 50 Hz output frequency.

THE ROTATING FIELD

In *Figure 17.3* the stator system of a three-phase alternator has been redrawn in a form which simply shows the appropriate coils spaced at 120° intervals around the stator frame. The rotor has been removed and we are interested only in the field generated within the stator when the coils are supplied with three-phase a.c. Just as the rotation of the magnetic field of the rotor in the alternator generates the three-phase supply, so when such a supply is applied to the stator system, we should expect a rotating field to be produced.

Let the stator coils of *Figure 17.3* be energized with three-phase a.c. as shown in *Figure 17.4*. This diagram illustrates the current in the coils, but the pattern can be interpreted as the flux waves set up by the stator windings. The total flux at any instant is the phasor sum of the separate fluxes at that instant. Consider the conditions at intervals of one-sixth of a cycle:

At instant 1: the field of poles AA' is zero, the fields of poles BB' and CC' are equal. Hence the resultant lies midway between poles BB' and CC'. At instant 2: the field of poles CC' is zero, the fields

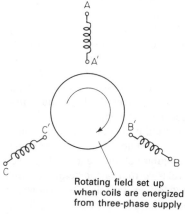

Rotating field set up when coils are energized from three-phase supply

Figure 17.3

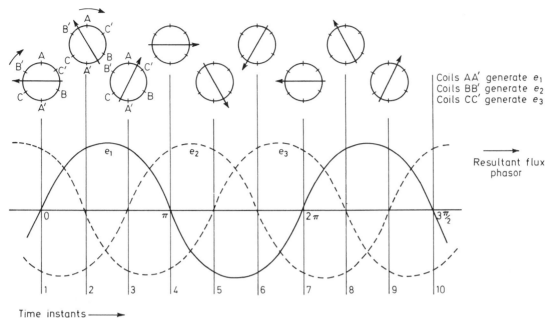

Coils AA′ generate e_1
Coils BB′ generate e_2
Coils CC′ generate e_3

Resultant flux phasor

Time instants ——→

Figure 17.4

of poles AA′ and BB′ are equal. Hence the resultant lies midway between poles AA′ and BB′.

At instant 3: the field of poles BB′ is zero, the fields of poles AA′ and CC′ are equal. Hence the resultant lies midway between poles AA′ and CC′.

And so on.

By taking intermediate instants of time to those shown, intermediate positions of the maximum field strength can be found which will lie between those marked. If you give the diagram a bit of thought, you should be able to deduce that in each case the resultant flux Φ is constant and equal in magnitude to 1.5Φ (the maximum flux due to any one phase), and that the field rotates at the same frequency as that of the applied alternating current.

If the supply frequency is f Hz then clearly the field rotates at a speed of $N = 60f/p$ rev/min where p is the number of pole pairs. Increasing the number of poles reduces the speed. The speed of field rotation is again known as the synchronous speed.

> (3)   In a manner similar to that used in *Figure 17.4*, draw phasor diagrams for the rotating field of a 2-pole, two-phase stator.
>
> Hence deduce that the resultant flux is equal to the maximum flux Φ due to any one phase. Write down an expression for the synchronous speed in this case.

THE INDUCTION MOTOR

The induction motor depends for its operation on a rotating field of the kind referred to above. The stator of the motor is basically the same as that of the alternator described previously and is wound for two, four, six poles, etc., depending on the speed required. The salient pole construction is, however, not used. Instead the stator is in the form of a cylinder, the coils being wound in slots cut in the cylinder walls. As always, the iron used is laminated to reduce iron losses and the windings are uniformly distributed, one-third of the stator surface being devoted to each phase. The windings are usually brought out to six terminals so that the machine may be either star or delta connected.

The most widely used form of rotor for low power machines is the so-called squirrel cage rotor. This is illustrated in *Figure 17.5*. It is built up from low hysteresis steel laminations and has longitudinal slots into each of which is placed a single bar of copper or aluminium conductor. Two thick conducting rings of copper are riveted to each end of the rotor and the conductor ends are brazed or welded to these, shorting all the conductors together and forming a cage from which the name of the rotor is derived.

The advantage of this kind of rotor is that there are no external connections to be made. Slip-rings, commutators or brushes are not required. When the rotor is centralized in the stator cylinder so that the air gap

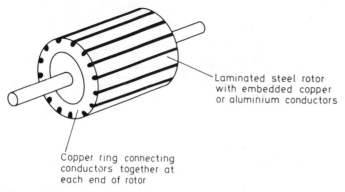

Laminated steel rotor with embedded copper or aluminium conductors

Copper ring connecting conductors together at each end of rotor

Figure 17.5

between the conductors and the poles of the stator is small, the assembly forms an induction motor, and the rotor will turn when the stator coils are energized from a three-phase supply.

Wound rotor

The squirrel cage motor is of value for comparatively small loads but it is not practicable when heavy loads have to be started up from rest.

For such operation, a wound rotor type is used. In this type the rotor is wound with a three-phase winding, brought out to three slip-rings and the starter used is a group of three resistances connected in star or delta and joined between the rings.

As the speed builds up, these resistances are cut out by centrifugal switches and the windings are short-circuited on themselves. The brushes are then also raised from the slip-rings and the machine then operates in the same manner as the squirrel cage.

ROTATION AND SLIP

Figure 17.6 shows one conductor on the rotor in the gap of an induction motor. As the rotating field passes,

the conductor is cut by the flux and an alternating e.m.f. is induced between its ends. As this conductor is connected to another conductor spaced one pole pitch away, a current will flow in this closed loop and the conductor will experience a force tending to move it out of the field and hence to turn the rotor. The direction of this turning force relative to the field will be opposite to that of the field. It is important to grasp this point and the diagram should help you do it.

In the diagram it has been assumed that the induced e.m.f. and the current are acting out of phase; therefore the direction of the force acting on the conductor will be such that the rotor will tend to turn clockwise (Fleming's Left-Hand Rule, remember?). The rotor will consequently be dragged in the direction of the field movement.

At switch-on, the frequency of the induced e.m.f. will be the same as that of the supply, but as the speed of the rotor increases the relative motion between field and rotor becomes less and the frequency of the e.m.f. decreases. The motor can clearly never run at the speed of rotation of the field (the synchronous speed) because the relative motion would then be zero and no e.m.f. would be induced in the conductors. For this reason the induction motor is known as an asynchronous machine.

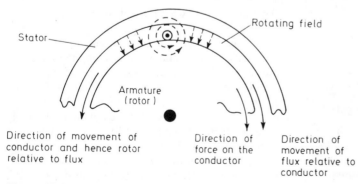

Stator

Rotating field

Armature (rotor)

Direction of movement of conductor and hence rotor relative to flux

Direction of force on the conductor

Direction of movement of flux relative to conductor

Figure 17.6

With no load on the rotor shaft, the torque required has only to be sufficient to overcome the bearing friction and the speed of rotation is very close to that of the field. The motor is then running at practically the synchronous speed. Let N rev/min be the synchronous speed of the field and let N_r be the actual speed of the rotor. Then

Slip speed $N_s = N - N_r$

The ratio $\dfrac{\text{synchronous speed} - \text{actual speed}}{\text{synchronous speed}}$

$$= \frac{N - N_r}{N} = \frac{N_s}{N}$$

is called the slip s and is usually expressed as a percentage.

The no-load slip will normally be of the order of 99 per cent of synchronous speed corresponding to a slip of 0.01 (1 per cent), and the full-load slip will be of the order of 0.05 (5 per cent). We can express the slip ratio in another way:

$$\frac{N_s}{N} = \frac{60f_s/p}{60f/p} = \frac{f_s}{f}$$

and here f_s is the slip frequency; it is the frequency of the rotor e.m.f. when running at slip s.

(4)   What is the slip at standstill?

(5)   What is the rotor frequency in terms of slip s and the supply frequency f?

Example (6)   The frequency of the supply to the stator of a four-pole induction motor is 50 Hz and the rotor frequency is 2 Hz. What is the slip and at what speed is the motor turning?

Solution   Slip $s = \dfrac{f_s}{f} = \dfrac{2}{50} = 0.04 \ (4\%)$

Synchronous speed $N = \dfrac{60f}{p} = \dfrac{60 \times 50}{2}$

$\qquad\qquad\qquad\qquad = 1500$ rev/min

But rotor speed $N_r = N(1 - s)$ since $s = \dfrac{N - N_r}{N}$

$\therefore\ N_r = 1500\,(1 - 0.04) = 1440$ rev/min

The speed of an induction motor tends to decrease as the load is increased in order that the increase in the speed of the rotor conductors relative to the field (the slip) can generate a current large enough to cope with the additional load, i.e. lead to an increased torque. The load may be increased until the slip is such that the maximum torque is reached; beyond this point the slip approaches unity and the machine stops. It can be proved that the slip (or speed) at which the maximum torque is developed depends on the rotor resistance.

This squirrel cage motor is ideally suited to drives where an approximately constant speed is required at relatively low power. For higher powers a wound rotor is employed. The rotor conductors here form a three-phase winding wired internally in star. The free ends are brought out to three slip-rings from which external contact is made by carbon brushes in the usual way.

Single-phase motor

The single-phase induction motor is widely used in all kinds of electrical appliances and is similar in its construction to the polyphase types we have been discussing. The field produced by the stator current, however, in not rotating but has its axis fixed relative to the body of the motor, its magnitude varying sinusoidally. We can look at this as a field which has two component fields of equal magnitude but rotating in opposite directions at equal speeds, each being half the maximum value of the alternating field. This is illustrated in the phasor diagram of *Figure 17.7*. An oscillating field is thereby established.

Once a squirrel cage rotor is turning in such a field, a pulsating torque is developed which tends to bring the rotor to nearly synchronous speed, and the velocity relative to the counterclockwise field component of *Figure 17.7* will be close to twice the synchronous speed. The motor will likewise rotate in the opposite sense if started in that direction. The problem in single-phase motor design is to get the rotor started in the first place.

There are several ways in which this can be done. In what is called the shaded pole motor, a single turn

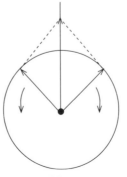

Figure 17.7

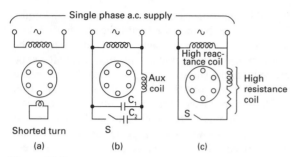

Single phase a.c. supply

High reactance coil

Aux coil

High resistance coil

Shorted turn

(a)   (b)   (c)

Figure 17.8

of heavy gauge copper wire is wound around half of each stator pole, usually embedded in a groove cut in the iron to accommodate it, see *Figure 17.8(a)*. This behaves at switch-on as a very low resistance shorted turn and the relatively heavy current which flows in it delays the otherwise rapid increase of magnetic flux in that region of the pole. The rotor then experiences the effect of a partially rotating field as the flux vector appears to shift as a function of time. This method is found on most low power motors.

An alternative to the shaded pole is found in capacitor start motors. Here a capacitor is used in series with an auxiliary starter winding to provide the necessary phase shift. A better performance is obtained when two capacitors are used as shown in *Figure 17.8(b)*. The larger capacitor provides a good starting torque to get the rotor almost up to speed and is then disconnected by a centrifugal switch S. The smaller capacitor is left in circuit where it maintains a good operating efficiency.

Figure 17.8(c) shows a split-phase starter system; here the main field winding is supplied through a resistor and the auxiliary winding has a much greater reactance–resistance ratio to that of the main coil. Currents through the two windings will be approximately in quadrature, hence the current in the high reactance coil reaches its maximum value behind that in the main winding and the rotor experiences a shift in the magnetic field which provides a starting torque. When the motor is run up to speed, the high resistance winding is cut out by a centrifugal switch.

SUMMARY

- The frequency generated by a three-phase generator $f = \dfrac{P\omega}{2\pi}$ Hz where ω is the speed in rad/s and P is the number of pole pairs.

- Synchronous speed $N = \dfrac{60f}{P}$ where f is the speed in revs/min. Increasing the number of poles reduces the synchronous speed.

- For synchronous speed N and rotation speed $N_r = \dfrac{N - N_r}{N_s}$ where N_s is the slip speed.

- Slip $(s) = \dfrac{N - N_r}{N} = \dfrac{N_s}{N} \times 100$ per cent also $\dfrac{N_s}{N} = \dfrac{f_s}{f}$ where f_s is the slip frequency.

- At standstill, the slip = 1 or 100 per cent and is zero at synchronous speed.

REVIEW QUESTIONS

1 Why is an induction motor so called?
2 Explain what you understand by 'synchronous speed'.
3 Explain the production of a rotating field by stationary coils.
4 Why must the rotor speed always be less than the synchronous speed?
5 What is meant by 'slip' in an induction motor?
6 What is the slip at synchronous speed? Would this be a practical possibility?
7 Why is a greater torque automatically produced when the load on an induction motor increases?
8 What is a shaded pole motor?
9 A single phase alternator has 12 poles and runs at 10 rev/s. What is its output frequency?
10 The rotor of a four-pole induction motor rotates at 23 rev/s and the slip is 8 per cent. What is the supply frequency?

EXERCISES AND PROBLEMS

(7) The synchronous speed of an induction motor is 750 rev/min. If the supply frequency is 50 Hz what is the number of stator poles?

(8)   A four-pole, three-phase induction motor is connected to a 50 Hz supply. If the slip is 4 per cent, what is the rotor speed?

(9)   The stator field of a four-pole, three-phase induction motor rotates at a speed of 24.9 rev/s. What is the supply frequency? If the frequency was increased to 50.4 Hz, what would be the new field speed?

(10)   The frequency of the e.m.f. in the stator of a four-pole induction motor is 50 Hz and that in the rotor is 2.5 Hz. What is the slip and at what speed is the motor turning?

(11)   A four-pole induction motor runs from a 50 Hz supply at a speed of 1450 rev/min. What is the frequency of the rotor current and the percentage slip?

(12)   A six-pole induction motor runs from 50 Hz supply at a speed of 970 rev/min. What is the percentage slip?

(13)   An eight-pole, three-phase induction motor has a full load slip of 0.025 when used on 50 Hz mains. Calculate (a) the synchronous speed, (b) the rotor speed, (c) the rotor frequency.

(14)   Show that the rotor e.m.f. is given by sE, where E is the e.m.f. in the stationary motor. Show also that the rotor frequency at slip s is given by sf, where f is the supply frequency.

(15)   A three-phase alternator has six poles and runs at 20 rev/s. Its output is connected to a four-pole induction motor which runs at a speed of 29 rev/s. What is the percentage slip in the motor?

(16)   A six-pole 50 Hz induction motor drives a pump which requires a speed of 40 rev/min for proper operation. If the motor slip is 3 per cent what gear ratio is necessary between the motor and the pump?

(17)   A three-phase, eight-pole, 50 Hz induction motor drives a haulage drum at a speed of 214 m/min. The effective diameter of the haulage drum is 1.83 m and the motor slip is 4 per cent. Calculate (a) the motor speed, (b) the drum speed in rev/min, (c) the gear ratio of the haulage.

Experiment E1

Characteristics of point contact and junction type diodes

Point contact diodes consisting of a pointed tungsten wire sprung against a small crystal of *n*-type germanium or silicon material preceded the junction type or *p-n* diode in development. In fact, the principle and form of the point contact diode goes back to the very early days of radio reception when the famous 'cat's whisker' tungsten wire was pressed against a crystal of galena in a hand-operated holder to provide the active component of the first 'crystal sets'.

Although the point contact diode does not have the power handling qualities of the junction sandwich, the very small area at the contact point results in a low self-capacitance and so enables this diode to operate as a demodulator up to very high radio frequencies.

This experiment investigates the characteristics of a germanium point contact diode and a silicon junction diode. The most apparent difference between these types is the forward voltage drop, that is, the voltage measured across the diode when it is conducting. For the point contact germanium this is of the order or 0.2–0.3 V while for the junction silicon it is about 0.6–0.7 V. This is the level of voltage which has to be applied to the diodes to get them into their conducting regions. This fact will be noted from the experimental results. This investigation also covers the characteristic of a typical zener diode which operates on its reverse characteristic unlike ordinary diodes which normally operate when forward biased. Suitable diodes are the germanium OA47 or OA91, and the silicon 1N4148.

Set up the circuit shown in *Figure E1.1(a)* and the diodes inserted in turn, increase the voltage applied in steps of 0.1 V up to about 1 V or until the current reaches about 10–15 mA. Record the voltage and current values for each diode type. With the circuit wired now to *Figure E1.1(b)*, measure the reverse current for each diode on a microammeter (50 μA f.s.d. or better) for reverse voltages up to about 10 V. It is quite probable that the reverse current through the silicon diode will be undetectable, but for the germanium diode it will rise fairly steadily as the

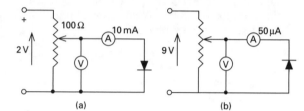

Figure E1.1

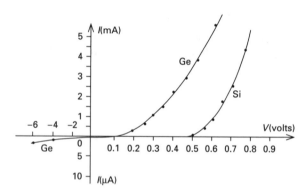

Figure E1.2

voltage is increased. Again, record the voltage and current values for each diode type.

Plot voltages horizontally against current vertically after the fashion of *Figure E1.2*. Notice the scaling for these graphs and the change which occurs at the origin for the reverse characteristics.

When the diodes are reverse biased, what happens if you squeeze each diode in turn between finger and thumb to increase its temperature? Compare the effect of a temperature rise on the germanium diode to that on the silicon diode.

Which type of diode would possibly be best as a detector in a radio receiver operating up to a few megahertz? As a power unit rectifier?

By drawing a suitable tangent to each of your graphs at an appropriate point on the curves estimate the dynamic resistance of each of the diodes.

Experiment E2

Slope resistance of a zener diode

As we have read in the main text, the zener diode acts as a voltage stabilizer when it is operated continuously in the reverse-bias mode beyond the breakdown point. The current through the diode under this condition must, of course, be limited, and a suitable series resistance must be introduced to prevent the power dissipation of the diode from exceeding its permissible maximum.

The zener breakdown voltage is usually marked on the diode casing and may be anything in a preferred range of values from 2.5 V to 100 V and more. The normal tolerance on the marked voltage is usually 5 per cent for general-purpose use.

You will want a zener of voltage working between 6.8 V and 8.2 V for this experiment and a nominal 400 mW power rating. This is because it is rather easier to measure the slope resistance of this rating than it is of higher ratings, although you may use a 1.3 W type if one is to hand.

Using the circuit of *Figure E2.1*, the selection of the series limiting resistance can be calculated knowing the zener power rating, its maximum current capability to keep within that power rating, and the greatest input voltage to the circuit, in this case 12 V. Once calculated, take the next higher preferred value. In any event, the diode reverse current will be restricted to no more than 20 mA in the interests of diode safety.

Using, if possible, digital meters for both voltage and current readings, set the potentiometer R_1 to its minimum value. Now carefully advance this control, noting the readings of both measuring instruments. As the breakdown voltage of the particular zener diode you are using is approached, the ammeter reading will slowly increase up, possibly, to the order of a milliampere, after which there will be a sudden increase in this level for very small increments in the voltage. This point of transition is known as the 'knee' of the characteristic; make a note of the *increase in voltage now necessary* to take the current to 20 mA; do not exceed this level.

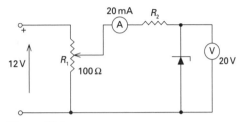

Figure E2.1

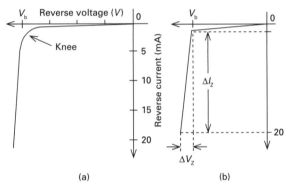

Figure E2.2

Plot reverse current against reverse voltage after the fashion illustrated in *Figure E2.2(a)*, from this determine the slope resistance $\Delta V_z / \Delta_z$ by the method shown in diagram (*b*).

Look up in the manufacturer's specification for the zener diode you have used to see if the slope resistances of typical examples are listed. If it does, how does your measurement compare with it? Is it within the tolerance which may also be quoted by the manufacturer? Within what error range would you expect your measurement to lie?

What would be the slope resistance of an ideal zener diode? Sketch a graph showing the characteristic of such an ideal diode.

Experiment E3

Investigation into constant-current sources

The output characteristics of a transistor in common-base configuration is similar in general appearance to that which was plotted in the earlier experiment for the common-emitter mode, but the common-base curves are almost perfectly horizontal. *Figure E3.1* shows a typical output characteristic for a bipolar transistor in common-base mode and this consequently demonstrates that the ratio $\Delta V_{CB}/\Delta I_c$ represents a very large output resistance. The consequence of a high resistance output is that any change in the resistance of the load has negligible effect on the current flowing in the load, which therefore remains constant at its selected value.

As we noted in the text, a constant-current source is designed to have a very high output resistance; it is therefore possible to derive such a source from a suitably designed circuit involving a common-base amplifier. However, in order to work along any of the characteristic curves, it is essential that the emitter current is held constant; under this condition the collector current I_c is constant and independent of the collector-base voltage which is the desired end of the design.

Make up the circuit of *Figure E3.2* which shows an experimental current source which can provide a constant current (for the purposes of this demonstration) up to 15 mA or so. The transistor, which is a small power type such as the BFY50 or 2N1893, should be provided with a crinkle-type push-on heat sink, and its collector current is monitored with a 10 mA (preferably digital) ammeter. The emitter current is maintained constant by being derived from a zener diode stabilized supply point; this holds the emitter-base voltage constant yet permits a constant level of I_E to be selected by potentiometer R_2. With the component values indicated and the load resistor set to zero ohms, the voltage across resistance R_1 is the V_{CE} on the transistor, that is, $15 - 9.1 = 5.9$ V. Hence the load current should remain constant until the drop across R_L (which is preferably a resistance box) reaches 5.9 V.

With R_L set at zero, load current I_L is adjusted by using potentiometer R_2 as a coarse control and the 470 Ω trimmer type variable as a fine control to

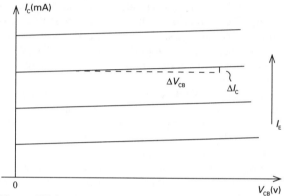

Figure E3.1

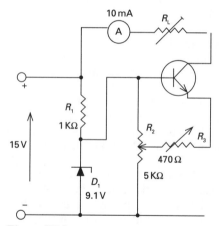

Figure E3.2

exactly 10 mA. R_L is then increased in small increments and I_L is recorded for each of these increments. This procedure is continued until I_L falls to about 9.5 mA. Now repeat this procedure with initial load currents of 5 mA, 1 mA and 0.1 mA.

Plot a graph of I_L against R_L and comment upon its appearance. From the graph obtain an estimate of the output resistance of the circuit.

Constant current for the purposes of charging Ni-Cad batteries can be obtained from integrated circuits designed for the purpose of providing constant voltage.

The 7805 is one such integrated system which gives a 5 V stabilized output up to a current of 1 A. A suitable circuit, which may be made the basis for a similar investigation to that above, is shown in *Figure E3.3*.

By connecting resistor R_1 from the output pin to the common pin of the voltage regulator package, the action of the circuit is then to try and maintain a fixed voltage across R_1. Hence the current through it will be constant. Varying R_1 varies this current to the desired level which can then be used for charging Ni-Cad at the required rate. If R_1 is 50 Ω and V is 5 V, a 50 mA

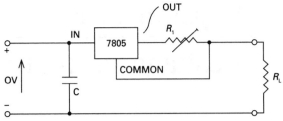

Figure E3.3

constant current is obtained. Build this circuit and try the effect of varying R on the load current R_L.

Experiment E4

Leakage current in a bipolar transistor

When a transistor is connected to a d.c. source and its emitter connection is left open-circulated with a reverse potential across its base-collector junction, the only current that can flow is the leakage current I_{cbo} brought about by the migration of minority carriers across the junction. This current is a function of the junction temperature, see *Figure E4.1(a)*.

The effect is most marked in germanium transistors where a leakage of 5–10 µA can take place at room temperature. For a silicon device, such a leakage is practically undetectable at room temperature but will become apparent if the temperature is increased sufficiently. For illustrative example, therefore, it is best to use germanium *p-n-p* transistors, particularly of the older small-signal varieties such as the OC71 series, and for these the battery polarity shown in the figure will be reversed.

When the transistor is connected so that the supply is wired between collector and emitter with the base open-circuited, the leakage is I_{ceo} and this is illustrated in *Figure E4.1(b)*. It might appear at first that this leakage current will be smaller than I_{cbo} since two junctions are now involved but this is not so. We have seen in the text that the transistor (effectually in the common-emitter mode) now amplifies I_{cbo} so that $I_{ceo} = h_{fe}I_{cbo}$ (or βI_{cbo}). Manufacturers quote the values of I_{cbo} and h_{fe} at room temperature (20°C) so that I_{ceo} may be easily calculated. Hence, if a germanium transistor has an I_{cbo} at 20°C of 5 µA, and $h_{fe} = 100$, I_{ceo} will be 500 µA.

It is necessary to raise the temperature of the test transistors to about 50°C for germanium and 100°C for silicon to show the effects of leakage; this heating is most easily done by fixing the transistor to a strip of insulating material (a piece of stripboard will do) so that it can be immersed below the surface of a small quantity of sunflower oil or any similar cooking oil, with the case of the transistor just immersed in the liquid, see *Figure E4.2*. The oil is then *gently* heated over a Bunsen or hot plate with its temperature recorded by an ordinary mercury-in-glass thermometer closely positioned to the transistor.

Measurements are made using germanium (OC71 etc.) and silicon (BC178 etc.) transistors, the circuit systems for I_{cbo} and I_{ceo} measurements being shown in

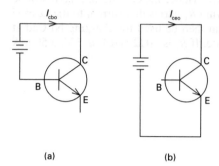

(a)   (b)

Figure E4.1

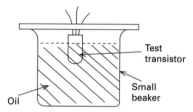

Figure E4.2

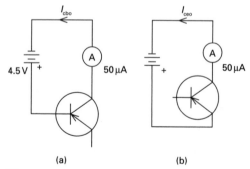

(a)   (b)

Figure E4.3

Figure E4.3. As the oil heating should be a relatively leisurely business, it is convenient to take the measurements alternately in the two arrangements, making a connection first to the base and then to the emitter as the temperature is recorded in 10°C steps. Do not take the temperature beyond 50°C for the germanium or 100°C for the silicon. Why is this?

Comparison of your results should enable you to obtain an estimation for the common-emitter gain

(h_{fe} or β) and from this a value for the common-base gain (h_{fb} or α) since $\beta = 1/(1 - \alpha)$. Plot graphs of I_{cbo}, I_{ceo} and α as functions of temperature. What can you deduce about the rate of change of leakage current with respect to temperature for each type of transistor? Why are common-emitter amplifiers less stable than common-emitter amplifiers, other things being equal, when a large temperature range may be experienced?

When some of the early transistorized radio receivers were left in strong sunlight or close to radiators they often stopped working. Why?

Experiment E5

Half- and full-wave rectification

This experiment will demonstrate either half-wave or full-wave rectification, and verify the relationship between peak and mean values of a rectified sinusoidal waveform.

Most power supplies for transistorized equipment provide d.c. voltages over the range 5 V to some 30 V; some of these supplies, particularly those for bench experimental work, are adjustable over such a range, with a current capability in many cases of up to several amperes. In fixed equipment such as domestic amplifiers, the voltage is constant and maintained at a level to suit that particular piece of equipment.

Their general purpose is to step down the 230 V mains supply by means of a transformer; then this smaller a.c. output is rectified and converted by appropriate filtering and smoothing systems into a steady d.c. supply. The simplest rectifier is the half-wave circuit using a single diode; for full-wave circuits, two diodes are necessary for the biphase arrangement and four for the bridge arrangement. We will examine here only the first and second of these systems.

Make up the circuit shown in *Figure E5.1*, using a step-down transformer of about 6 VA rating having a centre-tapped secondary winding. A 6 + 6 V type is illustrated, but anything up to 12 + 12 V is suitable. The primary winding (which must be protected against accidental contact) goes via the usual fuse to the mains supply. For the diodes, any 50 V 1 A types will be suitable, and the load R_L should be adjustable between 100 Ω and 5 kΩ and a resistance box is most conveniently used here. An oscilloscope is connected across the load so that the various waveforms may be observed.

In each part of the work sketch the voltage waveforms and the current waveforms; the current waveform can be observed by connecting the oscilloscope between the points A and B, since the voltage across resistance R_1 will be proportional to the diode current.

Half-wave rectification

The circuit has switch S open for this part of the experiment so that only a single diode is used. Set R_L to 5 kΩ and observe the voltage waveform across it. Measure this voltage with a d.c. ammeter connected across R_L. What does this reading (reading 1) represent? Now connect 'smoothing' capacitor C which is a 10 μF 25 V type where indicated and again observe the output waveform and measure the output voltage (reading 2). What does this voltage represent? Is the proportion between voltage readings 1 and 2 in accordance with theory? What should this proportion be?

Switch the value of R_L down to 500 Ω; what effect, if any, does this have on the output waveform and the output voltage? Account for any changes which may occur.

Look now at the current waveform across points A and B and account for its appearance. If you have a two-beam oscilloscope, compare the current waveform against the waveform developed across one half of the second winding; what conclusions do you draw?

Full-wave rectification

The same circuit is used but starting out with switch S closed, capacitor C removed and the load restored to 5 kΩ resistance.

Repeat the previous observations and measurements and answer the same questions.

If time is available, connect an ammeter in series with the load and for both half-wave circuits with C connected, plot regulation curves of output voltage against output current for a range of load values from 1 kΩ down to 100 Ω.

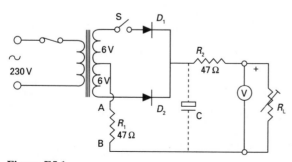

Figure E5.1

Experiment E6

Output and transfer characteristics of a bipolar transistor

This experiment examines the form of the output characteristic and the transfer characteristic of a small-signal transistor such as the BC107, and from these characteristics the output resistance and current gain of the transistor can be evaluated.

Make up the circuit shown in *Figure E6.1*. This is similar to that shown in the general text. The base current ammeter should be scaled 0–50 µA although a 0–100 µA may be used. The base voltmeter is not strictly required for this experiment but if one is used it should be of high resistance and preferably a digital type. The other meters measure collector current I_c and collector voltage V_1, respectively. The battery supplies are often the most convenient for this kind of experiment but these may be replaced by low voltage power units, and it these have their own output adjustments the potentiometers R_1 and R_2 may be omitted.

Start by having both potentiometers set to their minimum output positions, that is, there is no base current and no collector current. By means of R_1 adjust the base current I_B to some small value, say 5 µA. Now, by adjusting R_2 apply a small voltage, say 1 V, to the collector as measured by V_C. Keeping I_B constant, increase V_C in 1 V steps, recording the collector current I_C for each increment. This will give you a range of readings of V_C against I_C for a fixed value of I_B. At the end of these readings return R_2 to its minimum output position.

Now adjust R_1 so that the base current is increased to, say, 15 µA. Then repeat the above procedure, keeping I_B constant as before. Carry on in this way up to a maximum base current of 50 µA or so; you should finish up with about five or six related values of V_C and I_C each for a particular setting of I_B. From these readings the output characteristic may be plotted after the fashion of *Figure E6.2*.

From the linear (substantially flat) portion of any of the curves the effective output resistance of the transistor can be found by taking the gradient $\Delta V_C / \Delta I_C$. This will vary slightly for each of the curves, but an average value will be found by taking one of the gradients about the centre of the collection.

Since there is a connection between I_B and the output resistance R_o we can find a statement for this by assuming that $R_o = k I_B^n$ where k is a constant. Can you see why such an assumption is justified?

Taking logs we get $\log R_o = n \cdot \log I_B + \log k$, hence a plot of $\log R_o$ against $\log I_B$ will produce a straight line from which two pairs of related values will enable n and k to be evaluated. The required relationship will then be determined.

The transfer characteristic shows the relationship between input I_B and output I_C. Plot your values after the fashion of *Figure E6.3* and from this obtain a value for the common-emitter static gain h_{FE} (or β). What do the curvatures at the extremes of this last graph signify?

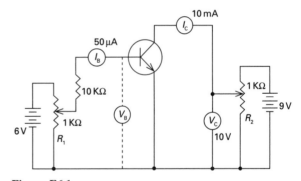

Figure E6.1

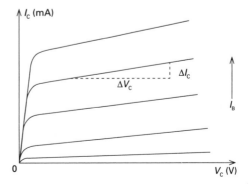

Figure E6.2

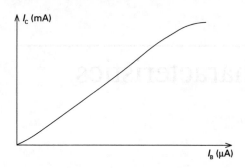

Figure E6.3

Experiment E7

Drain and transfer characteristics of a field-effect transistor

In general, a field-effect transistor (FET) is used as a low power device in the first stages of amplifiers to obtain low noise and a very high input resistance, simulating in many ways the characteristics of thermionic valves.

Set up the circuit of *Figure E7.1* which shows the FET connected in common source configuration. Voltmeter V measures the gate-source voltage V_{GS} and the milliammeter in the drain output circuit measures the drain current I_D. Set V_{GS} to -3 V by adjustment of the 1 kΩ potentiometer R_1 and record the drain current I_D for a series of values of drain-to-source voltage V_{DS} derived from the 15 V supply and potentiometer R_2. As a FET used is an *n*-channel type, make a note that the gate is maintained at a negative voltage and the drain at a positive voltage with respect to the source. Repeat the above procedure with settings of V_{GS} at -2.5 V, -2.0 V, -1.5 V, -0.5 V and 0 V. You can now plot the drain (or output) characteristic I_D against V_D for a range of values of V_{GS}.

There are two regions of interest in this characteristic: the ohmic region for small values of V_{DS} and the main region beyond the ohmic range where the transistor behaves as a constant-current generator where I_D is substantially independent of V_{DS}. It is in this region that the FET normally operates when it is used as an amplifier or as a switch. You will find all the details of this in the main text. Can you obtain an

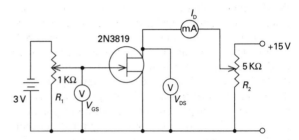

Figure E7.1

approximate figure for the output resistance of the FET from these curves?

Try, after the temperature rise of the FET case has stabilized, touching it with a small piece of ice to cool it, watching the effect on I_D while you do this. What happens to I_D and can you account for the effect? Keep in mind that the only carriers in the *n*-channel FET are electrons.

Using the same circuit, plot a curve of the transfer characteristic. Set V_{DS} to the full 15 V and measure I_D as a function of V_{GS} for values of V_{GS} between 0 and (about) -4 V. Identify, from your graph, the I_{DSS} value, which is the drain level for $V_{GS} = 0$, and the pinch-off voltage V_p. By drawing a suitable tangent to the transfer characteristic, calculate the curve gradient and hence find the mutual conductance g_m of the FET at the selected point.

Experiment E8

Characteristics of a series regulator

This experiment demonstrates the action of a series regulator when (a) the load current varies, (b) when the d.c. input voltage varies. The input to the circuit is a 20 V d.c. supply capable of delivering a current of at least 0.25 A. The circuit is shown in *Figure E8.1*.

Both transistors are used in the emitter-follower mode and a reference voltage is provided by zener diode D_1. This 12 V reference will be developed across the emitter load of transistor T_1 at approximately the same level and, depending on the setting of potentiometer R_2, a fraction of this will be applied between the base and emitter of the series control transistor T_2. This voltage in turn will be developed across the load resistor R_L and should remain at a constant level irrespective of a change in the value of R_L or a change in the input voltage V_L.

Make up the circuit of *Figure E8.1*, mounting transistor T_2 on to a small heat sink which can be made from a piece of 18 SWG aluminium measuring about 2 inches square, and selecting the power rating of the zener to suit these conditions. With V_i set to 20 V, measure the voltage across the zener diode and also across the 1 kΩ emitter load. These should be approximately equal; why? What is the probable tolerance rating of your zener diode?

Set R_L to 500 Ω and adjust R_2 to give a maximum output voltage V_o. Record this voltage and the output load current I_o, and repeat this for a number of settings of R_L, not allowing I_o to exceed 0.25 A and checking each time that the input voltage remains constant. Now adjust R_2 to give a maximum output V_o

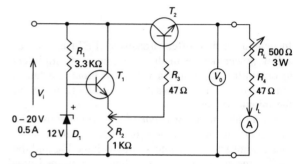

Figure E8.1

of 9 V and again with V_o at 6 V repeat the previous procedures.

Plot graphs of V_o against I_o and find by any suitable means the gradient of each curve. The value of this gradient gives us the output resistance R_o of the regulator, where $R_o = \Delta V_o / \Delta I_o$ with V_i constant. Should this resistance be large or small? What would be the theoretical ideal value?

Now recheck that V_i is set to 20 V and again set the output voltage to 10 V using potentiometer R_2. Adjust R_L to give a load current of 0.1 A and record V_o. Decrease the input in steps of 1 V down to about 15 V, maintaining I_o at its 0.1 A level. Meansure V_o for each of these settings of V_i. Plot a curve of V_o against V_i and hence determine the regulation or stabilization factor by finding the gradient $\Delta V_o / \Delta V_i$ with I_o constant. Should the figure be small or large? What would be the ideal regulation factor?

Experiment E9

Investigation into the operation of an emitter-follower

This experiment investigates the behaviour of an emitter-follower not only as a feedback amplifier but also as an impedance transformer.

Connect up the circuit of *Figure E9.1*, using a general-purpose transistor such as a BC107 or BC108. The voltages indicated are measured with a suitable high impedance a.c. millivoltmeter.

Set the frequency of the signal generator to, say, 500 Hz and adjust its output amplitude so that the input voltage v_1 is at some definite level, say 100 mV. Without the load resistor R_L connected, record the output voltage v_0; then connect R_L and readjust the signal generator output to restore voltage v to its original value. Record voltages v_1^* and v_0^* where the asterisk indicates that the load is connected.

Retune the signal generator to 25 Hz and repeat the above procedure for a number of spot frequencies within the range 25 Hz to about 25 kHz. The input resistance $R_i = 1000\, V_1/(v_s - v_1)$ without the load and $1000v_1/(v_s^* = v_0^*)$ with the load in place. In the

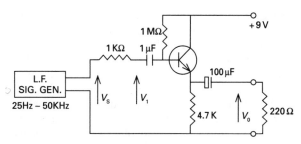

Figure E9.1

same way, the output resistance $R_o = 220(v_0 - v_0^*)/V_o^*$.

From your readings plot a graph of input resistance against log frequency (without the load) and on the same axes the input and output resistances when the load is attached.

What is the voltage gain of this amplifier? Deduce the current gain. What errors are likely to affect the results of this experiment?

Experiments E10

Measurement of iron and copper losses in a power transformer

If the secondary of a transformer is on open-circuit, the power taken by the primary will be equivalent to the power dissipated in the iron losses plus the very much smaller I^2R or copper loss. Neglecting this latter, a measurement of the total iron loss can be made using a wattmeter in the primary circuit.

A suitable transformer is one designed for valve-operated equipment, where the primary connects to the normal 230 V mains supply and the secondary provides from its main winding (neglecting any small subsidiary windings) probably up to 200–250 V at a design current of 100–150 mA. *Figure E10.1(a)* shows the general circuit arrangement; here a.c. voltmeters are connected across primary and secondary coils and the wattmeter is wired with its current coil in series with the primary. By this arrangement, the wattmeter ammeter reads only the primary current and not that in the voltage coil of the wattmeter.

The wattmeter therefore reads the power consumed, ignoring the very slight loss in the current coil itself. Record the primary volts V_1, primary current I_1, secondary volts V_2 and the wattmeter power. The latter reading will indicate the total transformer iron losses; these losses are substantially constant for all loadings. Now calculate the power factor and angle of lag of the primary no-load current, noting that $I_1V_1 \cos \phi =$ watts. What would $I_1V_1 \sin \phi$ represent in this case?

In order to evaluate the copper losses, a short-circuit test is applied to the transformer. In this, the secondary winding is short-circuited with a direct link of wire, but the input primary voltage must now be restricted so that the secondary current reaches its normal working full load value and no further. The secondary ammeter (which will have a very small self-resistance here) will record this; for the type of transformer suggested earlier, this could be assumed to be of the order of 100 mA.

With the circuit arrangement of *Figure E10.1(b)*, apply an input voltage from a Variac until the secon-

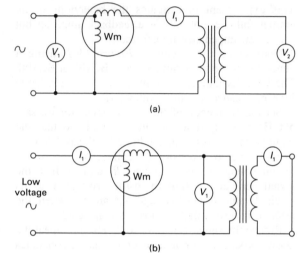

(a)

(b)

Figure E10.1

dary current I_2 reaches its working value. This time the wattmeter voltage coil records the actual primary voltage and its reading, which is the power consumed, will be large in comparison with the power loss in the voltage coil, which may therefore be neglected. As the primary input will now be very small, the flux density and hence the iron losses will be small and negligible in comparison with the copper losses. The wattmeter reading will this time indicate the total copper losses.

Record the primary voltage V_1, the primary current I_1 and the wattmeter power. From these readings derive an equivalent resistance R_C for the I^2R copper loss, where $R_C = W/I_1^2$. Find also the equivalent impedance $Z_e = V_1/I$, and the equivalent reactance $X_e = \sqrt{(Z_e^2 - R_C^2)}$.

Copper losses are proportional to the square of the current and also to the square of the VA output of the transformer.

Experiment E11

Determination of inductance by a voltmeter method

Although in general terms the measurement of inductance calls for a relatively expensive inductance bridge, particularly for small values of inductance such as radio-frequency coils, if the inductance concerned is of the order of 0.1 H or more, a simple a.c. voltage measurement gives not only the inductance value but the effective a.c. resistance as well.

The principle is that the voltage across an inductance L with a self-resistance R is the resultant of two phasors mutually at right angles, one of which is the resistive component in phase with the inductor current, and the other the reactive component leading the current by 90°. From a scaled measurement of these two components both L and R can be determined.

You will need an inductor of 0.1 H or above; this can be wound for the job or, alternately, one of the windings on a small iron-cored transformer such as is used for interstage coupling purposes may be satisfactory. In addition there needs to be a source of low voltage 50 Hz a.c. of the order of 15–25 V. This can be obtained from the mains supply by way of a small step-down transformer.

Make up the circuit of *Figure E11.1*; the resistance R_1 (which may be a resistance box) may normally be set to 50–100 Ω, and the three a.c. voltage measurements indicated are made in turn, using a digital meter.

Record the measurements V_1, V_2 and V_3. From these observations we can construct a voltage phasor diagram as shown in *Figure E11.2*. Here voltage V_1 is the reference phasor; voltage V_2 will lead V_1 by some angle ϕ and V_3 will be resultant. Only one triangle is possible knowing the lengths of the three sides, and this can be set out using a ruler and a pair of compasses to an appropriate scaling. We now need to break the V_2 phasor into its two mutually perpendicular components; the component in phase with V_1 being the in-phase voltage across the coil due to its self-resistance, and the 90° component being the out-of-phase voltage due to the coil reactance.

To do this, drop a perpendicular from B to the

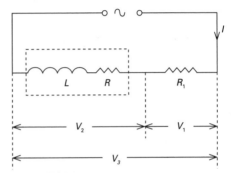

Figure E11.1

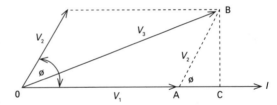

Figure E11.2

extension of OA at C. Now if the circuit current is I, then $V_1 = IR_1 = \text{OA}$; and if the self-resistance of the coil is R, then the in-phase voltage due to this is $IR_1 = \text{AC}$. Hence from these two relationships $R = (\text{AC}/\text{OA})R_1$. Side BC of the triangle now equals the out-of-phase voltage due to the coil reactance $X_L = 2\pi f L \cdot I$, and so $L = \text{BC} \cdot R_1/(2\pi f \cdot \text{OA}) = \text{BC} \cdot R_1/(100\pi \cdot \text{OA})$, since the mains frequency is 50 Hz.

If in this experiment your phasor diagram proves 'awkward' to use, that is, if ϕ happens to be very large or very small, so making the sketching difficult, adjust the series R_1 to some other value (which way should you go?) or try some other inductor.

Do you think this method leads to highly accurate results? If not, point out the reasons you think might upset things. If you have a good inductance bridge, check your inductor against that, using the same frequency if available.

Solutions to the exercises

CHAPTER 1

(1) (a) Energy; (b) power; (c) current; (d) quantity or charge
(2) 1 Ω
(4) 125C; 13×10^{18} electrons; 5.28 Ω; less
(5) 2.2 V; 0.1 Ω
(6) 1.3 Ω; 1 A (10 Ω), 5 A (2 Ω), 1.67 A (4 Ω), 1.11 A (3 Ω), 0.56 A (6 Ω)
(7) 180 W; 83.3 Ω, 55.6 Ω
(8) 30 Ω
(9) 200 Ω
(10) 0.92 A; 71.5 Ω; 140 W
(12) 1500 µJ or 1.5 mJ
(13) 2 mH
(14) 200 V; 50 µF

CHAPTER 2

(3) 2.38 Ω, 8.27 Ω
(4) 36 Ω
(5) 2.2 V; 0.1 Ω
(7) $I_1 = 40$ A, $I_2 = 80$ A, $I_3 = 150$ A
(12) (a) 1.29 A; 0.95 A; (b) Both 3.42 V (c) 1.17 W
(13) 26.9 mA, 38.6 Ω
(15) 5 A
(16) 281 Ω
(22) 26.8 Ω
(32) $I_1 = -3$ A, $I_2 = 2$ A, $I_3 = 3$ A
(33) $I_1 = 0.7$ A, $I_2 = 0.4$ A
(34) 2 A; 4 A
(35) 6 Ω
(36) 0.618 A
(37) 15.48 A
(38) 2 A
(39) 2.14 A, −2.93 A; 1.8 Ω
(40) 12 V, 800 Ω; 15 MA, 800 Ω; 45 MW
(41) 24 mΩ (milliohms)
(42) 102 µV, 52.3 Ω; 51 µW
(43) 61 mA
(44) $E = 30$ V, $E_2 = 10$ V
(46) 3.2 Ω

CHAPTER 3

(4) 500 Ω, 60 mA
(5) (a) 100 V, (b) 40 V, (c) 0
(7) 80 kΩ. (a) 1.25 mA, (b) 1.75 V/s
(17) 0.1 s
(22) 40 mA, 100 V
(23) 250 Ω, 40 mA
(24) (a) 1 s, (b) 300 kΩ, (c) 33.3 kΩ
(25) (a) 2 s, (b) 1.5 mA, (c) 0
(26) 62.5 kΩ, 3.125 MΩ
(27) (a) 100 V, (b) 39.4 V; 4 mA
(28) (a) 0 V, (b) 100 V, (c) 86.5 V
(29) (a) 0 V, (b) 0 V, (c) 86.5 V
(30) 125 ms
(31) (a) 33.3 ms, (b) 24 ms, (c) 15 ms, (d) 0.1 ms, (e) 250 µs
(32) 200 A/s, 2 A, 10 ms, 0.2 J
(34) 6.32 A
(35) 1.9 H, 380 Ω, 63 A/s
(36) 4.6 ms, 0.4 J
(37) 99.75 per cent
(38) (a) 2.5 A, (b) 937.5 J, (c) 1.56 J
(39) (a) 25 ms, (b) 6.32 A, (c) 8.65 A, (d) 5 H
   (a) $i = 10e^{-40t}$, (b) 400 A/s, (c) 3.68 A
(40) (a) o, (b) 0, (c) 10 V
(41) (a) 28.6 V, (b) 71.4 V, (c) 714 µC, (d) 25.5 mJ, (e) 71.4 µA; 19.67 s
(42) 6.67 A in R; 20 A in L; 2.71 A
(44) (a) 52.3 ms, (b) 23.2 ms

CHAPTER 4

(3) 200 kHz
(8) Zero. A moving-coil meter does not respond to a.c. signals
(17) 0.3 ms (milliseconds)
(18) 316 m
(19) 7.5 V, 19.06 V
(20) 0.8 V, 0.13 A
(22) 1.6 A, 4 A; 9600 J
(23) 5.42 V
(24) 13.75 V
(25) 16.5 V, 15.9 V
(28) 10 V, 7.07 V, 6.37 V; 40 Hz

(29)   20 . sin (628*t*) mA;   340 . sin (314*t*) V;   20 . sin (30 × 10³π*t*) V

(30)   17.8 A; 54°

(33)   *v* = 80 . sin (ω*t* + 51°), *v* = 80 . cos (ω*t* − 39°)

(34)   *i* = 15.5 . sin (ω*t* + 14°)

(35)   34 . cos (ω*t* + $\frac{\pi}{6}$)

(36)   2 ms

(37)   *i* = 4.48 . sin (314*t*) A, 1.56 ms, 4.76 A

(38)   0.424 . sin (2π10⁴*t*) + 0.212 . sin (6π10⁴*t*) A

(39)   (a) 5 . sin 942*t* V, (b) 0.625 . sin 1570*t* V, 7th

CHAPTER 5

(1)   314.2 Ω

(2)   2500 Hz or 2.5 kHz

(3)   7957 Ω, 6 mA

(4)   69.2 µF

(7)   *R* = 38.8 Ω, *L* = 154 mH; 953 Hz

(10)   306 Hz

(13) $Q = \dfrac{\omega L}{R}$. Put $\omega = \dfrac{1}{\sqrt{LC}}$, then $Q = \dfrac{L}{R\sqrt{LC}}$

$= \dfrac{1}{R}\sqrt{\dfrac{L}{C}}$. For a high *Q*-factor, *R* must be small and the ratio *L/C* must be large

(16)   (a) 12.5 A, (b) 37° lagging

(17)   150 mA, (b) 667 Ω, (c) 212 mH

(19)   30 mA, 24.8° leading

(20)   1.415 A, 2.06 A

(23)   795 kHz, 33.3 kΩ

(25)   (a) 31.4 Ω, (b) 50.3 Ω, (c) 628.3 Ω

(26)   0.5 H

(27)   600 Hz

(28)   (a) 6366 Ω, (b) 398 Ω, (c) 159 Ω

(29)   88.4 µF

(30)   8000 Hz or 8 kHz

(31)   13 Ω

(32)   10 Ω

(33)   (a) 4 Ω, (b) 3 Ω, (c) 4.8 mH

(34)   (a) 1.57 A, (b) 63 mA

(35)   306 Hz

(36)   (a) 250 Ω, (b) 1.215 H

(37)   23.1 Ω, 120 mH, (c) 56.5 lagging

(38)   (a) 500 Hz, (b) 5.66 A, (c) *R* = 42 Ω, *L* = 5.44 mH

(39)   0.084 Ω

(40)   50

(41)   1006 Hz, 100 Ω, 636 V

(42)   33.3 µH

(43)   25 000 Ω or 25 kΩ

(44)   1.887 to 0.909 MHz

(45)   800 kHz

CHAPTER 6

(5)   (a) 3.03 A, (b) 91.8 W, (c) 0.3

(13)   80 kHz

(14)   (a) unity, (b) capacitive, (c) unity

(15)   19.3°

(16)   63.2°, 0.45

(17)   0.47

(18)   (a) 173 Ω, (b) 93.3 W

(19)   (a) 5 V, (b) 32 mW, (c) 0.895, (d) 0.064 µF in parallel

(20)   1.275 kW

(21) I_L = 0.71 A, I_c = 1.0 A; *I* = 1.137 A

(22)   1.319 kW; 0.65 kW

(23)   (a) 77.6, (b) 8.84 kHz

CHAPTER 7

(2)   (a) 1125, (b) 2.25 A, 3 A, (c) 225 W

(6)   3360 V, 1.4 A

(12)   0.088 A or 88 mA, 0.023 A or 23 mA

(13)   (a) 48 V, (b) 320 mA, (c) 64 mA, (d) 15.4 W

(14)   (a) 200 lamps, (b) 0.417 A

(15)   (a) 5.33:1, (b) 328 turns, (c) 0.17 A, 0.9 A, (d) 40.5 W

(16)   28 125 Ω or 28.125 kΩ

(17)   3.53:1. Use a higher ratio; deduce a reason from example (10)

(18)   (a) 3.375 Ω, (b) 0.135 Ω

(19)   28 mW, 4.47:1, 156 mW

(20)   (a) 300 V, (b) 271 V

(21)   2.5 µF. Increased impedance means a smaller capacitance

(22)   (a) 158 W, (b) 4 per cent, (c) 95 per cent

(23)   130 W

(24)   1:3.16, 0.025 µW

(25)   AC = 4 A, BC = 0.545 A

(26)   480, 880; 8 A

(27)   (a) 21.9 W, (b) 93 per cent

CHAPTER 8

(5)   433 V, 250 V; unbalanced 250 V, 250 V, 433 V

(7)   Delta loads take three times the power as the same load in star

(14)   21.56 kVA

(15)   50.5 kVA

(16)   45 A, 430 V

(18)   (i) 288 V, (ii) 5 A, 8.66 A

(19)   Star 2.5 kW, delta 7.5 kW

(20) I_L = I_p = 54.3 A; I_p = 31.35 A

(21)   27.2 A

(23)   34.6:1

(24)   (a) About 9 kW, (b) 0.65

(25)   1.61 Ω, 10 kW

(26)   (a) 1100 V, (b) 635 V, (c) 1905 V, (d) 1100 V

CHAPTER 9

(4)   (a) 11.5 mA, (b) 2.95 V, (c) 3.55 V

(7)   107.5 V; about 120 V

(9) $I = \dfrac{\hat{V}}{\sqrt{2}(R_L + R_F)}$; $I = \dfrac{\hat{V}}{\sqrt{2}(R_L + 2R_F)}$

(19)   Carbon and WW resistors, capacitors and inductors are linear

(20)   4 V, 0 V

(21)   3.5 V, 2.82 V

(22)   (a) 0 V, (b) +10 V

(24)   33.7 per cent

(25)   14.5 mA, (b) 43.5 V, (c) 0.63 W, (d) 31.5 V

CHAPTER 10

(2) p-n-p

(3)   Common-emitter

(4)   0.07 mA or 70 µA

(5)   0.97, 37.6

(6)   0.997

(7)   65.7

(8)   0.993, 149

(11)   About 100

(12)   0.972

(13)   0.714 mA

(14)   A very large V_{CC} would be necessary; this would appear on the collector when the transistor was off and cause damage

(15)   No

(17)   Because $V_{CC} = V_{CE} + I_c R_L$ at all points

(18)   About 37, the ratio of 5.6 mA to 150 µA

(19)   About 55, the ratio of 8.25 V to 0.15 V

(20)   180 kΩ

(21)   2 V

(24)   Quarter watt would be adequate throughout

(26)   (a) 57.8, (b) 39, (c) 25.3, (d) 0.972, (e) 0.987, (f) 0.995

(27)   (a) 0.988, (b) 82.3

(28)   6000 Ω or 6 kΩ

(29)   200 kΩ

(30)   1230 Ω or 1.23 kΩ

(31)   (a) (i) False, it depends only on B_{BE}; (ii) False, it depends only on I_E; (iii) True, for V_{CB} about 0.2 V. (b) False, a leakage current flows from collector to emitter. (C) False, if they were the collector would not collect them. (d) False

(32) R_B = 0.104 kΩ, R_L = 500 Ω

(33)   46 kΩ

(34)   About 1000 Ω

(36) A_v = 90, A_i = 75, A_p = 6750

CHAPTER 11

(17)   16 mA, 1.6 V; 26 mW; yes

(21)   (a) 990 Ω, (b) –58.5

(22)   (i) About 600 Ω, very large; (ii) 100 kΩ, 12 Ω

(23) i_1 = 0.4 µA, i_2 = 11 µA, V_2 = 55 mV

(24) A_i ≃ 32,5, A_p ≃ 33

(25)   About 2.5:1

The solutions given to the above will be slightly at odds with what you may obtain. This comes from the degree of approximation made in working from an equivalent circuit. You should, however, be within 10 per cent of the solutions given.

CHAPTER 12

(2)   Doing the substitution we get $g_{fs} = g_{fso}$ [1 – V_{GS}/V_p]. I_{DSS} and V_p are easily found by direct measurement, hence g_{fso} can be found. Then g_{fs} for any V_{GS} can be calculated

(5)   12.5 kΩ

(7)   The gradient at V_{GS} = 0 is $2I_{DSS}/V_p$. From *Figure 3.11* it is at once clear that the tangent of the angle of gradient is $I_{DSS}/\frac{1}{2}V_p$ which gives the required proof

(8) r_d = 6.25 kΩ; g_{fs} = 6 mS, μ = 37.5

(13)   3 mA, assuming V_{DS} remains constant

(14)   66.7 kΩ

(15) μ = 44; g_{fs} = 3.3 mS, r_d = 13.3 k

(16)   (a) thermionic valves; (b) depletion; (c) n- or p-channel con-struction; (d) zero; (e) source to drain

(17) μ = 10; A_v = 16.7

(18)   About 15 kΩ

(19)   (a) 0.5 V; (b) 1.48 mS; (c) 470 Ω; (d) 6750 Ω

(20) r_d ≃ 6.67 kΩ; g_{fs} ≃ 1.7 mS

(21) R_1 = 3 MΩ; R_2 = 1 MΩ. A_v ≃ 7.5

(22)   About 2.0 mS

CHAPTER 13

(8)   Transistor T_2 would turn off

(9)   To limit base current of T_2

(14)   Approximately 0.94 A
(18)   100 mA, 23.4 V
(20)   0.034 Ω
(22)   300 Ω
(23)   27.5 V minimum; 51.5 V maximum
(24)   3 W
(25)   46 Ω, 350 mW; 47 Ω, 400 mW would serve
(26)   250 Ω, 4 mW; use 270 Ω, 0.25 W
(28) $R_1 = 3600$ Ω, $R_2 = 16\,040$ Ω
(29)   About 30 mV
(30)   68 Ω; (a) 3 W, (b) 1.4 V
(31)   A change of 1 V in V_L results from a change of 11 V in V_i (approx.)
(32)   (a) 51 V, 31 V; (b) 1.4%; (c) 0, 22 V

CHAPTER 14

(4)   2 = 16; F = A(B + C) + D
(12)   (a) Coincidence, (b) series, (c) 0
(13)   Logical 1 would appear in the 2nd, 3rd and 5th rows only
(14)   (a) 4–5 ms, 8–9 ms; (b) at all times except 7–8 ms; (c) 23 ms, 6–7 ms, 9–10 ms
(15)   Forward
(20)   (c) can (d)
(21)   (b)
(24)   (a) 1, (b) exclusive OR, (c) Nand, NOR
(25)   AND, OR
(28)   The first circuit will fail because the output will follow the B input irrespective of the condition at A
(29)   (a) When A or B are 10 V high, (b) about zero, (c) NOR
(32)   Red on when $A(\overline{B} + C)$; green on when $AB\overline{CD}$
(34) $\overline{A}CD$
(35) $AB + \overline{AB}$
(36) $\overline{C}(A + \overline{B})$

CHAPTER 15

(1)   Short the collector of the OFF transistor to earth
(3)   See table:

In	Q	$\overline{Q}$
0	0	1
0	1	0
1	1	0
1	0	1
0	0	1

(6) $2^9 = 512$, therefore use 9 stages
(7)   (a) Q outputs on B and C, $\overline{Q}$ outputs on A and D
       (b) Q outputs on A, B and D; $\overline{Q}$ on C

(8)   (a) A,B; (b) V_{CC}, V; (c) A,C; (d) B,C (e) A,D
       (f) C,D
(11)   250 Hz square wave
(17)   See table:

Decimal	a	b	c	d	e	f	g
0	1	1	1	1	1	1	0
1	0	1	1	0	0	0	0
2	1	1	0	1	1	0	1
3	1	1	1	1	0	0	1
4	0	1	1	0	0	1	1
5	1	0	1	1	0	1	1
6	1	0	1	1	1	1	1
7	1	1	1	0	0	0	0
8	1	1	1	1	1	1	1
9	1	1	1	1	0	1	1

(18)   Connect 1 to 2 as input A; connect 3 to 4 as input B; connect 6 to 9, 10 to 12 and take the output from 8.
(19)   Triple AND gate (In practice a quad AND gate would be used.)

CHAPTER 16

(3)   As B increases the relative permeability of the iron falls, hence the reluctance of the iron increases and the e.m.f. ceases to rise linearly
(7)   220 V; 250 V
(8)   The machine would not excite
(9)   203.4 V
(10)   See *Figure A*. The field flux is maintained constant by the external supply, so the generated e.m.f. is also constant. The slight drop on the output voltage is due to armature resistance reducing the output at high load currents

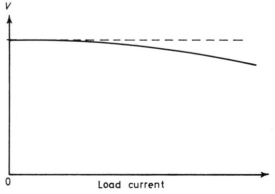

Figure A

(11) $E = V_B + R_A(I_L + I_F)$ where $V_B = V + R_F(I_L + I_F)$ this time

(12)   114 rad/s
(14)   Because of field weakening by magnetic distortion resulting from armature reaction
(19)   (a) Electrical, mechanical, (b) output/input, (c) poles, (d) two, (e) flux Φ
(20)   17.86 mWb
(21)   300 conductors
(22)   385 V
(23)   280 V
(24)   −215 V
(25)   178 rad/s
(26)   About 40 per cent
(27)   164 kW
(28)   (a) Flux Φ, (b) flux Φ, (c) opposes, (d) gross
(29)   (a) 272 Ω, (b) 230 V
(30)   62.5 N-m
(31)   491 rpm. Because of the high power loss in the series resistor
(32)   52.2 rad/s
(33)   62.5 per cent
(34)   One-sixth, 14.3 A, 5.71 kW

CHAPTER 17

(1)   Less iron and copper is required at higher frequencies
(2)   (a) 3000 rpm, 314 rad/s; (b) 1500 rpm, 157 rad/s; (c) 1000 rpm, 105 rad/s
(3) $N = 60f$
(4)   100 per cent. Standstill here means that the rotor is held stationary though power is applied to the stator
(5)   Slip times frequency
(7)   8 poles
(8)   1432 rev/min
(9)   49.8 Hz, 1512 rev/m
(10)   5 per cent; 1425 rev/m
(11)   1.67 Hz; 3.3 per cent
(12)   3 per cent
(13)   (a) 750 rpm, (b) 731 rpm, (c) 1.25 Hz
(15)   4 per cent
(16)   1:24.25
(17)   (a) 720 rpm, (b) 37.4 rpm, (c) 1:19.3

Index